VOLUME ONE HUNDRED AND SIXTY SEVEN

INTERNATIONAL REVIEW OF NEUROBIOLOGY

Microbiome in Neurological Disease

INTERNATIONAL REVIEW OF NEUROBIOLOGY

VOLUME 167

SERIES EDITOR

PATRICIA JANAK
Janak Lab, Department of Neuroscience
Dunning Hall, John Hopkins University
Baltimore, MD, USA

PETER JENNER
Division of Pharmacology and Therapeutics
GKT School of Biomedical Sciences
King's College, London, UK

EDITORIAL BOARD

VOLUME ONE HUNDRED AND SIXTY SEVEN

INTERNATIONAL REVIEW OF NEUROBIOLOGY

Microbiome in Neurological Disease

Edited by

TIMOTHY R. SAMPSON
Department of Cell Biology,
Emory University School of Medicine,
Atlanta, GA, United States

ELSEVIER

ACADEMIC PRESS
An imprint of Elsevier

Academic Press is an imprint of Elsevier
50 Hampshire Street, 5th Floor, Cambridge, MA 02139, United States
525 B Street, Suite 1650, San Diego, CA 92101, United States
The Boulevard, Langford Lane, Kidlington, Oxford OX5 1GB, United Kingdom
125 London Wall, London, EC2Y 5AS, United Kingdom

First edition 2022

Notices

Knowledge and best practice in this field are constantly changing. As new research and experience broaden our understanding, changes in research methods, professional practices, or medical treatment may become necessary.

Practitioners and researchers must always rely on their own experience and knowledge in evaluating and using any information, methods, compounds, or experiments described herein. In using such information or methods they should be mindful of their own safety and the safety of others, including parties for whom they have a professional responsibility.

To the fullest extent of the law, neither the Publisher nor the authors, contributors, or editors, assume any liability for any injury and/or damage to persons or property as a matter of products liability, negligence or otherwise, or from any use or operation of any methods, products, instructions, or ideas contained in the material herein.

ISBN: 978-0-323-99176-6
ISSN: 0074-7742

For information on all Academic Press publications
visit our website at https://www.elsevier.com/books-and-journals

Publisher: Zoe Kruze
Acquisitions Editor: Sam Mahfoudh
Developmental Editor: Federico Paulo Mendoza
Production Project Manager: Abdulla Sait
Cover Designer: Mark Rogers

Typeset by STRAIVE, India

Contents

Contributors

Laura M. Cox
Ann Romney Center for Neurologic Diseases, Brigham & Women's Hospital, Harvard Medical School, Boston, MA, United States

Millicent N. Ekwudo
Ann Romney Center for Neurologic Diseases, Brigham & Women's Hospital, Harvard Medical School, Boston, MA, United States

Sudeep Ghimire
Department of Pathology, University of Iowa; Iowa City VA Health System, Iowa City, IA, United States

Christopher G. Goetz
Department of Neurological Sciences, Rush University Medical Center, Chicago, IL, United States

Carolina Gubert
Florey Institute of Neuroscience and Mental Health, Melbourne Brain Centre, University of Melbourne, Parkville, Victoria, Australia

Adam M. Hamilton
Department of Cell Biology, Emory University School of Medicine, Atlanta, GA, United States

Anthony J. Hannan
Florey Institute of Neuroscience and Mental Health, Melbourne Brain Centre; Department of Anatomy and Neuroscience, University of Melbourne, Parkville, Victoria, Australia

Elaine Y. Hsiao
Department of Integrative Biology & Physiology, University of California, Los Angeles, Los Angeles, CA, United States

Ali Keshavarzian
Department of Internal Medicine, Division of Digestive Diseases and Nutrition; Rush Center for Integrated Microbiome and Chronobiology Research, Rush University Medical Center, Chicago, IL, United States

Chloe J. Love
Florey Institute of Neuroscience and Mental Health, Melbourne Brain Centre, University of Melbourne, Parkville, Victoria, Australia

Gregory R. Lum
Department of Integrative Biology & Physiology, University of California, Los Angeles, Los Angeles, CA, United States

Richard A. Manfready
Department of Internal Medicine, Division of Digestive Diseases and Nutrition, Rush University Medical Center, Chicago, IL, United States

Ashutosh K. Mangalam
Department of Pathology, University of Iowa; Iowa City VA Health System, Iowa City, IA, United States

Bethany A. Masson
Florey Institute of Neuroscience and Mental Health, Melbourne Brain Centre, University of Melbourne, Parkville, Victoria, Australia

Kelly B. Menees
Department of Neuroscience, Center for Translational Research in Neurodegenerative Disease, College of Medicine, University of Florida, Gainesville, FL, United States

Brittney A. Otero
Department of Neuroscience, Center for Translational Research in Neurodegenerative Disease, College of Medicine, University of Florida, Gainesville, FL, United States

Ezgi Özcan
Department of Integrative Biology & Physiology, University of California, Los Angeles, Los Angeles, CA, United States

Haydeh Payami
Professor of Genetics and Neurology, Department of Neurology, University of Alabama at Birmingham, Birmingham, AL; Aligning Science Across Parkinson's (ASAP) Collaborative Research Network, Chevy Chase, MD, United States

Martin R. Profant
Department of Immunology, University of Toronto, Toronto, ON, Canada

Timothy R. Sampson
Department of Cell Biology, Emory University School of Medicine, Atlanta, GA, United States

Shailesh K. Shahi
Department of Pathology, University of Iowa; Iowa City VA Health System, Iowa City, IA, United States

Ella Simonsen
Ann Romney Center for Neurologic Diseases, Brigham & Women's Hospital, Harvard Medical School, Boston; University of Massachusetts, Amherst, MA, United States

Malú Gámez Tansey
Department of Neuroscience, Center for Translational Research in Neurodegenerative Disease, College of Medicine, University of Florida; Norman Fixel Institute for Neurological Diseases, University of Florida Health, Gainesville, FL, United States

Helen E. Vuong
Department of Pediatrics, Division of Neonatology, University of Minnesota Twin Cities, Minneapolis, MN, United States

Caroline Wasén
Ann Romney Center for Neurologic Diseases, Brigham & Women's Hospital, Harvard Medical School, Boston, MA, United States; Department of Biology and Biological Engineering, Chalmers University of Technology, Gothenburg, Sweden

Meeta Yadav
Department of Pathology, University of Iowa; Iowa City VA Health System, Iowa City, IA, United States

Introduction: Unraveling the complex contributions of indigenous microbes to neurological health and disease

Timothy R. Sampson
Department of Cell Biology, Emory University School of Medicine, Atlanta, GA, United States

Abstract

The complex interactions between the human body and its indigenous microbes have come into focus as key mediators of neurological health. With both established and emerging association studies, alterations to the gut microbiome are observed to co-occur with many neurological diseases. Whether these associations are due to microbiome-mediated contributions to human health or an effect of the neurological disease itself is largely unknown across conditions. Here, we have collected contributions from a broad group of experts that highlight gut microbiome impacts across numerous neurological conditions. Ranging from neurodevelopmental disorders, to Parkinson's disease, Alzheimer's disease, epilepsy, traumatic injury, and amyotrophic lateral sclerosis, among others, we hope to provide a clearer picture of how our indigenous microbes impact neurological health. The study of these indigenous microbes will continue to reveal critical mechanisms that may 1 day be exploited for therapeutic benefits against these recalcitrant diseases.

We are not alone

Hundreds of unique species inhabit the human body. Spanning from tiny arthropods residing on our hair follicles, to numerous fungi, prokaryotes and their bacteriophage predators, and other viruses even engrained within our own DNA, our bodies are host to a legion of indigenous microorganisms from our heads to toes, inside and out. This complex and dynamic community of microorganisms inhabits every environmentally exposed surface beginning from the moment of birth and lasts through our lives. Humans and other animals have been interacting and coevolving with their associated microbial symbionts for millennia and together have drastically shaped their respective physiologies and evolutionary trajectories (McFall-Ngai et al., 2013).

When we consider the interactions between ourselves and microbes, largely these thoughts turn to microbial pathogens—organisms whose colonization of our body sites or tissues results in injury and significant detriment to our health. In the midst of an ongoing global viral pandemic, microbial pathogenesis is undoubtedly important to understand and consider (Moran, 2006). However, such detrimental pathogenic interactions are only one type of connectivity between species. In comparison to other types of symbioses, pathogenesis itself is exceedingly rare (and in some cases, it may be argued, an accidental occurrence) (Casadevall & Pirofski, 2019). Mutualistic interactions between animals and their indigenous microbes are dramatically more frequent.

The indigenous human microbiota are exceedingly vast. Estimates place the number of bacterial cells in the human body at ~39 trillion, more than the number of human cells in an adult (Sender, Fuchs, & Milo, 2016). This doesn't include bacteriophages and other viruses, estimated to be ~10× more than their replicative prokaryotic hosts (Mushegian, 2020). By sheer numbers, we are inhabited by more microbes than we are constructed of human cells. This fact becomes more astounding when considering the genetic repertoire encoded by the indigenous microbiota. Given modern metagenomic sequencing techniques, the catalog of unique microbially encoded genes within the human microbiome is estimated to be 10–100× more than encoded by the human genome (Qin et al., 2010). This huge genetic pool is an indicator of its immense capacity to touch upon all aspects of physiology of the host, rather than acting as simple bystanders or solely commensals to human physiology.

Direct mutualistic interactions between animals and microbes begin at birth, when microbes derived from the maternal environment (i.e., vaginal canal, skin) colonize the skin, nasal/oral cavity, and gut. The indigenous microbial community is seeded first by these maternal microbes and is subsequently shaped by diet and environmental, genetic, and immune factors. The resultant microbiota are relatively stable in their overall structure and during health are resilient to perturbation (Donaldson, Lee, & Mazmanian, 2016). Indirect symbiotic interactions, on the other hand, begin well before birth. The fetal developmental environment is intimately shaped by maternally derived microbes, through modulation of maternal immune and metabolic factors with capacity to cross or signal through the placental barrier (Pronovost & Hsiao, 2019).

Animal hosts are thus inextricably linked to their indigenous microbes from embryonic development through birth, growth, and death. Indeed, modern appreciation of these symbiotic connections has renewed the older concept of the holobiont or holo-organism (Rosenberg & Zilber-Rosenberg, 2016). This is a conceptual view of the host organism (i.e., humans) that takes into account the entirety of those indigenous microorganisms also stably residing within that host, together experiencing evolutionary pressures that guide their development. This is further detailed by the concept of the hologenome—whereby a given organism's genetic repertoire is defined not solely by what is encoded within their own genome, but by the entirety of the genes encoded by the (micro)organisms inhabiting them (Rosenberg & Zilber-Rosenberg, 2016). Thus, in order to understand the total physiology of the host, one must also understand the deep contributions of those other (micro)organisms to fully encapsulate all the genetic and metabolic factors that link their physiology.

Taking a metabolic perspective, it is fairly intuitive to envision mechanisms by which the consortium of indigenous microbes can broadly modify human physiology. On a foundational level, the fermentation of dietary fiber to short-chain fatty acids (SCFAs—acetate, propionate, and butyrate) by groups of gut microbes is a critical component of how our body receives carbon-derived energy. In fact, production of SCFAs by fermenting microbes is required for many ruminates as their primary carbon/energy source (Bergman, 1990). This is, of course, a very broad stroke. We understand now that gut microbial metabolism is critical for production of a plethora of nutrients and generation of essential vitamins and cofactors (Bishai & Palm, 2021; Sharon et al., 2014), and plays important roles in the metabolism of xenobiotics such as pharmaceuticals and environmental pollutants (Zimmermann, Zimmermann-Kogadeeva, Wegmann, & Goodman, 2019a, 2019b).

Given such life-long interactions and their impacts not only on metabolism but immune development as well, we now understand that microbial interactions shape a number of critical neurological functions through development, health, and disease (Sharon, Sampson, Geschwind, & Mazmanian, 2016). This is not an entirely new notion. Historically, Illya Metchnikov posited that microbes residing within us may be contributing to poor health and dementia, due to particular toxic factors that they produce, and that restoring the microbiome community might provide benefit for such age-related disorders (Bested, Logan, & Selhub, 2013). Parasitologists have long known that microorganisms can influence immense aspects of neurological function and behaviors. Now, these broader concepts of microbial impacts on brain health are beginning to be tested across both human observations and experimental models of disease.

With the advent of high-throughput sequencing and robust computational pipelines, cataloging microbial communities (through 16S rRNA sequencing, for instance) has become significantly more attainable for many researchers. Through this, we now have an understanding that our indigenous microbial communities—particularly those within the gut—have measurable variations that co-occur with neurological disease state and which are not fully explained by age or diet. Although highly intriguing, these studies are often conflated in press and popular media as demonstrating causality of the microbiome to the disease of interest. We, as a field, must be careful to not overinterpret such observations. As crucial as such studies are, demonstrating a concurrent gut microbiome alteration with a neurological disease, or with a beneficial pharmacological treatment, does not equate with microbiome contributions or causality of the disease or treatment outcome. Observing an impact on microbiome structure does not mean that the treatment is acting through microbiome activity, nor that the microbiome is causing the said disease. Systemic effects of many neurological illnesses can undoubtedly lead to both direct effects on the gut environment (e.g., intestinal inflammation, motility) or indirect effects (e.g., dietary preferences or caloric intake) that shape the gut microbiome community. However, the large body of observational studies that exist, which demonstrates these co-occurrences, provides a strong and critical foundation for testable hypotheses regarding the potential contributions of microbiome alterations to the brain. Many of these are presented in detail in the chapters that follow.

The question of whether neurological disease-associated gut microbiomes are epiphenomenon to disease, or whether they have some aspects of causation, triggering, or contribution is emerging as a prime subject of research. Highly controlled gnotobiotic animal models (across the span of model organisms) have allowed fine-tuning of the gut microbiome to explore the physiological consequences of microbiome absence (i.e., germ-free), enrichment/depletion of select disease-associated taxa, and the generation of both humanized and simplified communities. Like all therapeutic and etiological studies, gnotobiotic experimentation is still certainly limited by the aspects of animal models. However, these studies continue to provide critical insights into the contributions of gut microbes to neurological disease—coercively beginning to fulfill aspects of Koch's postulates for microbiome dysregulation.

Arguably, the experimental data to date in many models of neurological disease that have explored probiotics, fecal microbiome transplant, and designer dietary interventions support microbiome contributions to aspects of some neurological diseases. Given the relative ease and safety of such microbiome-modifying therapies, in conjunction with the significant need for new therapeutic pipelines, the translation of these microbiome-based (or informed) treatments into human trials is undoubtedly warranted for some neurological diseases. Although these still carry some health risk (albeit relatively low), there certainly exists potential for lack of efficacy in the human condition.

Current and future studies using metagenomic shotgun sequencing will also greatly enhance our knowledge of the microbiome communities associated with neurological disease as these technologies become less cost-prohibitive. Such identification of specific genes and metabolic pathways encoded by the microbiome may shed light on how these microbiome disruptions occur initially and which metabolic pathways might be shaping or associated with disease outcomes. As neurological diseases, particularly age-related neurodegenerative diseases, are slowly progressing, long-term longitudinal studies of how members of the microbiome shift through disease stages are desperately needed. Compounding the difficulty of the long-time scale is a need for more refined predictors of disease. Association of preclinical markers and genetic risk factors may help to identify those individuals to follow up longitudinally for microbiome compositional analysis to begin answering this "chicken-and-egg" problem. Given these difficulties, we are inherently struggling to understand whether microbiome differences occur first for many diseases, or instead arise later after the disease has firmly been diagnosed. These findings would be critical to delineate microbiome alterations as components of etiology, progression, or epiphenomenon to disease. The translation of microbiome-derived therapeutics to human incidences would therefore also shed light on this—as it will broadly test whether alterations to the human gut microbiome facilitate therapeutic changes, or if instead such microbial alterations are useful as early biomarkers of disease.

This volume has brought together a broad group of experts to touch upon many discrete and overlapping aspects of whether and how indigenous microbes contribute to neurological disease. It was my goal to highlight both established and emerging areas of interest in this complex space. Each chapter seeks to highlight either a specific human disease or a potential microbiome-mediated mechanism with a neurological impact. Together, these expound on what is known to date regarding the microbiome alterations that occur in neurological diseases, those experimental studies indicating specific contributions to such diseases, and the immense potential for various types of microbiome-modifying therapeutics to provide benefit against these recalcitrant diseases. Similar to how the study of pathogens has led to a deeper understanding of immune function, the study of our indigenous microbes will continue to reveal relevant interactions with our nervous system. Together we paint the vibrant picture of human physiology not as a sole species, but as a true holobiont.

References

Bergman, E. N. (1990). Energy contributions of volatile fatty acids from the gastrointestinal tract in various species. *Physiological Reviews*, *70*, 567–590.

Bested, A. C., Logan, A. C., & Selhub, E. M. (2013). Intestinal microbiota, probiotics and mental health: From Metchnikoff to modern advances: Part I—Autointoxication revisited. *Gut Pathogens*, *5*, 5.

Bishai, J. D., & Palm, N. W. (2021). Small molecule metabolites at the host-microbiota Interface. *Journal of Immunology*, *207*, 1725–1733.

Casadevall, A., & Pirofski, L. A. (2019). Benefits and costs of animal virulence for microbes. *mBio*, *10*, e00863-19.

Donaldson, G. P., Lee, S. M., & Mazmanian, S. K. (2016). Gut biogeography of the bacterial microbiota. *Nature Reviews. Microbiology*, *14*, 20–32.

McFall-Ngai, M., Hadfield, M. G., Bosch, T. C., Carey, H. V., Domazet-Loso, T., Douglas, A. E., et al. (2013). Animals in a bacterial world, a new imperative for the life sciences. *Proceedings of the National Academy of Sciences of the United States of America*, *110*, 3229–3236.

Moran, N. A. (2006). Symbiosis. *Current Biology*, *16*, R866–R871.

Mushegian, A. R. (2020). Are there 10(31) virus particles on earth, or more, or fewer? *Journal of Bacteriology*, *202*, e00052–00020.

Pronovost, G. N., & Hsiao, E. Y. (2019). Perinatal interactions between the microbiome, immunity, and neurodevelopment. *Immunity*, *50*, 18–36.

Qin, J., Li, R., Raes, J., Arumugam, M., Burgdorf, K. S., Manichanh, C., et al. (2010). A human gut microbial gene catalogue established by metagenomic sequencing. *Nature*, *464*, 59–65.

Rosenberg, E., & Zilber-Rosenberg, I. (2016). Microbes drive evolution of animals and plants: The hologenome concept. *mBio*, 7, e01395.

Sender, R., Fuchs, S., & Milo, R. (2016). Revised estimates for the number of human and bacteria cells in the body. *PLoS Biology*, *14*, e1002533.

Sharon, G., Garg, N., Debelius, J., Knight, R., Dorrestein, P. C., & Mazmanian, S. K. (2014). Specialized metabolites from the microbiome in health and disease. *Cell Metabolism*, *20*, 719–730.

Sharon, G., Sampson, T. R., Geschwind, D. H., & Mazmanian, S. K. (2016). The central nervous system and the gut microbiome. *Cell*, *167*, 915–932.

Zimmermann, M., Zimmermann-Kogadeeva, M., Wegmann, R., & Goodman, A. L. (2019a). Mapping human microbiome drug metabolism by gut bacteria and their genes. *Nature*, *570*, 462–467.

Zimmermann, M., Zimmermann-Kogadeeva, M., Wegmann, R., & Goodman, A. L. (2019b). Separating host and microbiome contributions to drug pharmacokinetics and toxicity. *Science*, *363*.

CHAPTER ONE

Intersections of the microbiome and early neurodevelopment

Helen E. Vuong*
Department of Pediatrics, Division of Neonatology, University of Minnesota Twin Cities, Minneapolis, MN, United States
*Corresponding author: e-mail address: hevuong@umn.edu

Contents

Abstract

Our resident microbes influence nearly all aspects of our biological systems. In particular, the maternal and early life microbiota is uniquely positioned to influence the development of the nervous system, and alterations to the gut microbiota, or dysbiosis, during this critical time in early life can have long-lasting negative effects on health. The question of how the maternal and early life microbiota shapes neurodevelopment is the topic of numerous investigations. Here, we discuss two possible, but not necessarily independent, hypotheses: (1) the maternal microbiota during pregnancy regulates the metabolites that are important for fetal development, (2) maternal microbiota seeded to offspring at birth and early postnatal days programs offspring immune and brain development, and regulates key molecules for postnatal brain development. In this chapter, we provide an overview of the impact of the microbiota on brain and behavior, introduce the maternal gut and vaginal microbiome during pregnancy, and discuss current understandings of microbiome in the context of developmental origins of health and disease.

We consider novel translational insights that harness the multitude of microbes and microbial metabolites for prevention or treatment of neurological disorders.

International Review of Neurobiology, Volume 167
ISSN 0074-7742
https://doi.org/10.1016/bs.irn.2022.06.004

1. Microbiota regulation of brain and behavior

The trillions of microorganisms that comprise the microbiota has emerged as a crucial component of host biology. The microbiome provides metabolic and immunological directives that aid in host physiology (here, the term "host" is used to refer to the organisms that harbor the microbiota, e.g., the human host or rodent host). Composition and function of the human microbiota is shaped by genetics and environmental factors including antibiotic exposure, diet, and exercise, varies from body sites, and is unique to each individual (Human Microbiome Project Consortium, 2012). Notably, the intimate relation with microbes begins before birth, with more direct interactions during birth, and continues throughout life, whereby these complex host–microbe interactions contribute to health and disease states.

The germ-free mouse model was developed in 1957 where mice are raised in gnotobiotic environments and are devoid of commensal microbes (Gustafsson, 1959; Gustafsson, Kahlson, & Rosengren, 1957). A study that showed germ-free mice are hyperresponsive to restraint stress and display exaggerated levels of plasma corticosterone and adrenocorticotrophic compared to conventional controls became the foundation for incorporating neuroscientists in the field of microbiome (Sudo et al., 2004). Several groups followed with findings that demonstrated germ-free mice show reduced anxiety-like behavior and decreased plasticity related genes in the hippocampus and amygdala (Clarke et al., 2013; Diaz Heijtz et al., 2011; Neufeld, Kang, Bienenstock, & Foster, 2011a; Neufeld, Kang, Bienenstock, & Foster, 2011b). Notably, conventionalization of germ-free microbiota was sufficient to reverse anxiety related behavior in germ-free mice (Clarke et al., 2013), suggesting the nuances of when colonization occurs is important for capturing effects on behaviors.

To better understand the role of the microbiota in clinically-relevant phenotypes and disease states, researchers have adapted gnotobiotics to transplant human microbiota into germ-free mice and evaluate the mechanistic contributions of the microbiota. One such study transplanted microbiota from twins discordant for obesity into adult germ-free mice, and successfully recapitulated increased body and fat mass, and obesity-associated metabolic phenotypes of the obese twin but not the lean twin (Ridaura et al., 2013). Colonization of alpha-synuclein overexpressing germ-free mice with microbiota from patients affected by Parkinson's elevated physical impairments compared to microbiota transplants from healthy human donors (Sampson et al., 2016). Germ-free mice that were colonized with microbiota from autism spectrum disorder compared to typically developing individuals show autistic-like behaviors and altered colonic and serum metabolome (Sharon et al., 2019). In a study aimed at discovering microbiota-directed complementary food prototypes that would best promote weaning phase microbiota and growth in childhood undernutrition, investigators colonized germ-free mice and piglets with a microbiota that appears younger than those of chronologically age-matched healthy individuals or specific bacterial taxa indicative of age- and growth-discriminatory taxa. One particular diet formulation sufficiently increased abundance of select bacteria and increased plasma protein that mediate growth, bone formation, neurodevelopment, and immune function in children with moderate acute malnutrition (Gehrig et al., 2019). Importantly, these studies were collected following clinical diagnosis and demonstrate that the microbiota is sufficient to maintain the disorders of obesity, Parkinson's or autism spectrum disorder. However, it remains unclear whether the microbiota is mechanistically responsible for onset and progression of disorders. A particular challenge in answering such causal questions involves coordinating human samples with the different stages of disease development to reveal development-specific contributions of the microbiome on disease outcomes.

2. Prenatal and early postnatal microbiota

While most microbiome studies have focused on the impact of the developed microbiota on health and disease states, researchers have begun to question the role of the maternal microbiota during pregnancy and the developing gut microbiota. The developing gut microbiota is interdependent in the concurrent maturation of the endocrine, immune, metabolic, and neurodevelopmental pathways during early life. Studies in the immunology field revealed that young children exposed to antibiotics during a critical window of immune system development and maturation exhibited increased susceptibility to several diseases. In 2013 and 2014, a series of epidemiological studies examined whether in utero antibiotic exposure could also have unintended consequences on the developing fetus and later in life health (Marild, Ludvigsson, Sanz, & Ludvigsson, 2014; Ortqvist et al., 2014; Stensballe, Simonsen, Jensen, Bonnelykke, & Bisgaard, 2013; Vidal et al., 2013). Therefore, it is crucial that we understand how the maternal and early-life microbiota may contribute to fetal and neonatal development.

Pregnancy is a remarkable biological process involving simultaneous changes in many physiological systems—hormonal changes, weight gain, immune system modulation—synchronized to support the health of mother and development of healthy offspring (Kumar & Magon, 2012). Observed changes in the maternal gut microbiota during pregnancy are associated with changes in maternal body weight, diet, metabolism. For example, the pregnant mom is physiologically presented as hyperglycemic, where the increased glucose availability is used to support the growing fetus. Consequently, there is increased abundance of bacterial species that can synthesize glycogen from the increased concentrations of glucose. In fact, the maternal microbiota during pregnancy shares a similar microbiome profile to that of a hyperglycemic microbiota. There are increased levels of secreted hormones, namely, progesterone and estrogen. In addition, there are obvious changes in immune responses, where first trimester is marked by degrees of immune suppression to accept the developing fetus which contains foreign genetic material that comes from the father (Abu-Raya, Michalski, Sadarangani, & Lavoie, 2020). However, strict immunity is maintained to protect the mother and fetus from infections.

Metabolic changes associated with pregnancy include weight gain, elevated fasting blood-glucose levels, insulin resistance, glucose intolerance, low grade inflammation, and changes to metabolic hormone levels (Lain & Catalano, 2007). While pregnancy-associated hormonal and metabolic changes are well documented, the concurring microbiome changes have only recently been investigated. During gestation, the maternal milieu is influenced by intrinsic and extrinsic factors, including the gut microbiome. Maternal perturbances in the form of stress, malnutrition, alcohol/drug exposure or immune activation during the perinatal period is associated with adverse health outcomes in the offspring that can have long lasting consequences (Dejong, Olyaei, & Lo, 2019; Dunkel Schetter & Tanner, 2012; Kourtis, Read, & Jamieson, 2014).

Efforts from clinical and preclinical researchers has led to the discovery of numerous mechanisms that mediate offspring risk, including changes in the maternal milieu, placental function, epigenetic modifications, and most recently, maternal intestinal and vaginal microbiota.

2.1 Maternal intestinal microbiota

During pregnancy in humans, the maternal intestinal microbiota changes dramatically between the first and third trimester of pregnancy. In the first trimester of pregnancy, the microbiota is similar to the microbiota of healthy non-pregnant women. From the first to third trimester, there is an expansion of diversity between mothers, an increase in Proteobacteria and Actinobacteria, reduced individual richness, alpha diversity, and an increase in bacterial load (Koren et al., 2012). There is an increase in lactic acid producing bacteria and a reduction in butyrate-producing bacteria. Analysis of third trimester stool samples reveal an inflammatory and energy loss profile. Using microbiota transplantation of third trimester stool into germ-free mice, the microbiota was sufficient to induce increased adiposity and insulin insensitivity compared to first trimester microbiota (Koren et al., 2012). This study demonstrates the microbial components actively shape host immunology and metabolism. Some mechanisms by which the microbiota can promote weight gain during pregnancy include enhanced absorption of glucose and fatty acids, increased fasting induced adipocyte factor secretion, induction of catabolic pathways, and stimulation of the immune system.

By the third trimester, shifts in bacterial composition include butyrate-producing bacteria that have anti-inflammatory functions decline, whereas bifidobacterial, proteobacteria, and lactic-acid producing bacteria increase (Koren et al., 2012; Mesa et al., 2020; Neuman & Koren, 2017; Nuriel-Ohayon, Neuman, & Koren, 2016). Further, in the third trimester there is increased between-subject diversity, beta diversity, which corresponds with weight gain, insulin sensitivity, and inflammation—higher levels of fecal cytokines.

While internal cues can influence the gut microbiota during pregnancy, exposure to different factors such as diet and antibiotics can also contribute to microbiota composition (Calatayud, Koren, & Collado, 2019; Maher et al., 2020). Particular maternal microbiota changes are associated with pregnancy complications. Antibiotics administered during pregnancy in rats dramatically altered their gut microbiome. Use of category B antibiotics (azithromycin, amoxicillin, and cefaclor) during pregnancy was shown to increase the fecal relative abundance of Proteobacteria and Enterobacter, while reducing the relative abundance of Firmicutes and Lactobacillus. Moreover, antibiotics during pregnancy significantly reduced bacterial diversity, and promoted weight gain. Another study found that maternal antibiotic treatment during pregnancy influenced development of type 1 diabetes in the offspring, suggesting a long-lasting effect of maternal microbiota disruptions (Hu et al., 2016). Most strikingly, non-absorbable antibiotics given to pregnant mice altered offspring behavior in a manner by which they exhibited low locomotor activity at 4 weeks of age, and less exploratory behavior in central regions at both 4 and 8 weeks of age. However, these behavioral phenotypes were reversed when pups were fostered by control mothers starting at postnatal day 1 (Tochitani et al., 2016).

2.2 The vaginal microbiota

The vaginal microbiota during pregnancy is characterized by low diversity and communities that is dominated by Lactobacillus that decreases the vaginal pH, (DiGiulio et al., 2015; Gajer et al., 2012). Vaginal epithelial cells will produce glycogen that is transformed into lactic acid, which induces a low vaginal pH (3.8–4.4) to create an unfavorable environment for the growth of pathogenic bacteria (Ravel et al., 2011). Environmental, diurnal and hormonal fluctuations determine the composition of the human vaginal microbiota (Anahtar, Gootenberg, Mitchell, & Kwon, 2018). Vaginal dysbiosis, which is associated with an inflammatory state, is associated with adverse pregnancy-related complications including increased risk for spontaneous pre-term birth, although the data is limited and the sample sizes are small (DiGiulio et al., 2015). Indeed, Elovitz et al. show reduced lactobacilli, increased diversity, and low vaginal levels of beta-defensin 2 is associated with women at risk for preterm births (Elovitz et al., 2019).

The vaginal microbiota is a strong determinant of offspring's initial microbiota, particularly if the newborn is birthed through vaginal delivery. The vaginal microbiota changes throughout a woman's reproductive life, where the healthy vaginal microbiota is dominated by Lactobacillus, Clostridiales, Bacteroidales, and Actinomycetales (Aagaard et al., 2012). Within the Lactobacillus genus, the most frequently isolated species are *Lactobacillus gasseri*, *Lactobacillus crispatus*, *Lactobacillus jensenii*, and *Lactobacillus iners*. Notably, interactions between vaginal microbiota and obstetric outcomes is population dependent—in that women of European descent are more likely to harbor a lactobacillus-dominated microbime, whereas African American women have a diverse microbiome profile. This microbiome profile is associated with African American women experiencing increased diagnosis of bacterial vaginosis and preterm birth compared to women of European descent (Fettweis et al., 2014).

The changes in hormone (progesterone and estrogen) levels during pregnancy corresponds with a shift in the vaginal microbiome diversity and stability. Further, changes in vaginal microbiota composition during pregnancy depends on the pre-existing community prior to pregnancy. Vaginal communities that are Lactobacillus dominated are more stable across gestation (Anahtar et al., 2018). One study shows a decrease in diversity, the community composition changes overtime, and there is an enrichment of Lactobacillus species (Aagaard et al., 2012). Importantly, loss of Lactobacillus during pregnancy are associated with adverse obstetric outcomes including preterm birth, preterm premature rupture of membranes, and bacterial vaginosis (Callahan et al., 2017; DiGiulio et al., 2015; Hyman et al., 2014). One study finds no significant alterations in the vaginal microbial composition throughout pregnancy, but, reported differences in the post-partum period, where Peptoniphilus, Prevotella, and Anaerococcus replaced the dominant Lactobacillus (DiGiulio et al., 2015). Whether the vaginal microbiota can be a source for diagnostics by defining predictive microbial signatures in complex environmental conditions like chronic stress during pregnancy remains to be determined.

3. Impact of microbiota on neurodevelopment: Considerations of the developmental origins of health and disease

The Barker's, or Developmental Origins of Health and Disease, hypothesis suggests that maternal pre-conception and pregnancy exposures can persistently alter physiology and behavior in ways that last into adulthood (Barker, 1995a, 1995b). Some examples of early life exposure include diet, stress, and infections. The proposed mechanisms include long lasting perturbations of the placenta, liver, pancreas, thyroid, and gastrointestinal tract through host gene and microbiome alterations. In particular, the gut microbiome can contribute small molecules and nutrients to the maternal circulation that can be actively or passively transferred across the placenta to the fetus.

Autism, obesity, and diabetes mellitus are prevalent neuropsychiatric and noncommunicable diseases. While these disorders are manifested in later life, they are affected by neurodevelopmental and metabolic programming that occur in early life and thus may be mitigated or averted via early correction of risk factors. Research in pre-clinical models that show microbiota mediated transmission of host disease phenotype provide some evidence that colonization history, developmental trajectories, and disturbances in the early gut microbiota are implicated in metabolic, immune, and neurological health later in life. Prospective studies in humans have shown links between changes in early life microbiota and pediatric overweight and obesity or autism spectrum disorder (Jian et al., 2021; Lou et al., 2021). Therefore, by understanding the ecological drivers and environmental factors contributing to shifts in microbial equilibrium during the critical window of development then, further disruptions of the normal development may be prevented.

One point of target is during the prenatal period, where metabolites, antibodies and substrates produced by the maternal microbiota may provide support for the molecular machinery involved in the formation of neural circuits. Second point of target is after birth, the offspring now seeded with its own microbiota can continue directing the development of key neural circuits that eventually mediate behaviors. Consistent with the Barker hypothesis, the maternal and early life microbiota can shape the development of the brain such that there is resilience to perturbations or yield vulnerabilities to challenges that may occur later in life.

In studies that compared germ-free mice to conventionally colonized mice, the absence of a maternal microbiota is associated with abnormal immune development, blood–brain barrier formation, programming of the HPA axis, microglia maturation, and regulation of hypothalamic circuits involved in appetite regulation and energy balance (Braniste et al., 2014; Erny et al., 2015; Frost et al., 2014; Gomez de Aguero et al., 2016; Matcovitch-Natan et al., 2016; Thion et al., 2018). These studies highlight that the complete absence of a maternal microbiota can have profound influences on the developing fetus, however, it remains unclear about the composition of the maternal microbiota that is implicated in different disease phenotypes. Recent work has aimed to describe the compositional differences of the maternal microbiota during pregnancy and lactation.

3.1 Maternal microbiota during pregnancy influences fetal development

Pregnancy is a time of tremendous growth and development for the fetus and placenta. There are important contributions from maternal immunity, metabolism, and nutrition. Nutrients, gases, waste, and biochemicals diffuse or are actively transported across the placenta to enter fetal circulation. Importantly, fetal nutrient requirements fluctuate over the course of pregnancy, requiring metabolic adaptations to support these demands (Nugent & Bale, 2015). Metabolic adaptations are particularly important for the normal development of the fetal brain, and the requirements for nutrients are regionally distributed based on which areas are rapidly developing at any given time, suggesting differing brain regions may present unique metabolic signatures (Georgieff, 2007; Kretchmer, Beard, & Carlson, 1996). The most striking transformation of the developing central nervous system occurs between 24 and 44 weeks after conception. Essential nutrients such as protein, energy, long-chain polyunsaturated fatty acids, iron, zinc, copper, iodine, selenium, vitamin A, choline, and folate support the rapidly developing brain through critical processes including neuronal growth and differentiation, neural migration, experience-dependent synapse formation, and myelination, all of which occur before term birth (Georgieff, 2007). Notably, the maternal microbiota also undergoes changes that are modified by hormonal and metabolic fluctuations during pregnancy. Conversely, as a regulator of immunity and metabolism, the microbiota has the capacity to modulate nearly all aspects of nutrition and metabolism physiology of pregnancy (Koren et al., 2012; Neuman & Koren, 2017; Nuriel-Ohayon et al., 2016). Humans are incapable of synthesizing most vitamins and consequently have to be obtained exogenously. The microbiome possesses the genetic code to influence metabolites that are not necessarily available to the host. Gut microbes harvest energy substrates from the diet by digesting and fermenting complex carbohydrates to generate short chain fatty acids and produce amino acids and vitamins (LeBlanc et al., 2013).

The maternal microbiota during pregnancy is a modifier of offspring health, contributing to development of the immune system or atopy and autoimmune phenotypes (Nyangahu & Jaspan, 2019). In a prospective study that examined the microbiota of normal ($n = 36$) and overweight ($n = 18$) women, the microbial counts increased from first to third trimester, and prepregnancy weight was associated with increased concentrations of Bacteroides, Clostridum, and Staphylococcus (Collado, Isolauri, Laitinen, & Salminen, 2008). In a follow-up study that examined the fecal microbiota of infants from normal ($<25\,kg/m^2$) or overweight ($>25\,kg/m^2$) mothers, researchers found correlations of increased Bacteroides and Staphylococcus concentrations in infants from overweight compared to normal weight mothers (Collado et al., 2008). Further, levels of metabolic hormones during pregnancy are associated with abundance of specific microbes (Gomez-Arango et al., 2016). For example, insulin and gastrointestinal polypeptide are positively correlated with the genus Collinsella and Coprococcus, respectively. Together, this suggests the microbiota composition and maternal weight can influence the microbiota of the infant and potentially infant's weight status in the future.

Maternal microbiome composition during pregnancy can impact offspring weight gain, immunity, and infant health. In a study of diet-induced obese rats, oligofructose prebiotic treatment during pregnancy and lactation altered the offspring cecal microbiome, prevented increased adiposity in dam and offspring (Paul, Bomhof, Vogel, & Reimer, 2016). In a separate study that targeted the maternal microbiota prior to birth, probiotics were administered to pregnant women 14 days before C-section, and this led to changes in the infant microbiota and expression of toll-like receptors genes in placenta and infant meconium, suggesting the maternal microbiota before birth can shape postnatal microbiota and infant transcriptome (Rautava, Collado, Salminen, & Isolauri, 2012). Microbial exposure during pregnancy can also be important for preventing allergic and autoimmune diseases in the offspring. These studies show immune-regulated epigenetic imprinting and bacterial translocation during pregnancy are the possible underlying mechanisms that serve as the initial exposure of bacterial molecular patterns to the fetus that prime the immune system to appropriately respond to pathogens and commensals after birth (Abrahamsson, Wu, & Jenmalm, 2015).

As the maternal microbiome emerges as a critical modifier of postnatal development, it is increasingly appreciated that the maternal microbiome during pregnancy also plays an important role in fetal development. Recent work from Gomez de Aguero and colleagues begins to dissect the impact of the maternal gut microbiota on newborn's gastrointestinal development and its influence on the development of the immune system. This study showed how transient colonization of female mice during pregnancy with a genetically modified *Escherichia coli* was sufficient to alter the levels of antibodies and metabolites in the mother and offspring (Gomez de Aguero et al., 2016). This strain of *E. coli* was genetically designed to be short-lived within the gastrointestinal tract. Offspring from dams that were transiently exposed to *E. coli* show a more mature innate immune system development and intestinal epithelial maturity compared to offspring from dams without *E. coli* exposure (Gomez de Aguero et al., 2016), suggesting microbial exposure during gestation can improve intestinal inflammatory response and shaping host-microbial symbiosis through maternal-fetal microbe interaction.

3.2 Early postnatal microbiota influences fetal development

Factors that shape the infant microbiome include mode of delivery, feeding, weaning, and exposure to antibiotics (Dominguez-Bello et al., 2010; Yang et al., 2016). During the birthing process, maternal fecal, skin, and vaginal microbes are vertically transmitted to the neonate to supports postnatal growth and development (McDonald & McCoy, 2019). Children born by vaginal delivery are colonized by vaginal and fecal microbes, while those born by Caesarean section are predominantly seeded by maternal skin microbes and sometimes hospital-acquired microbes, which then develops and diversify throughout the first 2 years of life until it is stabilized (Backhed et al., 2015; Dominguez-Bello et al., 2010; Koenig et al., 2011; Yatsunenko et al., 2012). The gut microbiota of vaginally born infants is first colonized by bacteria from the maternal vagina, which is characterized by enrichment in Prevotella, Sneathia, and Lactobacillus genera, and bacteria from the maternal gut (Dominguez-Bello et al., 2010; Mackie, Sghir, & Gaskins, 1999). In contrast, the gut microbiota of C-section born infants resemble that of maternal skin and oral microbiota, which are dominated by Propionibacterium, Corynebacterium, and Streptococcus (Backhed et al., 2015). Notably, the differences in species diversity between delivery modes are decreased after 4 months, and disappear by 12 months (Backhed et al., 2015).

There have only been a handful of studies that have considered the effects of C-section vs vaginal delivery on neurodevelopment and cognition. In one study of 3666C-section and vaginally-born children, C-section born children perform below vaginally-born children in Australian national numeracy test scores at age 8–9 years old (Polidano, Zhu, & Bornstein, 2017). In a meta-analysis of 61 studies that comprise 20 million deliveries, C-section was associated with autism spectrum disorder and attention-deficit/hyperactivity disorder (Zhang et al., 2019). However, in a study of 306 healthy children that were divided into three cohorts of 2-week-old, 8-year-old, and 3 months to 5-year-old children, there were in consistent effects on white matter development, functional connectivity, and gray matter volume in C-section compared to vaginally-born children (Deoni et al., 2019). Together, with confounding effects such as pregnancy complications and postnatal care and diet, it remains unclear whether delivery-mode can influence neurological outcomes, and detailed mechanisms will require further investigation.

Rapidly following birth, assembly of gut bacterial communities occurs as a dynamic process that is influenced by age-associated events such as transition from breast- or formula-fed to solid foods, antibiotic treatment and malnutrition (Koenig et al., 2011; Subramanian et al., 2015).

Neonatal bacteria develop across the newborn surfaces to prime the development of the gut microbiota, immune, metabolic, and nervous system. Much of the bacteria acquired postnatally aid to process indigestible molecules in the diet like breast milk, formula or solid foods, hence changing the infant gut microbiota as a reflection of dietary changes. For example, neonates are colonized with lactic acid bacteria that digest lactose, and other bacteria that process milk glycans such as oligosaccharides (human milk oligosaccharides; HMOs). The maturation of the gut microbiome is marked by enrichment of species that are able to utilize nutrients such as carbohydrates, and are involved in vitamin biosynthesis and xenobiotic degradation (Backhed et al., 2015). Moreover, microbial genes that encode for metabolic and enzymatic pathways support the transformation of substrates into energy sources for brain development (Goyal, Venkatesh, Milbrandt, Gordon, & Raichle, 2015). For example, the microbial metabolite butyrate which is used as a fuel source in mitochondria to sustain NADH/NAD+ balance and energy homeostasis (Donohoe et al., 2011). Signatures of gut microbiota maturation can provide a readout of healthy growth and development in infancy and childhood, whereas deviations from this program may be associated with negative health outcomes (Subramanian et al., 2014). Together, maternally acquired microbiota, microbial capacity to harvest energy and nutrients, and the metabolic demand of the developing brain during this early-in-life period may contribute to lasting effects on brain and behavior.

The following are some examples of maternal exposures during pregnancy and impact on offspring development:

Maternal diet has tremendous influence during pregnancy that contributes to maternal and fetal health. Maternal diet has been studied in the context of reduced sugar/carbohydrate, high-fat, or restricted protein intake. In a study that examined the effect of non-nutritive sweetener consumption during pregnancy, postnatal day 19 pups were found to have metabolic and hepatic detoxification changes, and exhibited a gut microbiome profile that was associated with human metabolic disease and obesity (Olivier-Van Stichelen, Rother, & Hanover, 2019). Pregnant mice fed a high-fiber diet compared to a regular- or no-fiber diet produced offspring that are more resistant to allergic airway disease, produced fewer eosinophils and proinflammatory cytokines. Authors propose the mechanism of high fiber diet effects could be mediated by increased acetate, as was observed in serum and feces (Nakajima et al., 2017; Thorburn et al., 2015). In a study where non-human primates were exposed to maternal high fat diet during pregnancy and lactation and then weaned onto a control diet for at least 6 months, the offspring had a distinct gut microbiome compared to peers that were only exposed to control diet (Ma et al., 2014).

Environmental chemicals like polycyclic aromatic hydrocarbons (PAH) are common maternal exposures. These noxious organic substances are by products of carbon combustion and cigarette smoke, and are associated with perinatal complications. Some adverse pregnancy outcomes associated with PAH exposure include preterm birth, low birth weight, child onset asthma, low cognitive assessment scores in children, neurodevelopmental delay, alteration of DNA methylation patterns, and fetal programming (Herbstman et al., 2012; Jedrychowski et al., 2015; Langlois et al., 2014; Lovasi et al., 2014; Miller et al., 2016). PAHs also altered maternal physiology including microbiome. Although the microbiome itself contributes to the transformation and metabolism of polycyclic aromatic hydrocarbon, ingested PAH can disrupt gut microbial enzymes that leads to a state of gut microbiota dysbiosis (Claus, Guillou, & Ellero-Simatos, 2016).

Finally, the intimate relationship between maternal adversities modeled by chronic stress is associated with abnormal immune function, dysfunctional hypothalamic pituitary adrenal stress axis, obstetric complications and negative offspring outcomes. Indeed, the effects of maternal adversities is wide-ranging, with implications in altered gut microbiota composition during pregnancy and disruption of inflammatory and gluococorticoid response to acute stress during pregnancy. While these studies provide the necessary frame work to begin to understand how maternal and childhood adversities impact maternal gut microbiota and inflammation, there are fundamental questions that remain regarding how these factors contribute to offspring development and behavioral outcomes. Stress during pregnancy is a risk factor for adverse maternal and fetal health. Stress can alter numerous biological pathways including the microbiome. Some studies that have examined maternal stress during pregnancy report offspring exhibit sleep disruption, atopic dermatitis, lower cognitive development scores, and increases in infectious and noncommunicable diseases. Jasarevic et al. demonstrated in a mouse model that maternal stress impacts the postnatal colonic microbiome from postnatal day 2–28. The disruption of the microbiome was observed in male mice, demonstrating a sex-dependent aspects of stress and microbiome interactions (Jasarevic, Howard, Misic, Beiting, & Bale, 2017). Further, work from Jasarevic et al. showed that the dams exposed to stress during the first week of pregnancy exhibited alterations in the maternal vaginal microbiota that is associated with alterations in hypothalamus transcriptome in the offspring (Jasarevic et al., 2018). Indeed, effects of other exposures during pregnancy such as diet changes or infection, on to the offspring can also be mediated by the maternal gut microbiome, altering the developing brain.

4. Potential mechanisms of microbiota influences on neurodevelopment

It is not surprising to microbiologists and immunologists that the microbiome is critically important for human health. However, some neuroscientists remain skeptical of how the microbiome can have such impact on neural circuits and complex behaviors. As such to understand mechanism we argue that these fields are not independent of each other, rather understanding the role of the microbiome in host physiology inherently suggests a multidisciplinary approach where fields of microbiology, immunology, neuroscience, psychiatry are all involved.

First example is the maternal immune activation (MIA) mouse model. In this model, pregnant dams are injected intraperitoneally with the synthetic double-stranded RNA (polyinosinic:polycytidylic acid, poly(I:C)), which mimics a viral infection (Smith, Li, Garbett, Mirnics, & Patterson, 2007). Offspring that are exposed to poly I:C exhibit altered microbiome immune response, metabolism, neurodevelopment, and behaviors (reviewed (Fung, Olson, & Hsiao, 2017; Pronovost & Hsiao, 2019)). Choi and colleagues discovered that poly I:C exposure activated the IL17a pathway and led to presence of patches of missing Satb2+ cells in the fetal cortex and ASD-like behaviors in the offspring (Choi et al., 2016). The embryonic cortical patches were associated with disruption of neural inhibition that contributes to abnormal MIA behaviors (Shin Yim et al., 2017). It was next discovered that the behavioral phenotypes that were caused by poly I:C induced maternal immune activation are dependent on a maternal gut commensal segmented filamentous bacteria (SFB), which drive differentiation and expansion of a T-cell subset known as T-helper 17 (TH17) cells that produced the inflammatory cytokine interleukin 17a (IL17a) (Kim et al., 2017; Shin Yim et al., 2017). To elucidate the role of SFB in mediating MIA-associated offspring phenotypes, C57 Bl/6 mice from Taconic Biosciences that are colonized by SFB were compared to C57Bl/6 mice from Jackson Laboratories that lack SFB (Kim et al., 2017). While offspring from poly I:C injected Taconic dams exhibited MIA-associated behavioral phenotypes, offspring from poly I:C injected Jax dams did not (Kim et al., 2017). Poly I:C treatment increased maternal gut levels of IL17a, maternal plasma levels of IL17a, and the increased IL17a promotes MIA-associated neurodevelopmental and behavioral abnormalities in offspring (Kim et al., 2017). Further, MIA offspring have an altered serum metabolomic profile, and treatment with a select gut bacterial metabolite reproduced certain abnormal behavioral phenotypes (Hsiao et al., 2013). These results shed light on the interactions across microbiome, immune system and neurobiology. Specifically, raises the question of what other interactions occur in the context of normal brain development. Together, these studies reveal mechanisms of neurodevelopmental programming that is mediated by specific microbes in the maternal gut.

The second case is in the context of diet, whereby the gut microbiota is a key interface for nutrition and brain development. Dietary substrates shape the microbial community composition and function, exerting significant metabolic consequences on the host. Dietary intervention studies during pregnancy and lactation reveal consumption of high-fat diet is associated with altered maternal gut microbiota composition and microbial metabolic pathways that are involved in lipid and glucose metabolism during pregnancy (Gohir, Ratcliffe, & Sloboda, 2015). Pregnant dams that were fed a high fat diet (HFD, 60% kcal from fat, 20% kcal from protein, and 40% kcal from carbohydrates) compared to a control diet (RD) consisting of 13.4% kcal from fat, 30% kcal from protein, and 57% kcal from carbohydrates) had significant shifts in their microbiome, which was also seen in the respective offspring (Buffington et al., 2016). Notably, the animals are consuming diets that are distinct in amount of fat and carbohydrates, making it difficult to distinguish whether the effects of the diet are due to the contribution of fat, fiber or both. To investigate whether maternal high-fat diet and microbiome changes can affect offspring brain and behavior, all offspring were weaned and placed on control diet and examined for behavioral differences. Offspring from maternal high-fat diet group exhibited social deficits including reduced social interaction with a mouse vs an empty arena and reduced social interaction with a novel mouse compared to a familiar mouse. The microbiome and social deficits were reversible by cohousing maternal high-fat diet offspring with maternal control diet offspring. The social deficits in maternal high-fat diet offspring corresponded with reduced oxytocin + parvalbumin neurons in the paraventricular nucleus of the hypothalamus and reduced long-term potentiation in dopamine neurons in the ventral tegmental area during social behavior assays as compared to maternal control diet offspring. Alterations in oxytocin expression, long-term potentiation, and social behaviors in the maternal high-fat diet offspring were corrected by colonization with a single bacterium, *Lactobacillus reuteri*. *Lactobacillus reuteri* has been shown to promote oxytocin (Poutahidis et al., 2013) and was identified to be reduced by more than ninefold in maternal high-fat diet offspring microbiome compared to control diet offspring. Interestingly, another bacterium *Lactobacillus johnsonii*, was also decreased in maternal high-fat diet compared to control diet offspring, however, *L. johnsonii* was not sufficient to correct behaviors. While it is clear that the interaction between maternal diet and microbiome determines social behaviors and particular neural activities, the precise mechanism of how microbes are carrying out such effects on the offspring remains unclear. Some hypothesis includes microbiome mediated vagal nerve signaling and/or gut-brain signaling through microbially mediated small molecules.

Finally, rodent studies have been helpful in providing mechanistic support of the role of the homeostatic maternal gut microbiota on offspring immune and neuroimmune development. In a proof-of-concept study, researchers generated germ-free pregnant female mice that were transiently colonized by inoculation with *E. coli* HA 107, a modified strain that is incapable of replicating in vivo, therefore allowing pregnant females to return to germ-free status (Gomez de Aguero et al., 2016). Pups that were born to dams that were transiently colonized displayed increased intestinal group 3 innate lymphoid cells and a distinct intestinal transcriptome including increased expression of genes encoding metabolism of microbial molecules (Gomez de Aguero et al., 2016). Using radiolabeled, 14C-labeled *E. coli* HA107, maternal antibodies and microbial molecules were determined to have crossed the placenta and reached the developing fetus, and were also present in postnatal day 14 offspring small intestine, liver, and spleen. In addition, this microbial exposure during pregnancy protected offspring from bacterial translocation during a challenge with *E. coli* JM83. While this study provides support for the maternal microbiota on offspring immune education, emerging research suggest microbes have far-reaching consequences. Epidemiological studies demonstrate an association between maternal diet and antibiotic exposure to the development of neurodevelopmental disorders in children. To directly investigate the homeostatic effects of the maternal microbiota in neurodevelopment, our group and others have compared conventionally colonized, germ-free, and antibiotic-treated mouse models. Thion and colleagues identified differences in microglial phenotype, abundance, and transcriptome at critical developmental stages that was microbiome-dependent and sex-specific (Thion et al., 2018). Our group showed colonization status of a pregnant mouse can influence the fetal brain transcriptome (Vuong et al., 2020).

Absence of the maternal gut microbiome decreased axons that project from the thalamus to the cortex in the fetal brain (Vuong et al., 2020).

Given that thalamocortical axons convey sensory information, we found that adult offspring from dams with depleted or deficient microbiota showed reduced forepaw and hindpaw tactile sensory behaviors (Vuong et al., 2020). The absence of the maternal microbiota decreased maternal serum and fetal brain levels of microbially-modulated metabolites, thus being the first study to demonstrate microbially-modulated metabolites can be regulated across compartments, particularly in the fetal brain.

Maternal supplementation of key metabolites in antibiotic treated pregnant dams was sufficient to prevent abnormal embryonic axon outgrowth and adult offspring tactile sensory behavioral deficits (Vuong et al., 2020).

The maternal microbiota's ability to break down food that we eat, secrete a wealth of metabolites that have wide-ranging—immune, growth, and neurodevelopmental functions. Some of these metabolites have been reported to pass from mother to child during gestation and breastfeeding. Short chain fatty acids (SCFAs) are products of dietary fiber fermentation by intestinal microbes. SCFAs can modulate the development of regulatory T cells, which teach our immune system to tolerate food and friendly bacteria, and are critical for protection against autoimmune disorders (Nakajima et al., 2017; Thorburn et al., 2015). In a recent study, Kimura and colleagues showed that propionate, a SCFA that is regulated by the maternal microbiota, could transfer across the placenta, act on embryonic G-protein coupled receptors (GPR41 and GPR43) and reduce susceptibility of the offspring to obesity and metabolic syndrome in response to a high-fat diet (Kimura et al., 2020). Together, these preclinical studies provide a framework supporting the impact of the maternal microbiota and microbial metabolites—modulating immune, metabolism, and neurodevelopment of the offspring, possibly even providing protection from immune, metabolic or neurological disorders later in life.

5. Future directions

Overall, the expansion of microbiome research across all fields, particularly the neurosciences has allowed us a better mechanistic understanding of how microbes can modulate brain function and complex behaviors. Until recently, microbe mediated effects on offspring development were thought to be initiated at birth. We discuss two general timeframes of microbial influences: (1) the maternal microbiota during gestation, and (2) the microbiota that is seeded at birth and develops over early-life. Several landmark studies highlighted in this chapter reveal contributions of earlier microbial communities that exist during gestation that modify the intrauterine milieu and influence various aspects of brain function and behaviors. Some proposed modifiers include microbially-modulated metabolites, chromatin remodeling, and signaling through G-protein coupled receptors. More work is needed to identify what aspects of brain development are dependent on the microbiota, and what are the critical windows for intervention. Future research is also needed to elucidate exact mechanisms of how microbially-modulated metabolites act on brain cells, and which brain region and/or cell type. Finally, while the general consensus is that there is no defined healthy or disease microbiota because there are a multitude of factors that determine one's health, it is important to understand how the microbiota may contribute to a healthy or disease phenotype. As such, additional methods that would allow targeted manipulation of the microbiota during development—pregnancy or in early postnatal life are needed. Addressing the fundamental gaps in our knowledge can promote translation of microbiota research toward better treatment and understanding of maternal-fetal health.

References

Aagaard, K., Riehle, K., Ma, J., Segata, N., Mistretta, T. A., Coarfa, C., et al. (2012). A metagenomic approach to characterization of the vaginal microbiome signature in pregnancy. *PLoS One*, 7, e36466.

Abrahamsson, T. R., Wu, R. Y., & Jenmalm, M. C. (2015). Gut microbiota and allergy: The importance of the pregnancy period. *Pediatric Research*, 77, 214–219.

Abu-Raya, B., Michalski, C., Sadarangani, M., & Lavoie, P. M. (2020). Maternal immunological adaptation during Normal pregnancy. *Frontiers in Immunology*, *11*, 575197.

Anahtar, M. N., Gootenberg, D. B., Mitchell, C. M., & Kwon, D. S. (2018). Cervicovaginal microbiota and reproductive health: The virtue of simplicity. *Cell Host & Microbe*, *23*, 159–168.

Backhed, F., Roswall, J., Peng, Y., Feng, Q., Jia, H., Kovatcheva-Datchary, P., et al. (2015). Dynamics and stabilization of the human gut microbiome during the first year of life. *Cell Host & Microbe*, *17*, 690–703.

Barker, D. J. (1995a). The fetal and infant origins of disease. *European Journal of Clinical Investigation*, *25*, 457–463.

Barker, D. J. (1995b). Fetal origins of coronary heart disease. *BMJ*, *311*, 171–174.

Braniste, V., Al-Asmakh, M., Kowal, C., Anuar, F., Abbaspour, A., Toth, M., et al. (2014). The gut microbiota influences blood-brain barrier permeability in mice. *Science Translational Medicine*, *6* (263ra158).

Buffington, S. A., Di Prisco, G. V., Auchtung, T. A., Ajami, N. J., Petrosino, J. F., & Costa-Mattioli, M. (2016). Microbial reconstitution reverses maternal diet-induced social and synaptic deficits in offspring. *Cell*, *165*, 1762–1775.

Calatayud, M., Koren, O., & Collado, M. C. (2019). Maternal microbiome and metabolic health program microbiome development and health of the offspring. *Trends in Endocrinology and Metabolism*, *30*, 735–744.

Callahan, B. J., DiGiulio, D. B., Goltsman, D. S. A., Sun, C. L., Costello, E. K., Jeganathan, P., et al. (2017). Replication and refinement of a vaginal microbial signature of preterm birth in two racially distinct cohorts of US women. *Proceedings of the National Academy of Sciences of the United States of America, 114*, 9966–9971.

Choi, G. B., Yim, Y. S., Wong, H., Kim, S., Kim, H., Kim, S. V., et al. (2016). The maternal interleukin-17a pathway in mice promotes autism-like phenotypes in offspring. *Science, 351*, 933–939.

Clarke, G., Grenham, S., Scully, P., Fitzgerald, P., Moloney, R. D., Shanahan, F., et al. (2013). The microbiome-gut-brain axis during early life regulates the hippocampal serotonergic system in a sex-dependent manner. *Molecular Psychiatry, 18*, 666–673.

Claus, S. P., Guillou, H., & Ellero-Simatos, S. (2016). The gut microbiota: A major player in the toxicity of environmental pollutants? *NPJ Biofilms Microbiomes, 2*, 16003.

Collado, M. C., Isolauri, E., Laitinen, K., & Salminen, S. (2008). Distinct composition of gut microbiota during pregnancy in overweight and normal-weight women. *The American Journal of Clinical Nutrition, 88*, 894–899.

Dejong, K., Olyaei, A., & Lo, J. O. (2019). Alcohol use in pregnancy. *Clinical Obstetrics and Gynecology, 62*, 142–155.

Deoni, S. C., Adams, S. H., Li, X., Badger, T. M., Pivik, R. T., Glasier, C. M., et al. (2019). Cesarean delivery impacts infant brain development. *AJNR. American Journal of Neuroradiology, 40*, 169–177.

Diaz Heijtz, R., Wang, S., Anuar, F., Qian, Y., Bjorkholm, B., Samuelsson, A., et al. (2011). Normal gut microbiota modulates brain development and behavior. *Proceedings of the National Academy of Sciences of the United States of America, 108*, 3047–3052.

DiGiulio, D. B., Callahan, B. J., McMurdie, P. J., Costello, E. K., Lyell, D. J., Robaczewska, A., et al. (2015). Temporal and spatial variation of the human microbiota during pregnancy. *Proceedings of the National Academy of Sciences of the United States of America, 112*, 11060–11065.

Dominguez-Bello, M. G., Costello, E. K., Contreras, M., Magris, M., Hidalgo, G., Fierer, N., et al. (2010). Delivery mode shapes the acquisition and structure of the initial microbiota across multiple body habitats in newborns. *Proceedings of the National Academy of Sciences of the United States of America, 107*, 11971–11975.

Donohoe, D. R., Garge, N., Zhang, X., Sun, W., O'Connell, T. M., Bunger, M. K., et al. (2011). The microbiome and butyrate regulate energy metabolism and autophagy in the mammalian colon. *Cell Metabolism, 13*, 517–526.

Dunkel Schetter, C., & Tanner, L. (2012). Anxiety, depression and stress in pregnancy: Implications for mothers, children, research, and practice. *Current Opinion in Psychiatry, 25*, 141–148.

Elovitz, M. A., Gajer, P., Riis, V., Brown, A. G., Humphrys, M. S., Holm, J. B., et al. (2019). Cervicovaginal microbiota and local immune response modulate the risk of spontaneous preterm delivery. *Nature Communications, 10*, 1305.

Erny, D., Hrabe de Angelis, A. L., Jaitin, D., Wieghofer, P., Staszewski, O., David, E., et al. (2015). Host microbiota constantly control maturation and function of microglia in the CNS. *Nature Neuroscience, 18*, 965–977.

Fettweis, J. M., Brooks, J. P., Serrano, M. G., Sheth, N. U., Girerd, P. H., Edwards, D. J., et al. (2014). Differences in vaginal microbiome in African American women versus women of European ancestry. *Microbiology (Reading), 160*, 2272–2282.

Frost, G., Sleeth, M. L., Sahuri-Arisoylu, M., Lizarbe, B., Cerdan, S., Brody, L., et al. (2014). The short-chain fatty acid acetate reduces appetite via a central homeostatic mechanism. *Nature Communications, 5*, 3611.

Fung, T. C., Olson, C. A., & Hsiao, E. Y. (2017). Interactions between the microbiota, immune and nervous systems in health and disease. *Nature Neuroscience, 20*, 145–155.

Gajer, P., Brotman, R. M., Bai, G., Sakamoto, J., Schutte, U. M., Zhong, X., et al. (2012). Temporal dynamics of the human vaginal microbiota. *Science Translational Medicine, 4*, 132ra152.

Gehrig, J. L., Venkatesh, S., Chang, H. W., Hibberd, M. C., Kung, V. L., Cheng, J., et al. (2019). Effects of microbiota-directed foods in gnotobiotic animals and undernourished children. *Science*, 365.

Georgieff, M. K. (2007). Nutrition and the developing brain: Nutrient priorities and measurement. *The American Journal of Clinical Nutrition*, *85*, 614S–620S.

Gohir, W., Ratcliffe, E. M., & Sloboda, D. M. (2015). Of the bugs that shape us: Maternal obesity, the gut microbiome, and long-term disease risk. *Pediatric Research*, 77, 196–204.

Gomez de Aguero, M., Ganal-Vonarburg, S. C., Fuhrer, T., Rupp, S., Uchimura, Y., Li, H., et al. (2016). The maternal microbiota drives early postnatal innate immune development. *Science*, *351*, 1296–1302.

Gomez-Arango, L. F., Barrett, H. L., McIntyre, H. D., Callaway, L. K., Morrison, M., Dekker Nitert, M., et al. (2016). Connections between the gut microbiome and metabolic hormones in early pregnancy in overweight and obese women. *Diabetes*, *65*, 2214–2223.

Goyal, M. S., Venkatesh, S., Milbrandt, J., Gordon, J. I., & Raichle, M. E. (2015). Feeding the brain and nurturing the mind: Linking nutrition and the gut microbiota to brain development. *Proceedings of the National Academy of Sciences of the United States of America*, *112*, 14105–14112.

Gustafsson, B. E. (1959). Lightweight stainless steel systems for rearing germfree animals. *Annals of the New York Academy of Sciences*, *78*, 17–28.

Gustafsson, B., Kahlson, G., & Rosengren, E. (1957). Biogenesis of histamine studied by its distribution and urinary excretion in germ free reared and not germ free rats fed a histamine free diet. *Acta Physiologica Scandinavica*, *41*, 217–228.

Herbstman, J. B., Tang, D., Zhu, D., Qu, L., Sjodin, A., Li, Z., et al. (2012). Prenatal exposure to polycyclic aromatic hydrocarbons, benzo[a]pyrene-DNA adducts, and genomic DNA methylation in cord blood. *Environmental Health Perspectives*, *120*, 733–738.

Hsiao, E. Y., McBride, S. W., Hsien, S., Sharon, G., Hyde, E. R., McCue, T., et al. (2013). Microbiota modulate behavioral and physiological abnormalities associated with neurodevelopmental disorders. *Cell*, *155*, 1451–1463.

Hu, Y., Jin, P., Peng, J., Zhang, X., Wong, F. S., & Wen, L. (2016). Different immunological responses to early-life antibiotic exposure affecting autoimmune diabetes development in NOD mice. *Journal of Autoimmunity*, 72, 47–56.

Human Microbiome Project Consortium. (2012). Structure, function and diversity of the healthy human microbiome. *Nature*, *486*, 207–214.

Hyman, R. W., Fukushima, M., Jiang, H., Fung, E., Rand, L., Johnson, B., et al. (2014). Diversity of the vaginal microbiome correlates with preterm birth. *Reproductive Sciences*, *21*, 32–40.

Jasarevic, E., Howard, C. D., Misic, A. M., Beiting, D. P., & Bale, T. L. (2017). Stress during pregnancy alters temporal and spatial dynamics of the maternal and offspring microbiome in a sex-specific manner. *Scientific Reports*, 7, 44182.

Jasarevic, E., Howard, C. D., Morrison, K., Misic, A., Weinkopff, T., Scott, P., et al. (2018). The maternal vaginal microbiome partially mediates the effects of prenatal stress on offspring gut and hypothalamus. *Nature Neuroscience*, *21*, 1061–1071.

Jedrychowski, W. A., Perera, F. P., Camann, D., Spengler, J., Butscher, M., Mroz, E., et al. (2015). Prenatal exposure to polycyclic aromatic hydrocarbons and cognitive dysfunction in children. *Environmental Science and Pollution Research International*, *22*, 3631–3639.

Jian, C., Carpen, N., Helve, O., de Vos, W. M., Korpela, K., & Salonen, A. (2021). Early-life gut microbiota and its connection to metabolic health in children: Perspective on ecological drivers and need for quantitative approach. *eBioMedicine*, *69*, 103475.

Kim, S., Kim, H., Yim, Y. S., Ha, S., Atarashi, K., Tan, T. G., et al. (2017). Maternal gut bacteria promote neurodevelopmental abnormalities in mouse offspring. *Nature*, *549*, 528–532.

Kimura, I., Miyamoto, J., Ohue-Kitano, R., Watanabe, K., Yamada, T., Onuki, M., et al. (2020). Maternal gut microbiota in pregnancy influences offspring metabolic phenotype in mice. *Science, 367*.

Koenig, J. E., Spor, A., Scalfone, N., Fricker, A. D., Stombaugh, J., Knight, R., et al. (2011). Succession of microbial consortia in the developing infant gut microbiome. *Proceedings of the National Academy of Sciences of the United States of America, 108*(Suppl 1), 4578–4585.

Koren, O., Goodrich, J. K., Cullender, T. C., Spor, A., Laitinen, K., Backhed, H. K., et al. (2012). Host remodeling of the gut microbiome and metabolic changes during pregnancy. *Cell, 150*, 470–480.

Kourtis, A. P., Read, J. S., & Jamieson, D. J. (2014). Pregnancy and infection. *The New England Journal of Medicine, 370*, 2211–2218.

Kretchmer, N., Beard, J. L., & Carlson, S. (1996). The role of nutrition in the development of normal cognition. *The American Journal of Clinical Nutrition, 63*, 997S–1001S.

Kumar, P., & Magon, N. (2012). Hormones in pregnancy. *Nigerian Medical Journal, 53*, 179–183.

Lain, K. Y., & Catalano, P. M. (2007). Metabolic changes in pregnancy. *Clinical Obstetrics and Gynecology, 50*, 938–948.

Langlois, P. H., Hoyt, A. T., Desrosiers, T. A., Lupo, P. J., Lawson, C. C., Waters, M. A., et al. (2014). Maternal occupational exposure to polycyclic aromatic hydrocarbons and small for gestational age offspring. *Occupational and Environmental Medicine, 71*, 529–535.

LeBlanc, J. G., Milani, C., de Giori, G. S., Sesma, F., van Sinderen, D., & Ventura, M. (2013). Bacteria as vitamin suppliers to their host: A gut microbiota perspective. *Current Opinion in Biotechnology, 24*, 160–168.

Lou, M., Cao, A., Jin, C., Mi, K., Xiong, X., Zeng, Z., et al. (2021). Deviated and early unsustainable stunted development of gut microbiota in children with autism spectrum disorder. *Gut*.

Lovasi, G. S., Eldred-Skemp, N., Quinn, J. W., Chang, H. W., Rauh, V. A., Rundle, A., et al. (2014). Neighborhood social context and individual polycyclic aromatic hydrocarbon exposures associated with child cognitive test scores. *Journal of Child and Family Studies, 23*, 785–799.

Ma, J., Prince, A. L., Bader, D., Hu, M., Ganu, R., Baquero, K., et al. (2014). High-fat maternal diet during pregnancy persistently alters the offspring microbiome in a primate model. *Nature Communications, 5*, 3889.

Mackie, R. I., Sghir, A., & Gaskins, H. R. (1999). Developmental microbial ecology of the neonatal gastrointestinal tract. *The American Journal of Clinical Nutrition, 69*, 1035S–1045S.

Maher, S. E., O'Brien, E. C., Moore, R. L., Byrne, D. F., Geraghty, A. A., Saldova, R., et al. (2020). The association between the maternal diet and the maternal and infant gut microbiome: A systematic review. *The British Journal of Nutrition*, 1–29.

Marild, K., Ludvigsson, J., Sanz, Y., & Ludvigsson, J. F. (2014). Antibiotic exposure in pregnancy and risk of coeliac disease in offspring: A cohort study. *BMC Gastroenterology, 14*, 75.

Matcovitch-Natan, O., Winter, D. R., Giladi, A., Vargas Aguilar, S., Spinrad, A., Sarrazin, S., et al. (2016). Microglia development follows a stepwise program to regulate brain homeostasis. *Science, 353*, aad8670.

McDonald, B., & McCoy, K. D. (2019). Maternal microbiota in pregnancy and early life. *Science, 365*, 984–985.

Mesa, M. D., Loureiro, B., Iglesia, I., Fernandez Gonzalez, S., Llurba Olive, E., Garcia Algar, O., et al. (2020). The evolving microbiome from pregnancy to early infancy: A comprehensive review. *Nutrients, 12*.

Miller, R. L., Yan, Z., Maher, C., Zhang, H., Gudsnuk, K., McDonald, J., et al. (2016). Impact of prenatal polycyclic aromatic hydrocarbon exposure on behavior, cortical gene expression and DNA methylation of the Bdnf gene. *Neuroepigenetics, 5*, 11–18.

Nakajima, A., Kaga, N., Nakanishi, Y., Ohno, H., Miyamoto, J., Kimura, I., et al. (2017). Maternal high fiber diet during pregnancy and lactation influences regulatory T cell differentiation in offspring in mice. *Journal of Immunology*, *199*, 3516–3524.
Neufeld, K. A., Kang, N., Bienenstock, J., & Foster, J. A. (2011a). Effects of intestinal microbiota on anxiety-like behavior. *Communicative & Integrative Biology*, *4*, 492–494.
Neufeld, K. M., Kang, N., Bienenstock, J., & Foster, J. A. (2011b). Reduced anxiety-like behavior and central neurochemical change in germ-free mice. *Neurogastroenterology and Motility*, *23*(255–264), e119.
Neuman, H., & Koren, O. (2017). The pregnancy microbiome. *Nestle Nutrition Institute Workshop Series*, *88*, 1–9.
Nugent, B. M., & Bale, T. L. (2015). The omniscient placenta: Metabolic and epigenetic regulation of fetal programming. *Frontiers in Neuroendocrinology*, *39*, 28–37.
Nuriel-Ohayon, M., Neuman, H., & Koren, O. (2016). Microbial changes during pregnancy, birth, and infancy. *Frontiers in Microbiology*, *7*, 1031.
Nyangahu, D. D., & Jaspan, H. B. (2019). Influence of maternal microbiota during pregnancy on infant immunity. *Clinical and Experimental Immunology*, *198*, 47–56.
Olivier-Van Stichelen, S., Rother, K. I., & Hanover, J. A. (2019). Maternal exposure to non-nutritive sweeteners impacts Progeny's metabolism and microbiome. *Frontiers in Microbiology*, *10*, 1360.
Ortqvist, A. K., Lundholm, C., Kieler, H., Ludvigsson, J. F., Fall, T., Ye, W., et al. (2014). Antibiotics in fetal and early life and subsequent childhood asthma: Nationwide population based study with sibling analysis. *BMJ*, *349*, g6979.
Paul, H. A., Bomhof, M. R., Vogel, H. J., & Reimer, R. A. (2016). Diet-induced changes in maternal gut microbiota and metabolomic profiles influence programming of offspring obesity risk in rats. *Scientific Reports*, *6*, 20683.
Polidano, C., Zhu, A., & Bornstein, J. C. (2017). The relation between cesarean birth and child cognitive development. *Scientific Reports*, *7*, 11483.
Poutahidis, T., Kearney, S. M., Levkovich, T., Qi, P., Varian, B. J., Lakritz, J. R., et al. (2013). Microbial symbionts accelerate wound healing via the neuropeptide hormone oxytocin. *PLoS One*, *8*, e78898.
Pronovost, G. N., & Hsiao, E. Y. (2019). Perinatal interactions between the microbiome, immunity, and neurodevelopment. *Immunity*, *50*, 18–36.
Rautava, S., Collado, M. C., Salminen, S., & Isolauri, E. (2012). Probiotics modulate host-microbe interaction in the placenta and fetal gut: A randomized, double-blind, placebo-controlled trial. *Neonatology*, *102*, 178–184.
Ravel, J., Gajer, P., Abdo, Z., Schneider, G. M., Koenig, S. S., McCulle, S. L., et al. (2011). Vaginal microbiome of reproductive-age women. *Proceedings of the National Academy of Sciences of the United States of America*, *108*(Suppl 1), 4680–4687.
Ridaura, V. K., Faith, J. J., Rey, F. E., Cheng, J., Duncan, A. E., Kau, A. L., et al. (2013). Gut microbiota from twins discordant for obesity modulate metabolism in mice. *Science*, *341*, 1241214.
Sampson, T. R., Debelius, J. W., Thron, T., Janssen, S., Shastri, G. G., Ilhan, Z. E., et al. (2016). Gut microbiota regulate motor deficits and neuroinflammation in a model of Parkinson's disease. *Cell*, *167*(1469–1480), e1412.
Sharon, G., Cruz, N. J., Kang, D. W., Gandal, M. J., Wang, B., Kim, Y. M., et al. (2019). Human gut microbiota from autism Spectrum disorder promote behavioral symptoms in mice. *Cell*, *177*(1600–1618), e1617.
Shin Yim, Y., Park, A., Berrios, J., Lafourcade, M., Pascual, L. M., Soares, N., et al. (2017). Reversing behavioural abnormalities in mice exposed to maternal inflammation. *Nature*, *549*, 482–487.

Smith, S. E., Li, J., Garbett, K., Mirnics, K., & Patterson, P. H. (2007). Maternal immune activation alters fetal brain development through interleukin-6. *The Journal of Neuroscience*, *27*, 10695–10702.

Stensballe, L. G., Simonsen, J., Jensen, S. M., Bonnelykke, K., & Bisgaard, H. (2013). Use of antibiotics during pregnancy increases the risk of asthma in early childhood. *The Journal of Pediatrics*, *162*, 832–838 (e833).

Subramanian, S., Blanton, L. V., Frese, S. A., Charbonneau, M., Mills, D. A., & Gordon, J. I. (2015). Cultivating healthy growth and nutrition through the gut microbiota. *Cell*, *161*, 36–48.

Subramanian, S., Huq, S., Yatsunenko, T., Haque, R., Mahfuz, M., Alam, M. A., et al. (2014). Persistent gut microbiota immaturity in malnourished Bangladeshi children. *Nature*, *510*, 417–421.

Sudo, N., Chida, Y., Aiba, Y., Sonoda, J., Oyama, N., Yu, X. N., et al. (2004). Postnatal microbial colonization programs the hypothalamic-pituitary-adrenal system for stress response in mice. *The Journal of Physiology*, *558*, 263–275.

Thion, M. S., Low, D., Silvin, A., Chen, J., Grisel, P., Schulte-Schrepping, J., et al. (2018). Microbiome influences prenatal and adult microglia in a sex-specific manner. *Cell*, *172*, 500–516 (e516).

Thorburn, A. N., McKenzie, C. I., Shen, S., Stanley, D., Macia, L., Mason, L. J., et al. (2015). Evidence that asthma is a developmental origin disease influenced by maternal diet and bacterial metabolites. *Nature Communications*, *6*, 7320.

Tochitani, S., Ikeno, T., Ito, T., Sakurai, A., Yamauchi, T., & Matsuzaki, H. (2016). Administration of non-absorbable antibiotics to pregnant mice to perturb the maternal gut microbiota is associated with alterations in offspring behavior. *PLoS One*, *11*, e0138293.

Vidal, A. C., Murphy, S. K., Murtha, A. P., Schildkraut, J. M., Soubry, A., Huang, Z., et al. (2013). Associations between antibiotic exposure during pregnancy, birth weight and aberrant methylation at imprinted genes among offspring. *International Journal of Obesity*, *37*, 907–913.

Vuong, H. E., Pronovost, G. N., Williams, D. W., Coley, E. J. L., Siegler, E. L., Qiu, A., et al. (2020). The maternal microbiome modulates fetal neurodevelopment in mice. *Nature*, *586*, 281–286.

Yang, I., Corwin, E. J., Brennan, P. A., Jordan, S., Murphy, J. R., & Dunlop, A. (2016). The infant microbiome: Implications for infant health and neurocognitive development. *Nursing Research*, *65*, 76–88.

Yatsunenko, T., Rey, F. E., Manary, M. J., Trehan, I., Dominguez-Bello, M. G., Contreras, M., et al. (2012). Human gut microbiome viewed across age and geography. *Nature*, *486*, 222–227.

Zhang, T., Sidorchuk, A., Sevilla-Cermeno, L., Vilaplana-Perez, A., Chang, Z., Larsson, H., et al. (2019). Association of Cesarean Delivery with Risk of neurodevelopmental and psychiatric disorders in the offspring: A systematic review and meta-analysis. *JAMA Network Open*, *2*, e1910236.

CHAPTER TWO

Microbiome influences on neuro-immune interactions in neurodegenerative disease

Kelly B. Menees[a], Brittney A. Otero[a], and Malú Gámez Tansey[a,b,*]
[a]Department of Neuroscience, Center for Translational Research in Neurodegenerative Disease, College of Medicine, University of Florida, Gainesville, FL, United States
[b]Norman Fixel Institute for Neurological Diseases, University of Florida Health, Gainesville, FL, United States
*Corresponding author: e-mail address: mgtansey@ufl.edu

Contents

Abstract

Mounting evidence points to a role for the gut microbiome in a wide range of central nervous system diseases and disorders including depression, multiple sclerosis, Alzheimer's disease, Parkinson's disease, and autism spectrum disorder. Moreover, immune system involvement has also been implicated in these diseases, specifically with inflammation being central to their pathogenesis. In addition to the reported changes in gut microbiome composition and altered immune states in many

International Review of Neurobiology, Volume 167
ISSN 0074-7742
https://doi.org/10.1016/bs.irn.2022.07.006

neurological diseases, how the microbiome and the immune system interact to influence disease onset and progression has recently garnered much attention. This chapter provides a review of the literature related to gut microbiome influences on neuro-immune interactions with a particular focus on neurological diseases. Gut microbiome-derived mediators, including short-chain fatty acids and other metabolites, lipopolysaccharide, and neurotransmitters, and their impact on neuro-immune interactions as well as routes by which these interactions may occur are also discussed.

1. Introduction

It is becoming increasingly evident that many neurological disorders and diseases, including autism spectrum disorder (ASD), Parkinson's disease (PD), Alzheimer's disease (AD), and multiple sclerosis (MS) are multifactorial and are not strictly disorders of the central nervous system (CNS). For example, gastrointestinal symptoms/dysfunction and gut microbiome dysbiosis have been implicated in multiple neurological disorders (Cryan et al., 2019) and specifically in autism (Vellingiri, Aishwarya, Benita Jancy, & Sriram Abhishek, 2022), PD (Dong, Sun, He, & Cheng, 2022; Wallen et al., 2020, 2022), AD (Giridharan, Barichello De Quevedo, & Petronilho, 2022), and MS (Noto & Miyake, 2022). The number of microorganisms estimated to inhabit the human gastrointestinal tract is upwards of 10^{14} (Thursby & Juge, 2017). Technological advances in the past decade have made available a number of tools to investigators to interrogate the complexity and function of gut microbiota, including biomarker profiling, metagenomics, metabolomics, large-scale transcriptomics and proteomics (Cryan et al., 2019). The human gut microbiome is primarily composed of bacteria from the phyla *Firmicutes*, *Bacteroidetes*, and *Actinobacteria* (Tap et al., 2009). Under homeostatic conditions, the human gut microbiome is crucial for metabolic function and development of the immune, endocrine, and nervous systems (Foster, 2016). Germ-free mice display a myriad of abnormalities in neurogenesis, myelination, microglia function, neurotransmission and synaptic plasticity, and blood-brain barrier permeability (Cryan et al., 2019).

In humans, approximately 70–80% of immune cells are found in the gut (Kagnoff, 1987) and the gut is directly connected to CNS via the vagal nerve (Fülling, Dinan, & Cryan, 2019), thus the gut microbiome can communicate directly with the CNS. The microbiome-gut-brain axis is a bidirectional pathway connecting the nervous system, immune system, endocrine system, and the gut microbiome (Cryan & Dinan, 2012). Gut microbiome dysbiosis has been linked to several neurological diseases in pre-clinical models of disease. As an example, stroke has been shown to induce gut microbiota dysbiosis in a mouse middle cerebral artery (MCA) occlusion model (Singh et al., 2016). Specifically, species diversity was reduced following stroke and these alterations were particularly noted in the three most abundant phyla *Firmicutes*, *Bacteroidetes*, and *Actinobacteria* (Singh et al., 2016). Thus, it is evident that gut microbiota influences the CNS and vice versa.

In this chapter we discuss the influence of the microbiome on the immune system and the nervous system, particularly in the context of neurological diseases. Gut microbiome influences on central and peripheral immune cells, microbiota-derived neuro-immune mediators, and routes of communication are discussed. We also discuss in broad strokes the likelihood that manipulating the gut-brain axis and gut microbiota could represent opportunities for therapeutic intervention and the challenges the field may face to get to that point.

2. Microbiome influences on microglia

Microglia are the resident innate immune cells of the central nervous system (CNS) and have a wide range of functions throughout the lifespan. Microglia are distinct from CNS resident and peripheral macrophages and are yolk-sac derived. Like other myeloid cells, microglia are phagocytic cells, cytokine producers, antigen presenters, and participate in the inflammatory response (Fung, Olson, & Hsiao, 2017; Nayak, Roth, & McGavern, 2014). Microglia serve a plethora of roles in development, homeostasis, and disease. During development microglia phagocytose excess neurons and secrete neurotrophic factors, such as nerve growth factor and brain derived neurotrophic factor, to support neurogenesis (Abdel-Haq, Schlachetzki, Glass, & Mazmanian, 2018).

In addition to serving important functions in development and homeostasis, microglia have been implicated in multiple CNS diseases. For example, microglial activation is observed in PD (Ferreira & Romero-Ramos, 2018). Recent positron emission tomography (PET) imaging studies have illustrated microglial activation in the brains of PD patients (Ferreira & Romero-Ramos, 2018). Furthermore, microglia can phagocytose and degrade α-synuclein aggregates, the protein implicated in PD. However, this leads to their activation resulting in production of pro-inflammatory cytokines and reactive oxygen species (ROS). Microglia activation is also implicated in AD (Hansen, Hanson, & Sheng, 2018; Leng & Edison, 2021; Rajesh & Kanneganti, 2022).

Recent work has demonstrated that microglia are influenced by metabolites and other factors produced by the gut microbiota. It has become apparent that the initial colonization of the microbiota at birth influences microglia development and function. Germ free (GF) mice were demonstrated to have distinct microglia mRNA profiles compared to specific pathogen free (SPF) mice (Erny et al., 2015). Furthermore, GF mice have increased CSF1R+, F4/80+, and CD31+ microglia indicating a more immature phenotype (Erny et al., 2015). Interestingly, when GF mice were colonized with *Bacteroides distasonis*, *Lactobacillus salivarius*, and *Clostridium cluster XIV* str. ASF356, abnormalities remained, suggesting the necessity of a complex (rather than simplified) microbiome. The gut microbiota can also significantly influence microglial responses. For example, in GF mice, microglia displayed reduced cytokine production following systemic or intracerebral injection of LPS compared to specific pathogen free mice (Erny et al., 2015). Microbiome effects on microglia have also been shown to be sex-specific. Following antibiotic treatment, male APPPS1-21 transgenic mice have reduced area of microglial cell bodies compared to vehicle treated controls while antibiotic treated female mice had significantly increased microglial cell body areas compared to vehicle treated female mice (Dodiya et al., 2019).

3. Microbiome influences on astrocytes

Astrocytes are the most abundant glial cells in the CNS and serve important roles in maintaining CNS homeostasis. Astrocytes have a wide range of functions including maintaining blood flow and the blood-brain barrier (Abbott, Rönnbäck, & Hansson, 2006), modulating synaptic plasticity, providing physical and metabolic support to neurons, and mediating axon pathfinding (Jha et al., 2018; Sofroniew & Vinters, 2010). While not typically considered part of the CNS immune cell repertoire, astrocytes have been shown to mediate immune responses in the brain (Colombo &

Farina, 2016), express pattern recognition receptors for detecting microbe associated molecular patterns, and produce cytokines and present antigens via the Major Histocompatibility Complex II (MHC II), also known as Human Leukocyte Antigen (HLA) proteins (Dong & Benveniste, 2001). Importantly, astrocytes have been implicated in the gut-brain axis in a variety of neurological diseases include major depressive disorder (MDD) (Tomova et al., 2020), autism spectrum disorder, AD, and PD. Astrocytes contribute to neuroinflammation through a process termed astrogliosis. Astrogliosis typically results in upregulation of glial fibrillary acidic protein (GFAP) and alterations in astrocyte morphology (Sofroniew & Vinters, 2010).

The gut microbiome influences astrocyte numbers and functions in a variety of pathological states. For example, microbiome dysbiosis has been associated with an increase in the number of S100β positive astrocytes in the brains of children with autism spectrum disorder (Tomova et al., 2020). However, astrocyte numbers were shown not to differ in GF and SPF mice (Erny et al., 2015). This discrepancy indicates that the gut microbiome alone may not be the sole factor influencing astrocyte numbers, but rather, microbiome dysbiosis in combination with other host factors may lead to changes in numbers of astrocytes. Astrocyte density in the hypothalamus is regulated by diet. Female mice fed a high-fat high-sugar diet displayed decreased astrocytic density in the hypothalamus and decreased *Bacteroidetes* and *Firmicutes* abundance in the gut (Daly et al., 2022).

Additionally, *Porphyromonas gingivalis* mediates activation of astrocytes via TLR4 thereby contributing to a proinflammatory response by astrocytes and inflammatory lesions (Boillot et al., 2015; Lalla & Papapanou, 2011; Zhang et al., 2018).

Microbial metabolites can interact with and modulate astrocyte function via aryl hydrocarbon receptors (AHR) (Rothhammer et al., 2016). The metabolism of dietary tryptophan by microbiota produces AHR ligands such as indole-3-aldehyde and indole-3-propionic acid. When human astrocytes were incubated with indoxyl-3-sulfate, a tryptophan metabolite, decreased expression of *NOS2*, *TNF*, *IL6*, and *CCL2*, genes encoding inflammatory mediators was observed (Rothhammer et al., 2016). *Lactobacillus reuteri* is known to produce indole-3-aldehyde (Rothhammer et al., 2016; Zelante et al., 2013), making it a prime candidate for further investigation into how specific microbes mediate astrocytic function.

4. Microbiome influences on other immune cells

Once thought to be a strictly peripheral immune cell type, T cells reside in the CNS and serve important functions in homeostasis and brain disease. The gut microbiome is crucial for normal T cell development. For example, GF mice completely lack Th17 cells and display altered Th1:Th2 ratios (Ivanov et al., 2009). *Bacteroides fragilis* promotes Th1 differentiation (Mazmanian, Liu, Tzianabos, & Kasper, 2005). Antibiotic treatment is an effective way to wipe out the gut microbiota to enable interrogation of the processes in which the latter is a critical mediator (Cryan et al., 2019). Recent work demonstrated that the CNS immune cell repertoire, particularly in aged mice, was affected by gut microbiome dysbiosis (Golomb et al., 2020). Specifically, young and aged mice received an antibiotic cocktail (metronidazole, vancomycin, neomycin, and ampicillin) for three weeks to deplete gut microbiota. This antibiotic paradigm shifted the predominant phyla from *Bacteroidetes* and *Firmicutes* to *Proteobacteria*. Interestingly, in young mice antibiotic treatment resulted in a slight increase in B cell frequency, which was not observed in aged mice. In the aged mice, the microbiota dysbiosis paradigm resulted in reduced ILCs and $Ly6C^{lo}$ monocytes and increased CD8+ T cells and $Ly6C^{hi}$ monocytes. These results further suggest the critical role of the gut microbiome in regulating the immune system and immune cell repertoire and highlights the importance of considering age as a critical factor in the outcomes of gut microbiome dysbiosis and its downstream effects. Importantly, T cells have been implicated in a host of neurological diseases including AD, PD, and MS (Dressman & Elyaman, 2021; Schetters, Gomez-Nicola, Garcia-Vallejo, & Van Kooyk, 2018). Thus, further understanding how the gut microbiome may impact the immune cell repertoire may shed light on potential therapeutic targets for these diseases.

Recent work in an experimental autoimmune encephalomyelitis (EAE) model of MS has demonstrated a role for gut-derived humoral immunity in mediating neuroinflammation (Rojas et al., 2019). IgA+ plasma cells that originated in the gut were shown to be present in the CNS of mice with EAE (Rojas et al., 2019). Furthermore, plasma cell deficient mice displayed exacerbated EAE. GF mice showed nearly a complete lack of IgA+ plasma cells in the meninges (Fitzpatrick et al., 2020). Re-colonization of GF mice with *C. rodentium* was sufficient to restore meningeal IgA+ plasma cell numbers (Fitzpatrick et al., 2020). Interestingly, when mono-colonization was limited to the skin, no difference in IgA+ plasma cells in the meninges was observed (Fitzpatrick et al., 2020), further indicating the role of the gut microbiome in the formation of this cell population.

Natural killer (NK) cells are cytotoxic lymphocytes with a wide range of functions including destroying cells that lack major histocompatibility complex (MHC) I (Camous, Pera, Solana, & Larbi, 2012) and resolving inflammation (Thorén et al., 2012; Waggoner & Kumar, 2012). NK cells are major producers of IFN-γ (Schoenborn & Wilson, 2007). The gut microbiome induces NK cell IFN-γ production (Ganal et al., 2012; Thiemann et al., 2017). A recent study demonstrated that IFN-γ producing NK cells present in the meninges, that were licensed by the microbiome, promote an anti-inflammatory astrocyte subset (Sanmarco et al., 2021). In an EAE model, this anti-inflammatory astrocyte subset limited CNS inflammation by driving T cell apoptosis (Sanmarco et al., 2021).

5. Gut microbiome studies in human populations

Interest in the gut microbiome and its connection to neurological disorders has prompted a wealth of cross-sectional studies in human populations in the last decade (see Table 1). While fewer studies have been conducted in subjects with AD and other dementias, substantial evidence is emerging to support the role of gut microbiome dysfunction in AD (see Table 2).

Cattaneo et al. (2017) reported that amyloid PET-positive patients harbored a rich source of pro-inflammatory bacteria and lower anti-inflammatory bacteria (e.g., *Eubacterium rectale, Eubacterium hallii, Faecalibacterium prausnitzii, Bacteroides fragilis*) in the gut relative to healthy controls. In addition, several pathogens including bacteria (spirochaete *Borrelia burgdorferi*), viruses (*herpes simplex virus type 1*), fungi, and protist parasites, suggesting that the Aβ peptide may be part of a generalized innate immune response that is not specific to a particular microorganism. Some have proposed that this defense mechanism of Aβ against invading microbes in the brain may turn deleterious in the aged brain, leading to inflammation and neurodegeneration when dysregulated (Kumar et al., 2016; Stilling et al., 2016).

Table 1 Fecal gut microbiome studies in Parkinson's disease (PD) patients and healthy controls (HC).

Reference	Scheperjans et al. (2015)	Hasegawa et al. (2015)	Keshavarzian et al. (2015)	Unger et al. (2016)	Bedarf et al. (2017)	Hill-Burns et al. (2017)	Hopfner et al. (2017)	Li et al. (2017)
n (PD/HC)	72/72	45/35	34/31	34/34	31/28	197/130	29/29	24/14
Mean age	65.3/64.5	68.9/68.4[a]	61.5/45.1	67.7/64.6	64.8/65.6	68.4/70.3	69.2/69.4	73.8/74.6
Sex % female	49/50	60/42	32/45	29/47	0/0	33/39	21/55	33/43
Country	Finland	Japan	United States	Germany	Germany	United States	Germany	China
Sample transport	DNA stabilizer (PSP®)	RNA stabilizer (RNAlater) at 4°C	BD Gaspak (no temp reported)	No stabilizer (no temp reported)	Transport not reported	Ambient temp (no stabilizer)	Ambient temp (no stabilizer)	Transport not reported
Methodology								
DNA extraction	PSP Spin Stool DNA Plus Kit	Custom protocol	FastDNA Spin Kit for Soil	QIAsym-phony® DSP Virus/ Pathogen Mini-Kit	NA	Earth microbiome project protocol	PowerSoil Kit	TIANamp stool DNA kit
Sequencing target	16S (V3-V4)	qPCR of selected taxa	16S (V4)	qPCR of selected taxa	Shotgun meta-genomics	16S (V4)	16S (V1-V2)	16S (V3-V5)

[a]Mean age of entire cohort, including participants that did not provide fecal samples. n, number of participants; NA, not available; qPCR, quantitative polymerase chain reaction; 16S, 16S rRNA gene sequencing.
[b]Median age.
[c]Mean age at baseline.
Table adapted from Boertien, J. M., Pereira, P. A. B., Aho, V. T. E., & Scheperjans, F. (2019). Increasing comparability and utility of gut microbiome studies in parkinson's disease: A systematic review. *Journal of Parkinson's Disease, 9*(s2): S297–S312.

Petrov et al. (2017)	Heintz-Buschart et al. (2018)	Lin et al. (2018)	Qian et al. (2018)	Aho et al. (2019)	Barichella et al. (2019)	Pietrucci et al. (2019)	Li et al. (2019)	Wallen et al. (2022)
89/66	26/38	75/45	45/45	64/64	193/113	80/72	10/10	490/234
67/63[b]	66.0/69.8	60.5/63.2	68.1/67.9	65.2/64.5[c]	67.6/65.9	66.2/62.6	79.5/76.5	68.7/65.8
N/A	34/27	35/49	51/49	48/50	40/58	35/50	30/50	37/70
Russia	Germany	China	China	Finland	Italy	Italy	China	United States
Transport not reported	Immediate freezing	Ambient temp (no stabilizer)	Transport on ice (no stabilizer)	DNA stabilizer (PSP®)	Immediate freezing	DNA stabilizer (PSP®)	Immediate freezing	Ambient temp
Custom protocol	Qiagen AllPrep	Huirui DNA kit	QIAamp DNA Stool Mini Kit	PSP Spin Stool DNA Plus Kit	QIAamp DNA Stool Mini Kit	PSP Spin Stool DNA Plus Kit	PSP Spin Stool DNA Plus Kit	Quiagen DNeasy PowerSoil Pro high throughput kit
16S (V3-V4)	16S (V4)	16S (V4)	16S (V3-V4)	16S (V3-V4)	16S (V3-V4)	16S (V3-V4)	16S (V3-V4)	Deep Shotgun meta-genomics

Table 2 Evidence supporting the role of microbiota in Alzheimer's disease.

Reference	Sample	Microbial profile	CNS pathology and phenotypes
Hammond et al., (2010)	5 AD; 5 controls (post-mortem brain samples)	*Chlamydia pneumoniae* infection	Promotes amyloid deposits, and neurofibrillary tangles aggravating the AD pathogenesis
Roubaud-Baudron, Krolak-Salmon, Quadrio, Mégraud and Salles (2012)	53 AD cases	Chronic *H. pylori* infection	Cerebrovascular lesions, neuroinflammation leading to cognitive impairment
Sparks Stein et al. (2012)	81 mild AD; 77 controls	*Fusobacterium nucleatum* and *Prevotella intermedia*	Cognitive Impairment
Poole, Singhrao, Kesavalu, Curtis, and Crean (2013)	10 AD; 10 controls	Periodontopathic *Porphyromonas gingivalis*	LPS in AD brain
Noble et al. (2014)	110 AD; 109 controls	High antibody titer for *A. naeslundii* In AD cases	Risk markers for AD, MCI
Vogt et al. (2017)	Fecal samples from 25 AD; 25 controls	A decrease in *Firmicutes*, increased *Bacteroidetes*, and decreased *Bifdobacterium*	CSF markers like Aβ42/Aβ40, phosphorylated tau (p-tau), and chitinase-3-like protein 1 (YKL-40) are elevated, revealing amyloid deposition in the brain
Zhao, Jaber, and Lukiw (2017)	10 AD; 8 controls	*Bacteroides fragilis* and *Escherichia coli*	LPS levels varied from a 2-fold increase in the neocortex to a 3- fold increase in the hippocampus in AD over age-matched controls
Haran et al. (2019)	51 No dementia; 24 AD; 33 other dementia types	AD pathogenic bacteria *O. splanchnicus*, *E. lenta*, and *K. neumoniae*, with decreased proportions of butyrate producing *B. hungatei*, *Blautia hansenii*, *E. eligens*, *R. hominis*, *Ruminococcus bicirculans*, and *F. prausnitzi*	AD pathological features

AD, Alzheimer's disease; CSF, cerebrospinal fluid; LPS, lipopolysaccharide; MCI, mild cognitive impairment.

Table adapted from Kesika, P., Suganthy, N., Sivamaruthi, B. S., & Chaiyasut, C. (2021). Role of gut-brain axis, gut microbial composition, and probiotic intervention in Alzheimer's disease. *Life Sciences, 264*, 118627.

6. Gut microbiome studies in model organisms of aging or neurological disease

Among the earliest and most elegant studies in models of neurological disease is the work from the Mazmanian group. Hsiao et al. demonstrated the importance of gut microbiota in a maternal immune activation (MIA) model of ASD (Hsiao et al., 2013). Sampson et al., interrogated the role of gut microbiota in contributing to the pathological effects of human alpha-synuclein in a transgenic mouse model of synucleinopathy; reporting that GF transgenic mice developed much less pathology compared to conventional mice and fecal microbiota transplants (FMTs) with gut microbes derived from PD patients accelerated alpha-synuclein pathology in transgenic mice relative to gut microbes derived from healthy control non-PD subjects (Sampson et al., 2016). In preclinical models of AD, several studies have demonstrated a critical role for gut microbiota (Bäuerl, Collado, Diaz Cuevas, Viña, & Pérez Martínez, 2018; Harach et al., 2017; Minter et al., 2016, 2017; Poole et al., 2015; Shen, Liu, & Ji, 2017). But few studies have gone far enough to identify bacterial taxa or microbial-derived molecules to which direct effects on the brain can be attributed. One such seminal and elegant study by Olson et al., identified the "good" gut microbes and the exact metabolites responsible for mitigation of epileptic seizures in mice mediated by a ketogenic diet (Olson et al., 2018).

In addition to the effects of manipulation of the gut microbiota on neurodegenerative pathology in pre-clinical models of disease, increased susceptibility to infection and decreased life-span has been reported in mutant human APP (amyloid precursor protein) transgenic mice infected with *Salmonella enterica* and transgenic *Caenorhabditis elegans* (human Aβ42 peptide) infected with *Candida albicans*. Some propose this might be due to the antimicrobial activity of Aβ peptide via which the heparin-binding motif of Aβ oligomers enable binding to the glycosyl group of carbohydrate moiety in the microbial cell wall preventing its adhesion to the host cell and inducing microbial agglutination (Kumar et al., 2016). In support of this hypothesized model for anti-microbial activity of Aβ peptides, proponents point out that in mouse models of AD brain infection often promotes formation of Aβ plaques much earlier than the normal timeline for development of plaques in uninfected mice. Following six weeks of oral infection of *P. gingivalis* in aged BALB/c mice, bacteria invaded the brain and brain levels of $A\beta_{1-42}$ were significantly increased (Dominy et al., 2019). Moreover, incubation of *P. gingivalis* with $A\beta_{1-42}$ resulted in increased bacterium death compared to controls, indicating $A\beta_{1-42}$ has antibacterial effects against *P. gingivalis* (Dominy et al., 2019). In recent work, the effects of a high-fat high-fructose diet on Aβ plaque deposition, immune profiles, and gut microbiome were examined in 5xFAD female mice. While the obesogenic diet had modest effect on plaque deposition, it had profound effects on immune cell profiles in the blood and brain, synaptic function, and gut microbiota (Macpherson et al., Under Revision). Interestingly, inhibition of soluble TNF half-way through the diet intervention normalized most of these outcomes, consistent with the role of pro-inflammatory cytokines like TNF and IL-6 in gut barrier permeability (Cryan et al., 2019) where blockade of IL-6 in newborn pups mitigated sociability deficits (Hsiao et al., 2013; Hsiao, McBride, Chow, Mazmanian, & Patterson, 2012) and blockade of soluble TNF mitigates inflammation and improves epithelial barrier functions, and colonic motility (Chandrasekharan et al., 2013).

7. Blood brain barrier (BBB)

The blood brain barrier (BBB) is a highly specialized physical barrier separating the brain parenchyma from the cerebral vasculature (Pandit, Chen, & Götz, 2020; Rice, Surian, & Chen, 2022). The BBB is essential for maintaining brain homeostasis. There are several cellular components that make up the BBB including endothelial cells, pericytes, a basement membrane, and astrocyte end feet (Rice et al., 2022). Disruption of the tight junctions can result in a leaky BBB, causing the brain to be exposed to a variety of factors in the circulation that might damage it or at the very least disrupt homeostasis. BBB permeability and dysfunction have been implicated in a variety of neurological diseases including MS, PD, and AD (Balasa, Barcutean, Mosora, & Manu, 2021; Gray & Woulfe, 2015; Sweeney, Sagare, & Zlokovic, 2018).

Metabolites produced by the gut microbiome have been shown to affect BBB integrity. For example, *Clostridium tyrobutyricum* produces high levels of butyrate, a short chain fatty acid (SCFA), and improved BBB integrity in GF mice (Braniste et al., 2014). FFAR3, a receptor for the SCFA propionate, is expressed on human brain endothelial cells (Hoyles et al., 2018). Development of the BBB is impacted by maternal gut microbiota. Expression of the tight junction protein occludin was significantly lower in brains of fetal mice from GF dams compared to those from pathogen-free dams (Braniste et al., 2014), indicating that maternal gut microbiome influences development of offspring BBB. Moreover, increased BBB permeability is observed in adult GF mice compared to pathogen free mice (Braniste et al., 2014). When mice were treated with antibiotics, inducing gut microbiome dysbiosis, decreased claudin 5 and occludin mRNA levels, both tight junction proteins, were observed in the hippocampus (Fröhlich et al., 2016) further indicating that the gut microbiome is important in maintaining BBB integrity. Brain endothelial cells express toll like receptors (TLRs) which enables them to respond to bacterial derived lipopolysaccharide (LPS) (Logsdon, Erickson, Rhea, Salameh, & Banks, 2018). Microbiome metabolites, including the SCFAs butyrate and propionate, help to promote BBB integrity (Braniste et al., 2014; Hoyles et al., 2018). In a collaboration with the Scheperjans team, we recently reported that bacterial butyrate and propionate were less abundant in PD stool compared to healthy controls; and in PD subjects, this deficiency is linked to epigenetic changes and depressive symptoms (Xie et al., 2022). The release of LPS and amyloidogenic proteins is enhanced by alterations in the gut microbial composition which may contribute to increased risk for neurodegenerative diseases, in particular, during aging when both the GI tract epithelium and BBB become more permeable. Therefore, it has been proposed that LPS and amyloids may directly pass through a breached GI tract barrier or BBB and/or signal through these protective physiological barriers via LPS/amyloid-triggered cytokines or other low molecular weight metabolites that are normally transited (Kesika, Suganthy, Sivamaruthi, & Chaiyasut, 2021).

8. Vagus nerve

The vagus nerve is part of the parasympathetic and sensory nervous systems that serves as a direct connection between the brain and enteric nervous system and is the main afferent pathway from abdomen to the brain (Forsythe, Bienenstock, & Kunze, 2014; Ma et al., 2019). The vagus nerve is a mixed nerve containing approximately 80% afferent and 20% efferent nerve fibers (Bonaz, Bazin, & Pellissier, 2018). Often referred to as the sixth sense, the vagus nerve is central to interoceptive awareness (Paciorek & Skora, 2020). The vagus nerve also has an immunomodulatory role. The inflammatory reflex is a neuro-immune circuit that involves the vagus nerve's modulation of TNF production in the spleen (Bernik et al., 2002). The inflammatory reflex is a mechanism by which the afferent signaling is related to efferent mediated output to regulate proinflammatory cytokine production and inflammation (Pavlov & Tracey, 2012). The efferent portion of the inflammatory reflex is often referred to as the "cholinergic anti-inflammatory pathway" due to the fact that acetylcholine is the primary vagal neurotransmitter (Borovikova et al., 2000; Johnston & Webster, 2009), the main purpose being to keep inflammatory responses in check.

While vagus nerve afferents are present throughout the different layers of the gastrointestinal mucosa and into the epithelial layer, they are not known to come in direct contact with the gastrointestinal lumen (Bonaz et al., 2018), and thus likely do not have direct contact with the microbiome. However, the vagus nerve can sense alterations in the gastrointestinal lumen through interaction with other cell types. Enteroendocrine cells (EECs) are epithelial cells within the intestine and directly interact with the vagus nerve (Latorre, Sternini, De Giorgio, & Greenwood-Van Meerveld, 2016). EECs express TLRs which allow for the detection of signals from gut microbiota such as LPS (Abreu, Fukata, & Arditi, 2005). LPS can also directly activate vagal afferent fibers at the nodose ganglia (Hosoi, Okuma, Matsuda, & Nomura, 2005).

In addition to acting as a pathway for gut microbiome metabolites to influence the brain, the vagus nerve also serves as a mechanism of protein propagation. Alpha-synuclein, a protein implicated in PD pathogenesis, is hypothesized to propagate from the gut via the vagal nerve to the brain (Hawkes, Del Tredici, & Braak, 2007). Recent work in mouse models of PD have demonstrated the propagation of pathologic alpha-synuclein from gut seeded alpha-synuclein preformed fibrils via the vagal nerve to the brain (Challis et al., 2020; Kim et al., 2019; Uemura et al., 2018). Moreover, it is well established that GI dysfunction is present in PD patients and that these symptoms may precede the onset of motor symptoms by decades. Further investigation into the mechanisms involved in how the microbiome may influence synuclein spread via the vagus is needed.

9. Mediators of gut/brain crosstalk

The gut microbiome influences neuro-immune interactions through a variety of mediators including short chain fatty acids, aryl hydrocarbon receptor ligands, lipopolysaccharide, secondary bile acids and neurotransmitters (Fig. 1).

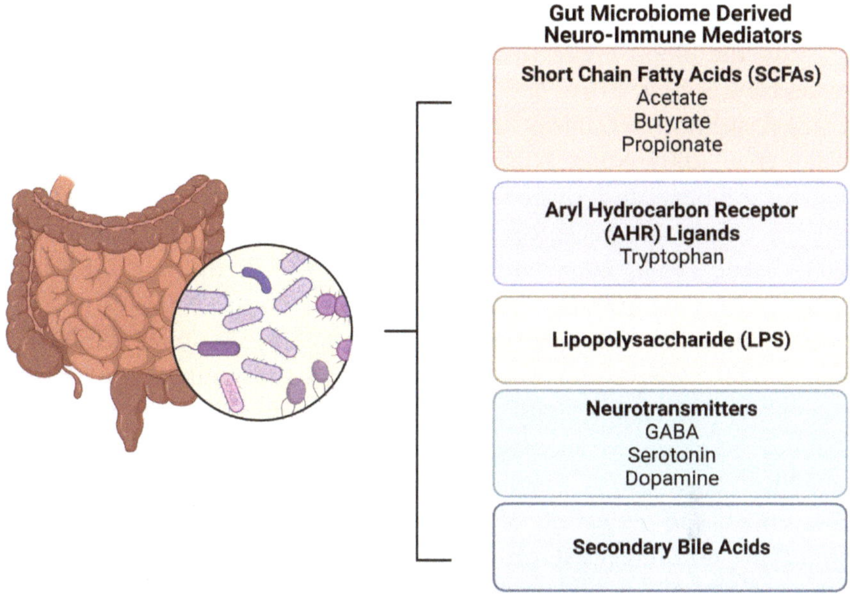

Fig. 1 Gut microbiome derived neuro-immune mediators. In addition to direct interactions with gut resident immune cells and the enteric nervous system, the gut microbiome produces mediators that influence neuro-immune interactions. Gut microbiome metabolism results in the production of short chain fatty acids (SCFAs) including acetate, butyrate and propionate, and aryl hydrocarbon receptor ligands (tryptophan). Lipopolysaccharide is derived from gram-negative bacteria and is a potent immune mediator. The gut microbiome is also able to produce several neurotransmitters that can have systemic effects. Created with BioRender.com

9.1 Short chain fatty acids

SCFAs are metabolites produced by the gut microbiota that are capable of influencing the brain. SCFAs are produced by gut microbiota through the anaerobic fermentation of dietary fiber and resistant starch (Silva, Bernardi, & Frozza, 2020). Acetate, propionate, and butyrate are the most abundant SCFAs in humans (Reyes, Zhang, Gao, & Asatryan, 2020; Silva et al., 2020). 500–600 mmol of SCFAs are produced in the gut daily, however, their amounts vary depending on dietary fiber content (Bergman, 1990). Following their production, SCFAs are rapidly absorbed by colonocytes. Butyrate is particularly important within the gut due to its role as an essential metabolic substrate for colonocyte proliferation and differentiation (Tan et al., 2014). Within the gut, SCFAs have a wide range of functions including helping to maintain intestinal barrier integrity, reducing intestinal inflammation, modulating mucous production, and influencing gut motility (Dalile, Van Oudenhove, Vervliet, & Verbeke, 2019). SCFAs can cross the blood brain barrier (BBB), however, various PET imaging studies have indicated that, despite the ability to cross the BBB, brain uptake of SCFAs is limited (Dalile et al., 2019; Frost et al., 2014; Kim et al., 2013; Seltzer et al., 2004; Song, Nielson, Banks, & Bradley, 2009). Acetate, propionate, and butyrate activate G-protein-coupled receptors 41 and 43 (also called free fatty acid receptors FFAR3 and 2 respectively) (Brown et al., 2003; Tazoe et al., 2008) and butyrate also signals through GPR109A (Thangaraju et al., 2009).

In addition to exerting effects in the gastrointestinal system, SCFAs have been shown to have systemic effects through the inhibition of histone deacetylase (HDAC) activity. Specifically, butyrate acts as an HDAC3 inhibitor resulting in increased antimicrobial activity of macrophages (Schulthess et al., 2019). Furthermore, SCFAs regulate the differentiation, recruitment, and activation of dendritic cells, macrophages, neutrophils, monocytes, and T cells (Corrêa-Oliveira, Fachi, Vieira, Sato, & Vinolo, 2016). SCFAs play a role in maintaining BBB integrity. Furthermore, SCFAs have been shown to be present in human cerebrospinal fluid (CSF). In humans, the FFAR3 is expressed in the colon (Tazoe et al., 2009), blood vessels, and sympathetic nervous system (Nøhr et al., 2015). In rats, FFAR3 is highly expressed in brain tissue (Bonini, Anderson, & Steiner, 1997).

Butyrate is known to exert anti-inflammatory effects within the gastrointestinal system (Siddiqui & Cresci, 2021). Butyrate can exert anti-inflammatory effects by inhibiting activation of NFκB and thereby modulating the production of pro-inflammatory cytokines (Siddiqui & Cresci, 2021). Moreover, butyrate has been shown to modulate neuroinflammation. Butyrate decreased LPS-induced inflammation in rat primary microglia, hippocampal slice cultures and co-cultures of rat cerebellar granule neurons, astrocytes, and microglia (Huuskonen, Suuronen, Nuutinen, Kyrylenko, & Salminen, 2004). Furthermore, in a mouse model of AD, butyrate supplementation was shown to improve synaptic plasticity which was associated with a reduction in TNF, IL-6, and IL-1β in the hippocampus and cortex (Jiang, Li, Li, Xu, & Yang, 2021). Microglial activation and Aβ accumulation were also reduced (Jiang et al., 2021). In addition to HDAC activity, SCFAs may also exert effects through their activation of G protein coupled receptors expressed in the large intestine, spleen and liver (Kimura et al., 2011).

SCFAs mediate microglial homeostasis (Erny et al., 2015). Microglia isolated from GF mice displayed an immature phenotype. However, following supplementation with SCFAs this phenotype was reversed. Furthermore, mice that were deficient for the SCFA receptor FFAR2 displayed similar microglial features as GF mice including altered dendrite lengths, branching points and cell volume (Erny et al., 2015). Collectively, these findings support an important role for gut microbiome metabolites in maintaining CNS homeostasis.

A recent study demonstrated that treating human THP-1 cells with a mixture of SCFAs (acetate, butyrate, propionate, formate, and valerate) in physiological ratios prior to stimulation with LPS and IFN-γ, significantly inhibited the secretion of IL-1β, MCP-1, and TNF (Wenzel, Gates, Ranger, & Klegeris, 2020). Similarly, a previous study reported that SCFAs inhibited inflammatory cytokine secretion (IFN-γ and TNF) in LPS-stimulated human PBMCs (Cox et al., 2009). SCFAs have been shown to mediate amyloid-β plaque formation in multiple GF mouse models of AD (APPPS1 and 3xTg) (Colombo et al., 2021). When GF APPPS1 AD mice were supplemented with SCFAs cerebral plaque load was doubled (Colombo et al., 2021). In PD patients, reduced abundance of *Bacteroidetes* and *Prevotellaceae* and increased abundance of *Enterobacteriaceae* were observed in combination with reduced fecal SCFA concentrations (Unger et al., 2016). A recent microbiome-wide association study demonstrated 15 genera that were associated with PD (Wallen et al., 2020). This work also revealed an overabundance of opportunistic pathogens in the PD microbiome. Furthermore, this study showed that SCFA producing bacteria were reduced in the PD microbiome.

9.2 Aryl hydrocarbon receptor (AHR) and AHR ligands

The aryl hydrocarbon receptor (AHR), an 848 amino acid protein, is a ligand-activated transcription factor (Hahn, Karchner, Shapiro, & Perera, 1997) that is expressed in both mice (Seldin, Howard, & D'Eustachio, 1989) and humans (Le Beau, Carver, Espinosa 3rd, Schmidt, & Bradfield, 1994). Expression of the transcription factor Ahr mediates colonic enteric neurons sensing the gut microbiota (Obata et al., 2020). AHR ligands, including derivatives of tryptophan, are produced by bacteria including *Peptostreptococcus russellii* and *Lactobacillus* species (Dong & Perdew, 2020; Zhao et al., 2021).

AHR signaling has been demonstrated to mediate neurological signaling and immune responses (Lamas, Natividad, & Sokol, 2018). Within the gut, a variety of immune cell types express AHR including T cells, dendritic cells, macrophages, neutrophils, and innate lymphoid cells (Lamas et al., 2018; Barroso, Mahler, Fonseca-Castro, & Quintana, 2021).

Type I interferon (IFN-I) treatment induces AHR expression in astrocytes (Rothhammer et al., 2016). Moreover, tryptophan metabolites were shown to suppress CNS inflammation through AHR signaling in astrocytes (Rothhammer et al., 2016).

When wild type mice are treated with antibiotics, *Ahr* expression is downregulated in enteric neurons (Obata et al., 2020), implicating the microbiome in its activity.

9.3 Lipopolysaccharide

LPS is a potent immune modulator derived from Gram-negative bacteria. Structurally, LPS is composed of lipid A, a core oligosaccharide, and an O antigen of oligo- or polysaccharide chain (Alexander & Rietschel, 2001; Lin et al., 2020). LPS is a pathogen-associated molecular pattern (PAMP) and elicits an immune response via TLR4 (Park & Lee, 2013). Once LPS is recognized by TLR4, downstream signaling through NFκB and IRF3 leads to the production of pro-inflammatory cytokines (Park & Lee, 2013).

In humans, 23% of the gut microbiota consists of bacteria from the phylum *Bacteroidetes*, with the majority of this population being gram-negative *Bacteroides* and *Prevotella* species (Lin et al., 2020). The endotoxicity of LPS produced by various gram-negative bacterial species can vary (Needham & Trent, 2013; Vatanen et al., 2016) and they can have different endotoxin structures due to differences in the O-antigen or lipid A (Needham & Trent, 2013). Lipid A is the primary immunostimulatory moiety of proinflammatory-LPS and is detected by TLR4 (Lin et al., 2020). For example, the endotoxicity of LPS produced by *Bacteroides* species is much lower than that produced by *E. coli* (Jacobson, Choudhury, & Fischbach, 2018). Proinflammatory-LPS derived from gut microbiota has been shown to contribute to the development of chronic inflammatory conditions and levels have been shown to be elevated in blood plasma of AD, PD, and amyotrophic lateral sclerosis (ALS) patients (Brown, 2019; Lin et al., 2020) and in patients with severe autism (Emanuele et al., 2010). Furthermore, LPS derived from *B. fragilis* has been shown to induce

NF-κB signaling in human neuron and glial co-cultures (Lukiw, 2016). Interestingly, there is approximately 1 g of endotoxin in the human gut (Sender, Fuchs, & Milo, 2016) indicating abundance to exert effects within the gastrointestinal system.

"Leaky gut" leads to increased translocation of LPS from the gut lumen to systemic circulation where it can elicit immune responses (Mohammad & Thiemermann, 2021). LPS contributes to the leaky gut phenomenon by altering the epithelial tight junction proteins, including ZO-1 and occludin (Mohammad & Thiemermann, 2021). The BBB is affected by circulating endotoxin levels. When plasma levels of endotoxin are high, the BBB permeability increases (Brown, 2019). This increase in permeability allows for circulating pathogens, cells, and gut microbiome derived mediators such as LPS to cross the BBB. Serum endotoxin levels are increased in patients with several neurological disorders including ASD (Emanuele et al., 2010), ALS, and AD (Zhang et al., 2009). Moreover, BBB disruption is observed in many neurodegenerative diseases including PD, AD, ALS, and MS (Knox, Aburto, Clarke, Cryan, & O'Driscoll, 2022). Thus, increased circulating endotoxin levels are likely to be associated with increased BBB permeability.

9.4 Secondary bile acids

Bile acids (BA) are molecules produced from cholesterol clearance in the liver (Martin, Osadchiy, Kalani, & Mayer, 2018). BA play an important role in absorption of dietary lipids and can influence GI functions, glucose homeostasis, inflammation and BBB integrity (Hegyi, Maléth, Walters, Hofmann, & Keely, 2018; Li et al., 2021; McMillin & DeMorrow, 2016; Wahlström, Sayin, Marschall, & Bäckhed, 2016). These primary BA are secreted into the bile and then into the intestinal lumen. In the ileum, up to 95% of BA are reabsorbed (Chiang, 2013). The BA that are not reabsorbed are transported to the colon where they can be metabolized by microbes to produce secondary BA (Mulak, 2021). This metabolic process is primarily mediated by microbiota in *Clostridium* clusters *XIVa* and *XI* and results in the production of deoxycholic acid (DCA), lithocholic acid (LCA), and ursodeoxycholic acid (UDCA) (Wahlström et al., 2016).

The secondary BAs isoallolithocholic acid (isoalloLCA) and isodeoxycholic acid (isoDCA) have previously been shown to mediate T_{reg} differentiation (Campbell et al., 2020; Hang et al., 2019; Song et al., 2020). A recent study also demonstrated that the human gut microbiome further metabolizes the secondary BA lithocholic acid into 3-oxolithocholic acid (3-oxoLCA) and isolithocholic acid (isoLCA), both of which inhibit T_H17 differentiation (Paik et al., 2022). Interestingly, increased levels of secondary BA have been associated with AD and cognitive decline (Baloni et al., 2020; Mahmoudian Dehkordi et al., 2019). A recent study reported an 18.7-fold increase in LCA and a 5.6-fold increase in DCA in the PD appendix compared to controls with concomitant alterations in microbiota (Li et al., 2021).

9.5 Neurotransmitters

The gut microbiome also produces neurotransmitters including glutamate, serotonin, dopamine, and gamma-aminobutyric acid (GABA) (Chen, Xu, & Chen, 2021). Microbiota are also able to indirectly mediate neurotransmitter production. For example, enterochromaffin cells are major producers of serotonin and it has been demonstrated that SCFAs, microbiome derived metabolites, are necessary for this production (Reigstad et al., 2015; Yano et al., 2015). In addition to producing these neurotransmitters in the gut, the microbiome can also influence the expression of neurotransmitter levels and receptors in the brain (Kennedy et al., 2016). In the human gut, GABA is produced by *Lactobacillus brevis* and *Bifidobacterium dentium* (Reyes et al., 2020). GABA produced by the gut microbiome can cross the BBB and exert effects in the CNS (Takanaga, Ohtsuki, Hosoya, & Terasaki, 2001). In GF rodents, GABA, serotonin, and norepinephrine levels in the gut are reduced (Foster, Baker, & Dursun, 2021). In conventionally colonized mice, treatment with *Lacotbacillus rhamnosus (JB-1)* resulted in alterations in GABA receptor mRNA expression, with $GABA_{A\alpha 2}$ decreased in the prefrontal cortex and amygdala and increased in the hippocampus and $GABA_{B1b}$ increased in the cortex and reduced in the hippocampus, amygdala, and locus coeruleus (Bravo et al., 2011).

Serotonin is derived from tryptophan (Gwak & Chang, 2021). The main source of serotonin is enterochromaffin cells in the gut, but it can also be produced in the brain (Mawe & Hoffman, 2013). Thus, it can serve as a mediator of communication between the gut and the CNS. GF mice have been shown to have reduced serotonin and $5\text{-}HT_{1A}$ receptor levels in the amygdala and hippocampus (Bercik et al., 2011; Clarke et al., 2013; Heijtz et al., 2011; Neufeld, Kang, Bienenstock, & Foster, 2011). Moreover, germ free mice have reduced serum and colonic levels of serotonin compared to conventionally colonized control mice (Wikoff et al., 2009). Some studies have reported sex-dependent effects on serotonin levels in the brains of GF mice. Clarke and colleagues reported higher serotonin and 5-HIAA levels in the hippocampus of GF male mice, but no difference in the hippocampus of GF female mice (Clarke et al., 2013).

Brain dopamine levels have also been shown to be modulated by the gut microbiome. Several studies have investigated the levels of dopamine, dopamine metabolites, and dopamine receptor expression in the brains of GF rodents (González-Arancibia et al., 2019). In GF mice, the DOPAC/DA ratio was shown to be higher in the striatum compared to specific pathogen-free mice (Heijtz et al., 2011). This indicates a higher turnover rate of dopamine. Models of antibiotic-induced dysbiosis have also revealed alterations in brain dopamine. Following a 42 day regiment of an antibiotic cocktail in rats, increased L-DOPA levels in the hippocampus and prefrontal cortex were observed and the HVA/DA ratio in the striatum and amygdala was reduced (Hoban et al., 2016). Recent work has strongly implicated the vagus nerve as the key route for dopamine gut-brain crosstalk (Han et al., 2018). In addition to influencing dopamine within the CNS, the gut microbiota can also produce dopamine including several *Bacillus* strains and *E. coli* (reviewed in (Strandwitz, 2018)).

10. Therapeutic potential of probiotics and prebiotics to treat neurodegenerative disease

Given the strong links between gut dysbiosis and risk for or exacerbation of progression in several neurodegenerative diseases, diet modification, the use of probiotics and prebiotics and fecal microbiota transplantation have begun to be explored as new paradigms for treating individuals with PD, AD, ALS, and other neurodegenerative diseases (Cryan et al., 2019; Kesika et al., 2021).

Several common factors are known to impinge on the microbiota-gut-brain axis (Cryan et al., 2019). These include heredity and epigenetic modifications, mode of delivery at birth, environmental exposures, medications, diet, and exercise. In addition, various behaviors are known to disrupt the microbiota-gut-brain axis, including cognitive and social behaviors, fear, stress and food intake. While the more holistic approach to maintain good gut health is likely to include a good diet (low carbohydrates/high fiber or Mediterranean-style), regular physical activity (daily for 20 min to break a sweat), stress-management (meditation or yoga), and adequate amounts of sleep (a minimum of 6–7 h), pro/pre-biotics and/or interventions such as FMTs may be necessary in some people to more rapidly assist in restoring gut microbiota to healthy levels, such as the use of FMTs for individuals with life-threatening *Clostridium difficile* infections. For a more in-depth discussion of pro/pre-biotics and FMTs, the reader is referred elsewhere (Cryan et al., 2019). Importantly, while there may be plenty of data supporting the fact that these interventions improve gut health and enable individuals to reduce constipation and other GI disturbances, to date there is no strong evidence that these interventions modify the course of any neurodegenerative disease in humans.

11. Future challenges/directions

It is evident that investigators aiming to unravel the causes of many neurological diseases have realized it was time to embrace the role of the gut brain axis. An overwhelming amount of data from cross-sectional studies implicate the microbiome as a determinant of risk and progression in neurological disorders, including AD, PD, ASD, MDD and stroke. Yet it's clear that longitudinal studies as well as systems biology and well-designed clinical trials will be needed to identify mechanisms and targets to identify subjects at risk for neurological disorders and for therapeutic intervention. Additional challenges thus far have been differences in sample collection and processing, differences in analytical methods and platforms, as well as differences in study populations and methodologies between gut microbiome studies in all disease areas.

As an example, in PD subjects, the nearly two dozen or so studies published thus far (Table 1) involve differences in PD subtypes, disease duration and PD medication regimen all of which will influence gut microbiome composition changes. Furthermore, increasing evidence from pre-clinical models suggests microbiome-immune interactions and microbiome influences on PD related pathology. Scheperjans and colleagues outline several important issues that need to be addressed regarding the gut-brain axis in PD (Scheperjans, Derkinderen, & Borghammer, 2018), including the need for correlational analyses to harmonize the multi-omics findings from multicenter studies with pre-clinical models which will aid the field to translate these findings, diagnose subjects at risk for neurological disorders and develop cutting-edge therapies.

Acknowledgments

The authors would like to acknowledge partial funding support for this work from the National Institutes of Health (7RF1AG057247), the Michael J. Fox Foundation for Parkinson's Research, and the Parkinson's Foundation.

Conflict of interests

Malú Gámez Tansey is a co-inventor on the XPro1595 patent, and is a consultant to and has stock ownership in INmune Bio, which has licensed XPro1595 for neurological indications. The remaining authors declare no competing interests.

References

Abbott, N. J., Rönnbäck, L., & Hansson, E. (2006). Astrocyte–endothelial interactions at the blood–brain barrier. *Nature Reviews Neuroscience*, 7(1), 41–53.

Abdel-Haq, R., Schlachetzki, J. C. M., Glass, C. K., & Mazmanian, S. K. (2018). Microbiome–microglia connections via the gut–brain axis. *Journal of Experimental Medicine*, *216*(1), 41–59.

Abreu, M. T., Fukata, M., & Arditi, M. (2005). TLR signaling in the gut in health and disease. *The Journal of Immunology*, *174*(8), 4453–4460.

Aho, V. T. E., Pereira, P. A. B., Voutilainen, S., Paulin, L., Pekkonen, E., Auvinen, P., et al. (2019). Gut microbiota in Parkinson's disease: Temporal stability and relations to disease progression. *eBioMedicine*, *44*, 691–707.

Alexander, C., & Rietschel, E. T. (2001). Invited review: Bacterial lipopolysaccharides and innate immunity. *Journal of Endotoxin Research*, 7(3), 167–202.

Balasa, R., Barcutean, L., Mosora, O., & Manu, D. (2021). Reviewing the significance of blood-brain barrier disruption in multiple sclerosis pathology and treatment. *International Journal of Molecular Sciences*, *22*(16).

Baloni, P., Funk, C. C., Yan, J., Yurkovich, J. T., Kueider-Paisley, A., Nho, K., et al. (2020). Metabolic network analysis reveals altered bile acid synthesis and metabolism in Alzheimer's disease. *Cell Reports Medicine*, *1*(8), 100138.

Barichella, M., Severgnini, M., Cilia, R., Cassani, E., Bolliri, C., Caronni, S., et al. (2019). Unraveling gut microbiota in Parkinson's disease and atypical parkinsonism. *Movement Disorders*, *34*(3), 396–405.

Barroso, A., Mahler, J. V., Fonseca-Castro, P. H., & Quintana, F. J. (2021). The aryl hydrocarbon receptor and the gut–brain axis. *Cellular & Molecular Immunology*, *18*(2), 259–268.

Bäuerl, C., Collado, M. C., Diaz Cuevas, A., Viña, J., & Pérez Martínez, G. (2018). Shifts in gut microbiota composition in an APP/PSS1 transgenic mouse model of Alzheimer's disease during lifespan. *Letters in Applied Microbiology*, *66*(6), 464–471.

Bedarf, J. R., Hildebrand, F., Coelho, L. P., Sunagawa, S., Bahram, M., Goeser, F., et al. (2017). Functional implications of microbial and viral gut metagenome changes in early stage L-DOPA-naïve Parkinson's disease patients. *Genome Medicine*, *9*(1), 39.

Bercik, P., Denou, E., Collins, J., Jackson, W., Lu, J., Jury, J., et al. (2011). The intestinal microbiota affect central levels of brain-derived neurotropic factor and behavior in mice. *Gastroenterology*, *141*(2), 599–609. e593.

Bergman, E. N. (1990). Energy contributions of volatile fatty acids from the gastrointestinal tract in various species. *Physiological Reviews*, *70*(2), 567–590.

Bernik, T. R., Friedman, S. G., Ochani, M., DiRaimo, R., Susarla, S., Czura, C. J., et al. (2002). Cholinergic antiinflammatory pathway inhibition of tumor necrosis factor during ischemia reperfusion. *Journal of Vascular Surgery*, *36*(6), 1231–1236.

Boillot, A., Demmer, R. T., Mallat, Z., Sacco, R. L., Jacobs, D. R., Benessiano, J., et al. (2015). Periodontal microbiota and phospholipases: the oral infections and vascular disease epidemiology study (INVEST). *Atherosclerosis*, *242*(2), 418–423.

Bonaz, B., Bazin, T., & Pellissier, S. (2018). The vagus nerve at the interface of the microbiota-gut-brain axis. *Frontiers in Neuroscience*, *12*.

Bonini, J. A., Anderson, S. M., & Steiner, D. F. (1997). Molecular cloning and tissue expression of a novel orphan G protein-coupled receptor from rat lung. *Biochemical and Biophysical Research Communications*, *234*(1), 190–193.

Borovikova, L. V., Ivanova, S., Zhang, M., Yang, H., Botchkina, G. I., Watkins, L. R., et al. (2000). Vagus nerve stimulation attenuates the systemic inflammatory response to endotoxin. *Nature*, *405*(6785), 458–462.

Braniste, V., Al-Asmakh, M., Kowal, C., Anuar, F., Abbaspour, A., Tóth, M., et al. (2014). The gut microbiota influences blood-brain barrier permeability in mice. *Science Translational Medicine*, *6*(263), 263ra158.

Bravo, J. A., Forsythe, P., Chew, M. V., Escaravage, E., Savignac, H. M., Dinan, T. G., et al. (2011). Ingestion of Lactobacillus strain regulates emotional behavior and central GABA receptor expression in a mouse via the vagus nerve. *Proceedings of the National Academy of Sciences*, *108*(38), 16050–16055.

Brown, G. C. (2019). The endotoxin hypothesis of neurodegeneration. *Journal of Neuroinflammation*, *16*(1), 1–10.

Brown, A. J., Goldsworthy, S. M., Barnes, A. A., Eilert, M. M., Tcheang, L., Daniels, D., et al. (2003). The Orphan G protein-coupled receptors GPR41 and GPR43 are activated by propionate and other short chain carboxylic acids. *Journal of Biological Chemistry*, *278*(13), 11312–11319.

Camous, X., Pera, A., Solana, R., & Larbi, A. (2012). NK cells in healthy aging and age-associated diseases. *Journal of Biomedicine & Biotechnology*, *2012*, 195956.

Campbell, C., McKenney, P. T., Konstantinovsky, D., Isaeva, O. I., Schizas, M., Verter, J., et al. (2020). Bacterial metabolism of bile acids promotes generation of peripheral regulatory T cells. *Nature*, *581*(7809), 475–479.

Cattaneo, A., Cattane, N., Galluzzi, S., Provasi, S., Lopizzo, N., Festari, C., et al. (2017). Association of brain amyloidosis with pro-inflammatory gut bacterial taxa and peripheral inflammation markers in cognitively impaired elderly. *Neurobiology of Aging*, *49*, 60–68.

Challis, C., Hori, A., Sampson, T. R., Yoo, B. B., Challis, R. C., Hamilton, A. M., et al. (2020). Gut-seeded α-synuclein fibrils promote gut dysfunction and brain pathology specifically in aged mice. *Nature Neuroscience*, *23*(3), 327–336.

Chandrasekharan, B., Jeppsson, S., Pienkowski, S., Belsham, D. D., Sitaraman, S. V., Merlin, D., et al. (2013). Tumor necrosis factor-neuropeptide Y cross talk regulates inflammation, epithelial barrier functions, and colonic motility. *Inflammatory Bowel Diseases*, *19*(12), 2535–2546.

Chen, Y., Xu, J., & Chen, Y. (2021). Regulation of neurotransmitters by the gut microbiota and effects on cognition in neurological disorders. *Nutrients*, *13*(6), 2099.

Chiang, J. Y. (2013). Bile acid metabolism and signaling. *Comprehensive Physiology*, *3*(3), 1191.

Clarke, G., Grenham, S., Scully, P., Fitzgerald, P., Moloney, R., Shanahan, F., et al. (2013). The microbiome-gut-brain axis during early life regulates the hippocampal serotonergic system in a sex-dependent manner. *Molecular Psychiatry*, *18*(6), 666–673.

Colombo, E., & Farina, C. (2016). Astrocytes: Key regulators of neuroinflammation. *Trends in Immunology*, *37*(9), 608–620.

Colombo, A. V., Sadler, R. K., Llovera, G., Singh, V., Roth, S., Heindl, S., et al. (2021). Microbiota-derived short chain fatty acids modulate microglia and promote Aβ plaque deposition. *eLife*, *10*, e59826.

Corrêa-Oliveira, R., Fachi, J. L., Vieira, A., Sato, F. T., & Vinolo, M. A. R. (2016). Regulation of immune cell function by short-chain fatty acids. *Clinical & Translational Immunology*, *5*(4), e73.

Cox, M. A., Jackson, J., Stanton, M., Rojas-Triana, A., Bober, L., Laverty, M., et al. (2009). Short-chain fatty acids act as antiinflammatory mediators by regulating prostaglandin E2 and cytokines. *World Journal of Gastroenterology: WJG*, *15*(44), 5549.

Cryan, J. F., & Dinan, T. G. (2012). Mind-altering microorganisms: The impact of the gut microbiota on brain and behaviour. *Nature Reviews Neuroscience*, *13*(10), 701–712.

Cryan, J. F., O'Riordan, K. J., Cowan, C. S. M., Sandhu, K. V., Bastiaanssen, T. F. S., Boehme, M., et al. (2019). The microbiota-gut-brain axis. *Physiological Reviews*, *99*(4), 1877–2013.

Dalile, B., Van Oudenhove, L., Vervliet, B., & Verbeke, K. (2019). The role of short-chain fatty acids in microbiota-gut-brain communication. *Nature Reviews. Gastroenterology & Hepatology*, *16*(8), 461–478.

Daly, C. M., Saxena, J., Singh, J., Bullard, M. R., Bondy, E. O., Saxena, A., et al. (2022). Sex differences in response to a high fat, high sucrose diet in both the gut microbiome and hypothalamic astrocytes and microglia. *Nutritional Neuroscience*, *25*(2), 321–335.

Dodiya, H. B., Kuntz, T., Shaik, S. M., Baufeld, C., Leibowitz, J., Zhang, X., et al. (2019). Sex-specific effects of microbiome perturbations on cerebral Aβ amyloidosis and microglia phenotypes. *Journal of Experimental Medicine*, *216*(7), 1542–1560.

Dominy, S. S., Lynch, C., Ermini, F., Benedyk, M., Marczyk, A., Konradi, A., et al. (2019). *Porphyromonas gingivalis* in Alzheimer's disease brains: Evidence for disease causation and treatment with small-molecule inhibitors. *Science Advances*, *5*(1), eaau3333.

Dong, Y., & Benveniste, E. N. (2001). Immune function of astrocytes. *Glia*, *36*(2), 180–190.

Dong, F., & Perdew, G. H. (2020). The aryl hydrocarbon receptor as a mediator of host-microbiota interplay. *Gut Microbes*, *12*(1), 1859812.

Dong, S., Sun, M., He, C., & Cheng, H. (2022). Brain-gut-microbiota axis in Parkinson's disease: A historical review and future perspective. *Brain Research Bulletin*, *183*, 84–93.

Dressman, D., & Elyaman, W. (2021). T cells: A growing universe of roles in neurodegenerative diseases. *The Neuroscientist*, *28*(4), 335–348.

Emanuele, E., Orsi, P., Boso, M., Broglia, D., Brondino, N., Barale, F., et al. (2010). Low-grade endotoxemia in patients with severe autism. *Neuroscience Letters*, *471*(3), 162–165.

Erny, D., Hrabě de Angelis, A. L., Jaitin, D., Wieghofer, P., Staszewski, O., David, E., et al. (2015). Host microbiota constantly control maturation and function of microglia in the CNS. *Nature Neuroscience*, *18*(7), 965–977.

Ferreira, S. A., & Romero-Ramos, M. (2018). Microglia response during Parkinson's disease: Alpha-synuclein intervention. *Frontiers in Cellular Neuroscience*, *12*.

Fitzpatrick, Z., Frazer, G., Ferro, A., Clare, S., Bouladoux, N., Ferdinand, J., et al. (2020). Gut-educated IgA plasma cells defend the meningeal venous sinuses. *Nature*, *587*(7834), 472–476.

Forsythe, P., Bienenstock, J., & Kunze, W. A. (2014). Vagal pathways for microbiome-brain-gut axis communication. In M. Lyte, & J. F. Cryan (Eds.), *Microbial endocrinology: The microbiota-gut-brain axis in health and disease* (pp. 115–133). New York, NY: Springer New York.

Foster, J. A. (2016). Gut microbiome and behavior: Focus on neuroimmune interactions. *International Review of Neurobiology*, *131*, 49–65.

Foster, J. A., Baker, G. B., & Dursun, S. M. (2021). The relationship between the gut microbiome-immune system-brain axis and major depressive disorder. *Frontiers in Neurology*, *12* (eCollection 2021).

Fröhlich, E. E., Farzi, A., Mayerhofer, R., Reichmann, F., Jačan, A., Wagner, B., et al. (2016). Cognitive impairment by antibiotic-induced gut dysbiosis: Analysis of gut microbiota-brain communication. *Brain, Behavior, and Immunity*, *56*, 140–155.

Frost, G., Sleeth, M. L., Sahuri-Arisoylu, M., Lizarbe, B., Cerdan, S., Brody, L., et al. (2014). The short-chain fatty acid acetate reduces appetite via a central homeostatic mechanism. *Nature Communications*, *5*, 3611.

Fülling, C., Dinan, T. G., & Cryan, J. F. (2019). Gut microbe to brain signaling: What happens in vagus.... *Neuron*, *101*(6), 998–1002.

Fung, T. C., Olson, C. A., & Hsiao, E. Y. (2017). Interactions between the microbiota, immune and nervous systems in health and disease. *Nature Neuroscience*, *20*(2), 145–155.

Ganal, S. C., Sanos, S. L., Kallfass, C., Oberle, K., Johner, C., Kirschning, C., et al. (2012). Priming of natural killer cells by nonmucosal mononuclear phagocytes requires instructive signals from commensal microbiota. *Immunity*, *37*(1), 171–186.

Giridharan, V. V., Barichello De Quevedo, C. E., & Petronilho, F. (2022). Microbiota-gut-brain axis in the Alzheimer's disease pathology—An overview. *Neuroscience Research*, *181*, 17–21.

Golomb, S. M., Guldner, I. H., Zhao, A., Wang, Q., Palakurthi, B., Aleksandrovic, E. A., et al. (2020). Multi-modal single-cell analysis reveals brain immune landscape plasticity during aging and gut microbiota dysbiosis. *Cell Reports*, *33*(9), 108438.

González-Arancibia, C., Urrutia-Piñones, J., Illanes-González, J., Martinez-Pinto, J., Sotomayor-Zárate, R., Julio-Pieper, M., et al. (2019). Do your gut microbes affect your brain dopamine? *Psychopharmacology*, *236*(5), 1611–1622.

Gray, M. T., & Woulfe, J. M. (2015). Striatal blood-brain barrier permeability in Parkinson's disease. *Journal of Cerebral Blood Flow and Metabolism*, *35*(5), 747–750.

Gwak, M.-G., & Chang, S.-Y. (2021). Gut-brain connection: microbiome, gut barrier, and environmental sensors. *Immune Network*, *21*(3).

Hahn, M. E., Karchner, S. I., Shapiro, M. A., & Perera, S. A. (1997). Molecular evolution of two vertebrate aryl hydrocarbon (dioxin) receptors (AHR1 and AHR2) and the PAS family. *Proceedings of the National Academy of Sciences of the United States of America*, *94*(25), 13743–13748.

Hammond, C. J., Hallock, L. R., Howanski, R. J., Appelt, D. M., Little, C. S., & Balin, B. J. (2010). Immunohistological detection of Chlamydia pneumoniae in the Alzheimer's disease brain. *BMC Neuroscience*, *11*, 121.

Han, W., Tellez, L. A., Perkins, M. H., Perez, I. O., Qu, T., Ferreira, J., et al. (2018). A neural circuit for gut-induced reward. *Cell*, *175*(3), 665–678.e623.

Hang, S., Paik, D., Yao, L., Kim, E., Trinath, J., Lu, J., et al. (2019). Bile acid metabolites control TH17 and Treg cell differentiation. *Nature*, *576*(7785), 143–148.

Hansen, D. V., Hanson, J. E., & Sheng, M. (2018). Microglia in Alzheimer's disease. *The Journal of Cell Biology*, *217*(2), 459–472.

Harach, T., Marungruang, N., Duthilleul, N., Cheatham, V., Mc Coy, K. D., Frisoni, G., et al. (2017). Reduction of Abeta amyloid pathology in APPPS1 transgenic mice in the absence of gut microbiota. *Scientific Reports*, 7(1), 41802.

Haran, J. P., Bhattarai, S. K., Foley, S. E., Dutta, P., Ward, D. V., Bucci, V., et al. (2019). Alzheimer's disease microbiome is associated with dysregulation of the anti-inflammatory P-glycoprotein pathway. *MBio*, *10*(3).

Hasegawa, S., Goto, S., Tsuji, H., Okuno, T., Asahara, T., Nomoto, K., et al. (2015). Intestinal dysbiosis and lowered serum lipopolysaccharide-binding protein in Parkinson's disease. *PLoS One*, *10*(11), e0142164.

Hawkes, C. H., Del Tredici, K., & Braak, H. (2007). Parkinson's disease: A dual-hit hypothesis. *Neuropathology and Applied Neurobiology*, *33*(6), 599–614.

Hegyi, P., Maléth, J., Walters, J. R., Hofmann, A. F., & Keely, S. J. (2018). Guts and gall: Bile acids in regulation of intestinal epithelial function in health and disease. *Physiological Reviews*, *98*(4), 1983–2023.

Heijtz, R. D., Wang, S., Anuar, F., Qian, Y., Björkholm, B., Samuelsson, A., et al. (2011). Normal gut microbiota modulates brain development and behavior. *Proceedings of the National Academy of Sciences*, *108*(7), 3047–3052.

Heintz-Buschart, A., Pandey, U., Wicke, T., Sixel-Döring, F., Janzen, A., Sittig-Wiegand, E., et al. (2018). The nasal and gut microbiome in Parkinson's disease and idiopathic rapid eye movement sleep behavior disorder. *Movement Disorders*, *33*(1), 88–98.

Hill-Burns, E. M., Debelius, J. W., Morton, J. T., Wissemann, W. T., Lewis, M. R., Wallen, Z. D., et al. (2017). Parkinson's disease and Parkinson's disease medications have distinct signatures of the gut microbiome. *Movement Disorders*, *32*(5), 739–749.

Hoban, A. E., Moloney, R. D., Golubeva, A. V., Neufeld, K. M., O'Sullivan, O., Patterson, E., et al. (2016). Behavioural and neurochemical consequences of chronic gut microbiota depletion during adulthood in the rat. *Neuroscience*, *339*, 463–477.

Hopfner, F., Künstner, A., Müller, S. H., Künzel, S., Zeuner, K. E., Margraf, N. G., et al. (2017). Gut microbiota in Parkinson disease in a northern German cohort. *Brain Research*, *1667*, 41–45.

Hosoi, T., Okuma, Y., Matsuda, T., & Nomura, Y. (2005). Novel pathway for LPS-induced afferent vagus nerve activation: Possible role of nodose ganglion. *Autonomic Neuroscience*, *120*(1–2), 104–107.

Hoyles, L., Snelling, T., Umlai, U.-K., Nicholson, J. K., Carding, S. R., Glen, R. C., et al. (2018). Microbiome–host systems interactions: Protective effects of propionate upon the blood–brain barrier. *Microbiome*, *6*(1), 55.

Hsiao, E. Y., McBride, S. W., Chow, J., Mazmanian, S. K., & Patterson, P. H. (2012). Modeling an autism risk factor in mice leads to permanent immune dysregulation. *Proceedings of the National Academy of Sciences of the United States of America*, *109*(31), 12776–12781.

Hsiao, E. Y., McBride, S. W., Hsien, S., Sharon, G., Hyde, E. R., McCue, T., et al. (2013). Microbiota modulate behavioral and physiological abnormalities associated with neurodevelopmental disorders. *Cell*, *155*(7), 1451–1463.

Huuskonen, J., Suuronen, T., Nuutinen, T., Kyrylenko, S., & Salminen, A. (2004). Regulation of microglial inflammatory response by sodium butyrate and short-chain fatty acids. *British Journal of Pharmacology*, *141*(5), 874–880.

Ivanov, I. I., Atarashi, K., Manel, N., Brodie, E. L., Shima, T., Karaoz, U., et al. (2009). Induction of intestinal Th17 cells by segmented filamentous bacteria. *Cell*, *139*(3), 485–498.

Jacobson, A. N., Choudhury, B. P., & Fischbach, M. A. (2018). The biosynthesis of lipooligosaccharide from Bacteroides thetaiotaomicron. *MBio*, *9*(2). e02289-02217.

Jha, M. K., Kim, J.-H., Song, G. J., Lee, W.-H., Lee, I.-K., Lee, H.-W., et al. (2018). Functional dissection of astrocyte-secreted proteins: Implications in brain health and diseases. *Progress in Neurobiology*, *162*, 37–69.

Jiang, Y., Li, K., Li, X., Xu, L., & Yang, Z. (2021). Sodium butyrate ameliorates the impairment of synaptic plasticity by inhibiting the neuroinflammation in 5XFAD mice. *Chemico-Biological Interactions*, *341*, 109452.

Johnston, G. R., & Webster, N. R. (2009). Cytokines and the immunomodulatory function of the vagus nerve. *British Journal of Anaesthesia*, *102*(4), 453–462.

Kagnoff, M. (1987). Immunology of the digestive system. In L. R. Johnson (Ed.), *Vol. 2. Physiology of the gastrointestinal tract* (pp. 1699–1728). New York: Raven Press.

Kennedy, P. J., Murphy, A. B., Cryan, J. F., Ross, P. R., Dinan, T. G., & Stanton, C. (2016). Microbiome in brain function and mental health. *Trends in Food Science & Technology*, *57*, 289–301.

Keshavarzian, A., Green, S. J., Engen, P. A., Voigt, R. M., Naqib, A., Forsyth, C. B., et al. (2015). Colonic bacterial composition in Parkinson's disease. *Movement Disorders*, *30*(10), 1351–1360.

Kesika, P., Suganthy, N., Sivamaruthi, B. S., & Chaiyasut, C. (2021). Role of gut-brain axis, gut microbial composition, and probiotic intervention in Alzheimer's disease. *Life Sciences*, *264*, 118627.

Kim, S. W., Hooker, J. M., Otto, N., Win, K., Muench, L., Shea, C., et al. (2013). Whole-body pharmacokinetics of HDAC inhibitor drugs, butyric acid, valproic acid and 4-phenylbutyric acid measured with carbon-11 labeled analogs by PET. *Nuclear Medicine and Biology*, *40*(7), 912–918.

Kim, S., Kwon, S.-H., Kam, T.-I., Panicker, N., Karuppagounder, S. S., Lee, S., et al. (2019). Transneuronal propagation of pathologic α-synuclein from the gut to the brain models Parkinson's disease. *Neuron*, *103*(4), 627–641.e627.

Kimura, I., Inoue, D., Maeda, T., Hara, T., Ichimura, A., Miyauchi, S., et al. (2011). Short-chain fatty acids and ketones directly regulate sympathetic nervous system via G protein-coupled receptor 41 (GPR41). *Proceedings of the National Academy of Sciences*, *108*(19), 8030–8035.

Knox, E. G., Aburto, M. R., Clarke, G., Cryan, J. F., & O'Driscoll, C. M. (2022). The blood-brain barrier in aging and neurodegeneration. *Molecular Psychiatry*, *27*(6), 2659–2673.

Kumar, D. K. V., Choi, S. H., Washicosky, K. J., Eimer, W. A., Tucker, S., Ghofrani, J., et al. (2016). Amyloid-β peptide protects against microbial infection in mouse and worm models of Alzheimer's disease. *Science Translational Medicine*, *8*(340), 340ra372.

Lalla, E., & Papapanou, P. N. (2011). Diabetes mellitus and periodontitis: A tale of two common interrelated diseases. *Nature Reviews Endocrinology*, 7(12), 738–748.

Lamas, B., Natividad, J. M., & Sokol, H. (2018). Aryl hydrocarbon receptor and intestinal immunity. *Mucosal Immunology*, *11*(4), 1024–1038.

Latorre, R., Sternini, C., De Giorgio, R., & Greenwood-Van Meerveld, B. (2016). Enteroendocrine cells: A review of their role in brain-gut communication. *Neurogastroenterology and motility: the official journal of the European Gastrointestinal Motility Society*, *28*(5), 620–630.

Le Beau, M. M., Carver, L. A., Espinosa, R., 3rd, Schmidt, J. V., & Bradfield, C. A. (1994). Chromosomal localization of the human AHR locus encoding the structural gene for the Ah receptor to 7p21 → p15. *Cytogenetics and Cell Genetics*, *66*(3), 172–176.

Leng, F., & Edison, P. (2021). Neuroinflammation and microglial activation in Alzheimer disease: Where do we go from here? *Nature Reviews Neurology*, *17*(3), 157–172.

Li, P., Killinger, B. A., Ensink, E., Beddows, I., Yilmaz, A., Lubben, N., et al. (2021). Gut microbiota dysbiosis is associated with elevated bile acids in Parkinson's disease. *Metabolites*, *11*(1), 29.

Li, F., Wang, P., Chen, Z., Sui, X., Xie, X., & Zhang, J. (2019). Alteration of the fecal microbiota in North-Eastern Han Chinese population with sporadic Parkinson's disease. *Neuroscience Letters*, *707*, 134297.

Li, W., Wu, X., Hu, X., Wang, T., Liang, S., Duan, Y., et al. (2017). Structural changes of gut microbiota in Parkinson's disease and its correlation with clinical features. *Science China Life Sciences*, *60*(11), 1223–1233.

Lin, T.-L., Shu, C.-C., Chen, Y.-M., Lu, J.-J., Wu, T.-S., Lai, W.-F., et al. (2020). Like cures like: Pharmacological activity of anti-inflammatory lipopolysaccharides from gut microbiome. *Frontiers in Pharmacology*, *11*, 554.

Lin, A., Zheng, W., He, Y., Tang, W., Wei, X., He, R., et al. (2018). Gut microbiota in patients with Parkinson's disease in southern China. *Parkinsonism & Related Disorders*, *53*, 82–88.

Logsdon, A. F., Erickson, M. A., Rhea, E. M., Salameh, T. S., & Banks, W. A. (2018). Gut reactions: How the blood–brain barrier connects the microbiome and the brain. *Experimental Biology and Medicine*, *243*(2), 159–165.

Lukiw, W. J. (2016). Bacteroides fragilis lipopolysaccharide and inflammatory signaling in Alzheimer's disease. *Frontiers in Microbiology*, 7, 1544.

Ma, Q., Xing, C., Long, W., Wang, H. Y., Liu, Q., & Wang, R.-F. (2019). Impact of microbiota on central nervous system and neurological diseases: The gut-brain axis. *Journal of Neuroinflammation*, *16*(1), 53.

Macpherson, K. P., Eidson, L. N., Weiss, B. E., Gollihue, J. L., Herrick, M., & Houser, M. C., et al. (n.d.). Soluble TNF mediates amyloid-independent, diet-induced alterations to immune and neuronal functions in an Alzheimer's disease mouse model. Frontiers in Cellular Neuroscience. Submitted for publication.

Mahmoudian Dehkordi, S., Arnold, M., Nho, K., Ahmad, S., Jia, W., Xie, G., et al. (2019). Altered bile acid profile associates with cognitive impairment in Alzheimer's disease—An emerging role for gut microbiome. *Alzheimer's & Dementia*, *15*(1), 76–92.

Martin, C. R., Osadchiy, V., Kalani, A., & Mayer, E. A. (2018). The brain-gut-microbiome axis. *Cellular and Molecular Gastroenterology and Hepatology*, *6*(2), 133–148.

Mawe, G. M., & Hoffman, J. M. (2013). Serotonin signalling in the gut—Functions, dysfunctions and therapeutic targets. *Nature Reviews. Gastroenterology & Hepatology*, *10*(8), 473–486.

Mazmanian, S. K., Liu, C. H., Tzianabos, A. O., & Kasper, D. L. (2005). An immunomodulatory molecule of symbiotic bacteria directs maturation of the host immune system. *Cell*, *122*(1), 107–118.

McMillin, M., & DeMorrow, S. (2016). Effects of bile acids on neurological function and disease. *The FASEB Journal*, *30*(11), 3658–3668.

Minter, M. R., Hinterleitner, R., Meisel, M., Zhang, C., Leone, V., Zhang, X., et al. (2017). Antibiotic-induced perturbations in microbial diversity during post-natal development alters amyloid pathology in an aged APPSWE/PS1ΔE9 murine model of Alzheimer's disease. *Scientific Reports*, 7(1), 10411.

Minter, M. R., Zhang, C., Leone, V., Ringus, D. L., Zhang, X., Oyler-Castrillo, P., et al. (2016). Antibiotic-induced perturbations in gut microbial diversity influences neuro-inflammation and amyloidosis in a murine model of Alzheimer's disease. *Scientific Reports*, *6*(1), 30028.

Mohammad, S., & Thiemermann, C. (2021). Role of metabolic endotoxemia in systemic inflammation and potential interventions. *Frontiers in Immunology*, *11*, 594150.

Mulak, A. (2021). Bile acids as key modulators of the brain-gut-microbiota axis in Alzheimer's disease. *Journal of Alzheimer's Disease*, *84*, 461–477.

Nayak, D., Roth, T. L., & McGavern, D. B. (2014). Microglia development and function. *Annual Review of Immunology*, *32*, 367–402.

Needham, B. D., & Trent, M. S. (2013). Fortifying the barrier: The impact of lipid A remodelling on bacterial pathogenesis. *Nature Reviews Microbiology*, *11*(7), 467–481.

Neufeld, K., Kang, N., Bienenstock, J., & Foster, J. A. (2011). Reduced anxiety-like behavior and central neurochemical change in germ-free mice. *Neurogastroenterology and Motility*, *23*(3), 255. e119.

Noble, J. M., Scarmeas, N., Celenti, R. S., Elkind, M. S., Wright, C. B., Schupf, N., et al. (2014). Serum IgG antibody levels to periodontal microbiota are associated with incident Alzheimer disease. *PLoS One*, *9*(12), e114959.

Nøhr, M. K., Egerod, K. L., Christiansen, S. H., Gille, A., Offermanns, S., Schwartz, T. W., et al. (2015). Expression of the short chain fatty acid receptor GPR41/FFAR3 in autonomic and somatic sensory ganglia. *Neuroscience*, *290*, 126–137.
Noto, D., & Miyake, S. (2022). Gut dysbiosis and multiple sclerosis. *Clinical Immunology*, *235*, 108380.
Obata, Y., Castaño, Á., Boeing, S., Bon-Frauches, A. C., Fung, C., Fallesen, T., et al. (2020). Neuronal programming by microbiota regulates intestinal physiology. *Nature*, *578*(7794), 284–289.
Olson, C. A., Vuong, H. E., Yano, J. M., Liang, Q. Y., Nusbaum, D. J., & Hsiao, E. Y. (2018). The gut microbiota mediates the anti-seizure effects of the ketogenic diet. *Cell*, *173*(7), 1728–1741.e1713.
Paciorek, A., & Skora, L. (2020). Vagus nerve stimulation as a gateway to interoception. *Frontiers in Psychology*, *11*, 1659.
Paik, D., Yao, L., Zhang, Y., Bae, S., D'Agostino, G. D., Zhang, M., et al. (2022). Human gut bacteria produce TH17-modulating bile acid metabolites. *Nature*, *603*(7903), 907–912.
Pandit, R., Chen, L., & Götz, J. (2020). The blood-brain barrier: Physiology and strategies for drug delivery. *Advanced Drug Delivery Reviews*, *165-166*, 1–14.
Park, B. S., & Lee, J.-O. (2013). Recognition of lipopolysaccharide pattern by TLR4 complexes. *Experimental & Molecular Medicine*, *45*(12), e66.
Pavlov, V. A., & Tracey, K. J. (2012). The vagus nerve and the inflammatory reflex—Linking immunity and metabolism. *Nature Reviews. Endocrinology*, *8*(12), 743–754.
Petrov, V. A., Saltykova, I. V., Zhukova, I. A., Alifirova, V. M., Zhukova, N. G., Dorofeeva, Y. B., et al. (2017). Analysis of gut microbiota in patients with Parkinson's disease. *Bulletin of Experimental Biology and Medicine*, *162*(6), 734–737.
Pietrucci, D., Cerroni, R., Unida, V., Farcomeni, A., Pierantozzi, M., Mercuri, N. B., et al. (2019). Dysbiosis of gut microbiota in a selected population of Parkinson's patients. *Parkinsonism & Related Disorders*, *65*, 124–130.
Poole, S., Singhrao, S. K., Chukkapalli, S., Rivera, M., Velsko, I., Kesavalu, L., et al. (2015). Active invasion of Porphyromonas gingivalis and infection-induced complement activation in ApoE-/- mice brains. *Journal of Alzheimer's Disease*, *43*(1), 67–80.
Poole, S., Singhrao, S. K., Kesavalu, L., Curtis, M. A., & Crean, S. (2013). Determining the presence of periodontopathic virulence factors in short-term postmortem Alzheimer's disease brain tissue. *Journal of Alzheimer's Disease*, *36*, 665–677.
Qian, Y., Yang, X., Xu, S., Wu, C., Song, Y., Qin, N., et al. (2018). Alteration of the fecal microbiota in Chinese patients with Parkinson's disease. *Brain, Behavior, and Immunity*, *70*, 194–202.
Rajesh, Y., & Kanneganti, T.-D. (2022). Innate immune cell death in neuroinflammation and Alzheimer's disease. *Cell*, *11*(12), 1885.
Reigstad, C. S., Salmonson, C. E., Rainey, J. F., III, Szurszewski, J. H., Linden, D. R., Sonnenburg, J. L., et al. (2015). Gut microbes promote colonic serotonin production through an effect of short-chain fatty acids on enterochromaffin cells. *The FASEB Journal*, *29*(4), 1395–1403.
Reyes, R. E. N., Zhang, Z., Gao, L., & Asatryan, L. (2020). Microbiome meets microglia in neuroinflammation and neurological disorders. *Neuroimmunology and Neuroinflammation*, 7(3), 215–233.
Rice, O., Surian, A., & Chen, Y. (2022). Modeling the blood-brain barrier for treatment of central nervous system (CNS) diseases. *Journal of Tissue Engineering*, *13*. 20417314221095997.
Rojas, O. L., Pröbstel, A.-K., Porfilio, E. A., Wang, A. A., Charabati, M., Sun, T., et al. (2019). Recirculating intestinal IgA-producing cells regulate neuroinflammation via IL-10. *Cell*, *176*(3), 610–624.e618.

Rothhammer, V., Mascanfroni, I. D., Bunse, L., Takenaka, M. C., Kenison, J. E., Mayo, L., et al. (2016). Type I interferons and microbial metabolites of tryptophan modulate astrocyte activity and central nervous system inflammation via the aryl hydrocarbon receptor. *Nature Medicine*, *22*(6), 586–597.

Roubaud-Baudron, C., Krolak-Salmon, P., Quadrio, I., Mégraud, F., & Salles, N. (2012). Impact of chronic Helicobacter pylori infection on Alzheimer's disease: Preliminary results. *Neurobiology of Aging*, *33*(5), 1009.e1011–1009.e1019.

Sampson, T. R., Debelius, J. W., Thron, T., Janssen, S., Shastri, G. G., Ilhan, Z. E., et al. (2016). Gut microbiota regulate motor deficits and neuroinflammation in a model of Parkinson's disease. *Cell*, *167*(6), 1469–1480. e1412.

Sanmarco, L. M., Wheeler, M. A., Gutiérrez-Vázquez, C., Polonio, C. M., Linnerbauer, M., Pinho-Ribeiro, F. A., et al. (2021). Gut-licensed IFNγ+ NK cells drive LAMP1 +TRAIL+ anti-inflammatory astrocytes. *Nature*, *590*(7846), 473–479.

Scheperjans, F., Aho, V., Pereira, P. A., Koskinen, K., Paulin, L., Pekkonen, E., et al. (2015). Gut microbiota are related to Parkinson's disease and clinical phenotype. *Movement Disorders*, *30*(3), 350–358.

Scheperjans, F., Derkinderen, P., & Borghammer, P. (2018). The gut and Parkinson's disease: Hype or hope? *Journal of Parkinson's Disease*, *8*(s1), S31–s39.

Schetters, S. T. T., Gomez-Nicola, D., Garcia-Vallejo, J. J., & Van Kooyk, Y. (2018). Neuroinflammation: Microglia and T cells get ready to tango. *Frontiers in Immunology*, *8* (eCollection 2017).

Schoenborn, J. R., & Wilson, C. B. (2007). Regulation of interferon-γ during innate and adaptive immune responses. *Advances in Immunology*, *96*, 41–101.

Schulthess, J., Pandey, S., Capitani, M., Rue-Albrecht, K. C., Arnold, I., Franchini, F., et al. (2019). The short chain fatty acid butyrate imprints an antimicrobial program in macrophages. *Immunity*, *50*(2), 432–445. e437.

Seldin, M. F., Howard, T. A., & D'Eustachio, P. (1989). Comparison of linkage maps of mouse chromosome 12 derived from laboratory strain intraspecific and Mus spretus interspecific backcrosses. *Genomics*, *5*(1), 24–28.

Seltzer, M. A., Jahan, S. A., Sparks, R., Stout, D. B., Satyamurthy, N., Dahlbom, M., et al. (2004). Radiation dose estimates in humans for (11)C-acetate whole-body PET. *Journal of Nuclear Medicine*, *45*(7), 1233–1236.

Sender, R., Fuchs, S., & Milo, R. (2016). Revised estimates for the number of human and bacteria cells in the body. *PLoS Biology*, *14*(8), e1002533.

Shen, L., Liu, L., & Ji, H. F. (2017). Alzheimer's disease histological and behavioral manifestations in transgenic mice correlate with specific gut microbiome state. *Journal of Alzheimer's Disease*, *56*(1), 385–390.

Siddiqui, M. T., & Cresci, G. A. M. (2021). The immunomodulatory functions of butyrate. *Journal of Inflammation Research*, *14*, 6025–6041.

Silva, Y. P., Bernardi, A., & Frozza, R. L. (2020). The role of short-chain fatty acids from gut microbiota in gut-brain communication. *Frontiers in Endocrinology*, *11*, 25.

Singh, V., Roth, S., Llovera, G., Sadler, R., Garzetti, D., Stecher, B., et al. (2016). Microbiota dysbiosis controls the neuroinflammatory response after stroke. *The Journal of Neuroscience*, *36*(28), 7428–7440.

Sofroniew, M. V., & Vinters, H. V. (2010). Astrocytes: Biology and pathology. *Acta Neuropathologica*, *119*(1), 7–35.

Song, W. S., Nielson, B. R., Banks, K. P., & Bradley, Y. C. (2009). Normal organ standard uptake values in carbon-11 acetate PET imaging. *Nuclear Medicine Communications*, *30*(6), 462–465.

Song, X., Sun, X., Oh, S. F., Wu, M., Zhang, Y., Zheng, W., et al. (2020). Microbial bile acid metabolites modulate gut RORγ+ regulatory T cell homeostasis. *Nature*, *577*(7790), 410–415.

Sparks Stein, P., Steffen, M. J., Smith, C., Jicha, G., Ebersole, J. L., Abner, E., et al. (2012). Serum antibodies to periodontal pathogens are a risk factor for Alzheimer's disease. *Alzheimer's & Dementia*, *8*(3), 196–203.

Stilling, R. M., van de Wouw, M., Clarke, G., Stanton, C., Dinan, T. G., & Cryan, J. F. (2016). The neuropharmacology of butyrate: The bread and butter of the microbiota-gut-brain axis? *Neurochemistry International*, *99*, 110–132.

Strandwitz, P. (2018). Neurotransmitter modulation by the gut microbiota. *Brain Research*, *1693*(Pt B), 128–133.

Sweeney, M. D., Sagare, A. P., & Zlokovic, B. V. (2018). Blood-brain barrier breakdown in Alzheimer disease and other neurodegenerative disorders. *Nature Reviews. Neurology*, *14*(3), 133–150.

Takanaga, H., Ohtsuki, S., Hosoya, K., & Terasaki, T. (2001). GAT2/BGT-1 as a system responsible for the transport of gamma-aminobutyric acid at the mouse blood-brain barrier. *Journal of Cerebral Blood Flow and Metabolism*, *21*(10), 1232–1239.

Tan, J., McKenzie, C., Potamitis, M., Thorburn, A. N., Mackay, C. R., & Macia, L. (2014). Chapter three—The role of short-chain fatty acids in health and disease. In F. W. Alt (Ed.), *121. Advances in immunology* (pp. 91–119). Academic Press.

Tap, J., Mondot, S., Levenez, F., Pelletier, E., Caron, C., Furet, J. P., et al. (2009). Towards the human intestinal microbiota phylogenetic core. *Environmental Microbiology*, *11*(10), 2574–2584.

Tazoe, H., Otomo, Y., Kaji, I., Tanaka, R., Karaki, S., & Kuwahara, A. (2008). Roles of short-chain fatty acids receptors, GPR41 and GPR43 on colonic functions. *Journal of Physiology and Pharmacology*, *59*(Suppl 2), 251–262.

Tazoe, H., Otomo, Y., Karaki, S., Kato, I., Fukami, Y., Terasaki, M., et al. (2009). Expression of short-chain fatty acid receptor GPR41 in the human colon. *Biomedical Research*, *30*(3), 149–156.

Thangaraju, M., Cresci, G. A., Liu, K., Ananth, S., Gnanaprakasam, J. P., Browning, D. D., et al. (2009). GPR109A is a G-protein–coupled receptor for the bacterial fermentation product butyrate and functions as a tumor suppressor in colon. *Cancer Research*, *69*(7), 2826–2832.

Thiemann, S., Smit, N., Roy, U., Lesker, T. R., Gálvez, E. J., Helmecke, J., et al. (2017). Enhancement of IFNγ production by distinct commensals ameliorates Salmonella-induced disease. *Cell Host & Microbe*, *21*(6), 682–694. e685.

Thorén, F. B., Riise, R. E., Ousbäck, J., Della Chiesa, M., Alsterholm, M., Marcenaro, E., et al. (2012). Human NK cells induce neutrophil apoptosis via an NKp46- and Fas-dependent mechanism. *Journal of Immunology*, *188*(4), 1668–1674.

Thursby, E., & Juge, N. (2017). Introduction to the human gut microbiota. *The Biochemical Journal*, *474*(11), 1823–1836.

Tomova, A., Soltys, K., Repiska, G., Palkova, L., Filcikova, D., Minarik, G., et al. (2020). Specificity of gut microbiota in children with autism spectrum disorder in Slovakia and its correlation with astrocytes activity marker and specific behavioural patterns. *Physiology & Behavior*, *214*, 112745.

Uemura, N., Yagi, H., Uemura, M. T., Hatanaka, Y., Yamakado, H., & Takahashi, R. (2018). Inoculation of α-synuclein preformed fibrils into the mouse gastrointestinal tract induces Lewy body-like aggregates in the brainstem via the vagus nerve. *Molecular Neurodegeneration*, *13*(1), 21.

Unger, M. M., Spiegel, J., Dillmann, K.-U., Grundmann, D., Philippeit, H., Bürmann, J., et al. (2016). Short chain fatty acids and gut microbiota differ between patients with Parkinson's disease and age-matched controls. *Parkinsonism & Related Disorders*, *32*, 66–72.

Vatanen, T., Kostic, A. D., d'Hennezel, E., Siljander, H., Franzosa, E. A., Yassour, M., et al. (2016). Variation in microbiome LPS immunogenicity contributes to autoimmunity in humans. *Cell*, *165*(4), 842–853.

Vellingiri, B., Aishwarya, S. Y., Benita Jancy, S., Sriram Abhishek, G., Winster Suresh Babu, H., Vijayakumar, P., et al. (2022). An anxious relationship between autism spectrum disorder and gut microbiota: A tangled chemistry? *Journal of Clinical Neuroscience*, *99*, 169–189.

Vogt, N. M., Kerby, R. L., Dill-McFarland, K. A., Harding, S. J., Merluzzi, A. P., Johnson, S. C., et al. (2017). Gut microbiome alterations in Alzheimer's disease. *Scientific Reports*, 7(1), 13537.

Waggoner, S. N., & Kumar, V. (2012). Evolving role of 2B4/CD244 in T and NK cell responses during virus infection. *Frontiers in Immunology*, *3*, 377.

Wahlström, A., Sayin, S. I., Marschall, H.-U., & Bäckhed, F. (2016). Intestinal crosstalk between bile acids and microbiota and its impact on host metabolism. *Cell Metabolism*, *24*(1), 41–50.

Wallen, Z. D., Appah, M., Dean, M. N., Sesler, C. L., Factor, S. A., Molho, E., et al. (2020). Characterizing dysbiosis of gut microbiome in PD: Evidence for overabundance of opportunistic pathogens. *NPJ Parkinsons Disease*, *6*, 11.

Wallen, Z. D., Demirkan, A., Twa, G., Cohen, G., Dean, M. N., Standaert, D. G., et al. (2022). Metagenomics of Parkinson's disease implicates the gut microbiome in multiple disease mechanisms. *bioRxiv*. 2022.2006.2008.495316.

Wenzel, T. J., Gates, E. J., Ranger, A. L., & Klegeris, A. (2020). Short-chain fatty acids (SCFAs) alone or in combination regulate select immune functions of microglia-like cells. *Molecular and Cellular Neuroscience*, *105*, 103493.

Wikoff, W. R., Anfora, A. T., Liu, J., Schultz, P. G., Lesley, S. A., Peters, E. C., et al. (2009). Metabolomics analysis reveals large effects of gut microflora on mammalian blood metabolites. *Proceedings of the National Academy of Sciences*, *106*(10), 3698–3703.

Xie, A., Ensink, E., Li, P., Gordevičius, J., Marshall, L. L., George, S., et al. (2022). Bacterial butyrate in Parkinson's disease is linked to epigenetic changes and depressive symptoms. *Movement Disorders*. https://doi.org/10.1002/mds.29128.

Yano, J. M., Yu, K., Donaldson, G. P., Shastri, G. G., Ann, P., Ma, L., et al. (2015). Indigenous bacteria from the gut microbiota regulate host serotonin biosynthesis. *Cell*, *161*(2), 264–276.

Zelante, T., Iannitti, R. G., Cunha, C., De Luca, A., Giovannini, G., Pieraccini, G., et al. (2013). Tryptophan catabolites from microbiota engage aryl hydrocarbon receptor and balance mucosal reactivity via interleukin-22. *Immunity*, *39*(2), 372–385.

Zhang, R., Miller, R. G., Gascon, R., Champion, S., Katz, J., Lancero, M., et al. (2009). Circulating endotoxin and systemic immune activation in sporadic amyotrophic lateral sclerosis (sALS). *Journal of Neuroimmunology*, *206*(1-2), 121–124.

Zhang, J., Yu, C., Zhang, X., Chen, H., Dong, J., Lu, W., et al. (2018). Porphyromonas gingivalis lipopolysaccharide induces cognitive dysfunction, mediated by neuronal inflammation via activation of the TLR4 signaling pathway in C57BL/6 mice. *Journal of Neuroinflammation*, *15*(1), 1–14.

Zhao, C., Hu, X., Bao, L., Wu, K., Feng, L., Qiu, M., et al. (2021). Aryl hydrocarbon receptor activation by Lactobacillus reuteri tryptophan metabolism alleviates Escherichia coli-induced mastitis in mice. *PLoS Pathogens*, *17*(7), e1009774.

Zhao, Y., Jaber, V., & Lukiw, W. J. (2017). Secretory products of the human GI tract microbiome and their potential impact on alzheimer's disease (AD): Detection of lipopolysaccharide (LPS) in AD hippocampus. *Frontiers in Cellular and Infection Microbiology*, 7, 318.

CHAPTER THREE

The many genomes of Parkinson's disease

Haydeh Payami*

[a]Professor of Genetics and Neurology, Department of Neurology, University of Alabama at Birmingham, Birmingham, AL, United States
[b]Aligning Science Across Parkinson's (ASAP) Collaborative Research Network, Chevy Chase, MD, United States
*Corresponding author: e-mail addresses: haydehpayami@yahoo.com; haydehpayami@uabmc.edu

Contents

Abstract

Genetic component of Parkinson's disease, once firmly believed non-existent, involves the human genome, mitochondrial genome, and the microbiome. Understanding the genomics of PD requires identification of PD-relevant genes and learning how they interact within the hologenome and with their environment. This chapter is an evidence-based perspective of a geneticist on how far we have come in this endeavor. The contemporary scientific society started with a naive and simplistic view of PD, evolved to accept that Parkinson's disease is probably the most complex disease there is, the progress we have made in discovering the genes and elucidating their functions, and now assembling the parts to create the whole.

International Review of Neurobiology, Volume 167
ISSN 0074-7742
https://doi.org/10.1016/bs.irn.2022.07.007

1. Holistic view

Aristotle originated the concept of holism. He is said to have believed that systems should be studied in their holistic entirety, concentrating on the interaction between the different components rather than on the individual parts. "The whole is greater than the sum of its parts". Easier said than done if you are not Aristotle.

1.1 Human holobiont

Humans are holobionts composed of a host and trillions of microorganisms whose collective functioning keeps the whole alive. At the core of each organism is its genome. The genome in the nucleus of human cells (henceforth referred to as human genome) is made of linear double stranded DNA packed in 23 pairs of chromosomes in each of the ~37 trillion human cells in a body. Human genome has ~60,000 genes, ~20,000 of them are protein-coding (Nurk et al., 2022; Schneider et al., 2017). Each human cell also contains one or many mitochondria, and each mitochondrion has varying numbers of circular DNA (Sharma & Sampath, 2019). There are ~39 trillion microorganism in human body (microbiota), each with its own unique genome. These include Bacteria and Archaea with circular DNA genomes and viruses having either DNA or RNA genomes in linear or circular forms. Microbiome refers to the collective genomes of the microbiota in their ecological environment (gut microbiome, skin microbiome, etc.). The genetic composition of human microbiome is highly diverse (Pasolli et al., 2019). The gut microbiome alone has an estimated 22 million non-redundant genes.

1.2 Complexities of PD

Not long ago, PD was thought to be a single disease that involved the substantia nigra in the brain and caused movement disorder. The etiology was believed to be purely environmental with no genetic component. Owing to investigators who dared to go outside this box and challenge the old dogmas, substantial progress was made towards a more holistic view of PD in a relatively short time. While some may still disagree hanging on tight to the classical definitions, many in the field have come to believe that PD is not a single disease, involves more than just the brain, and causes more problems than movement disorder (Obeso et al., 2017). The cause of PD is not purely environmental. In fact, the only proven causes of PD so far are genes (plural: even the rare genetic forms of PD have different causes) (Klein & Westenberger, 2012). PD affects nearly every system in the body, including the brain and gastrointestinal (GI) tract as the potential sites of origin and with a heavy influence from the immune system. The earliest manifestations are constipation, olfactory dysfunction, sleep disorder, depression, and mood disorder (Postuma & Berg, 2016). The cardinal motor signs, which are still the diagnostic requirements, develop decades after the GI symptoms. As disease progresses, so do the motor and non-motor symptoms including GI, cognitive, and psychiatric disorders (Weintraub & Mamikonyan, 2019). In later stages of disease an unproportionally high number of persons with PD develop dementia (Aarsland et al., 2001). To solve a disease this complex, it would behoove us to maintain a holistic perspective in our daily research, whether we are experimenting with molecules or running a clinical trial.

2. Hologenome

Genomics revolutionized PD research three times in the past three decades. The first was in the 1990's, a paradigm shift brought on by family-based genetic studies that challenged the dogma: *PD is a purely environmental disease*, and ushered geneticists into PD research. Second was the discovery of the genes that cause PD in the 2000's and the identification of 90 genetic risk factors in the 2010's, each of which opened a new window to PD pathobiology. The third is happening now in the 2020's, with the emerging field of metagenomics pivoting and expanding the genomics of PD to include the microbiome.

Understanding the full genetic contribution to PD would entail identifying all the individual components and studying their interactions within the hologenome (i.e., human genome, mitochondrial genome, and microbiome), and understanding how the hologenome responds to and interacts with the elements in its environment. That is the ideal we strive for, as there is compelling evidence that human genome, mitochondria, gut microbiome and environment all contribute to PD.

2.1 Human genome

Up until early 1990's, the medical and scientific community firmly believed that PD had no genetic component. A family-based study conducted in the

1990's unearthed evidence for a genetic component in PD (Payami, Larsen, Bernard, & Nutt, 1994), not only in familial and early onset forms of disease which is typical of genetic disorders, but also in sporadic and late-onset PD. Replicated world-wide, this paradigm shift generated mounting interest from geneticists to engage in PD research. Geneticists began in the 2000's with identifying the genes that cause familial forms of PD for which tools were already perfected (Klein & Westenberger, 2012). Gene discovery for non-familial forms of PD was having limited success with candidate gene approach, having identified robust associations with *SNCA* (which encodes alpha-synuclein) and *MAPT* (which encodes tau), but also a mounting literature of irreproducible candidate genes. Genetics of idiopathic PD took off with the advent of genome-wide association studies (GWAS). GWAS was made possible by three groundbreaking and well-orchestrated advances in the 2000's: The Human Genome Project provided the map, companies manufactured genotyping microarrays, and analysts developed the statistical tools. Coupled with large cohorts that had been assembled in anticipation, the path was paved to the identification of genetic risk factors for sporadic and late onset PD. The first two GWAS which identified novel loci for PD (in addition to confirming *SNCA* and *MAPT*) were published in 2009 by Satake et al. who discovered association of PD risk with *BSL1* variants, and in 2010 by Hamza et al. who discovered association of PD risk with *HLA* variants. These were the first novel PD-associated variants that met the newly established standards of rigor: a genome-wide multiple testing corrected $P < 5E\text{-}8$ and replication in a second dataset. While it may have felt jarringly prohibitive at the time, the standards of rigors adopted by geneticist and enforced by top tier journals, did prohibit excessive publication of unreproducible associations with candidate genes and small sample sizes, and as a result, transformed genetic associations into a robust body of data. This is a salient point for this chapter because microbiome association studies today are operating without these safeguards, and the field is as inconsistent as genetics associations were before multiple testing correction and independent replication were required. To meet these requirements effectively means having either a large and very clean data set (Satake et al., and Hamza et al) or an extremely large albeit noisy data set (meta-analyses that ensued (Nalls et al., 2019)). As of this writing, over 100 genetic risk factors have been identified for PD, which range from rare but highly penetrant causative mutations to common variants that increase disease susceptibility. Each gene has opened a new window to disease pathology. The discoveries of *SNCA*, *PINK1*, *PRKN*, *LRRK2*, and *GBA* each began a new and vigorous field of investigation and have revealed central involvement of protein aggregation, and dysregulation of endosomal, lysosomal, and mitochondrial pathways in PD (Hardy, 2010; Vazquez-Velez & Zoghbi, 2021). Similarly, the discovery of an association with *HLA* ignited the interest in the role of immune system in PD, which is now understood to be integral to disease pathogenesis (Schonhoff, Williams, Wallen, Standaert, & Harms, 2020).

2.2 Mitochondrial genome

Mitochondrion is known first and foremost for being the powerhouse of the cell, although its functions extend well beyond energy production. Mitochondria are fully integrated with other eukaryotic cellular compartments in regulating oxidative stress response, apoptosis, autophagy, and communication between cells and tissues (Nunnari & Suomalainen, 2012). An essential organelle of eukaryotic cells, mitochondrion was once a free-living bacterium (Sagan, 1967) that integrated with an archaeon (Roger, Munoz-Gomez, & Kamikawa, 2017). The transformation from endosymbiosis to permanent eukaryotic organelle involved incremental yet drastic evolutionary changes, including integration of metabolism and transfer of genes between endosymbiotic bacteria and the archaea host (Gogoi et al., 2022). Today's mitochondrial proteome is 40% bacterial, 60% archaeal/eukaryotic. Bacteria and archaea have jointly created an organelle that eukaryotes, including humans, cannot live without. Mitochondrial dysfunction halts energy production, disrupts calcium buffering, and increases oxidative stress. Loss of mitochondrial function causes cell death, including those neurons that are lost in PD. Mitochondrial involvement in PD was first noted in the 1980's via exposure to the neurotoxin, MPTP (Langston, Ballard, Tetrud, & Irwin, 1983) and has gained strong and increasing support over the years. The discovery and elucidation of the function of *PRKN* and *PINK1* as causal PD genes solidified the central role of mitochondria in autosomal recessive PD, and *SNCA* and *LRRK2* mutations showed the relevance of mitochondria to autosomal dominant PD (Borsche, Pereira, Klein, & Grunewald, 2021). Mitochondrial dysfunction is a common feature in idiopathic as well as familial PD. While toxicant exposure and genetic variations are well-establish as causes of mitochondrial deficiency, the role of the mitochondria genome itself has remained elusive (Muller-Nedebock et al., 2019). Unlike the human chromosomes that can be traced faithfully from parent to child (23 from mother, 23 from father), and from cell to cell (the same 23 pairs in practically every cell), mitochondrial DNA (mtDNA) is difficult to trace. Each cell can have many mitochondria, and each mitochondrion may have thousands of mtDNA depending on the energy requirements of the cell. mtDNA accumulates somatic mutations at a much higher rate than nuclear DNA. One reason for the high mutation rate is that mtDNA replication is error prone. Another is because mtDNA is physically close to DNA-damaging oxidative products that mitochondria generate. With age, these somatic mutations accumulate. An individual who may have started life with a homogenous mitochondrial genome, may acquire variation in mtDNA (heteroplasmy) in some but not all cells, and to varying degrees. This unpredictable variation makes mtDNA alterations difficult to pin to disease in a meaningful way. That being said, there is a consistent pattern in PD of deletions in mtDNA and overall depletion of mitochondrial genomes (Muller-Nedebock et al., 2019). In addition to genes involved in energy production and metabolic necessities of mitochondria, mitochondrial genome encodes peptides that regulate nuclear genes. Thus, common age-related deletions of segments of mtDNA, could not only deplete critical protein coding genes for mitochondrial function, but it can also disrupt mtDNA-nuclear genome cross-regulation. It has also been shown that the number of mtDNA genomes, a critical threshold of which is required to meet the energy demands for survival, is depleted in blood, cerebrospinal fluid, and vulnerable regions of substantia nigra in PD (Pyle et al., 2016). It is worth noting that the depletion of mtDNA in PD has been linked to both human genome (variations in nuclear encoded mitochondrial polymerase POLG genes affects the rate and efficacy of mtDNA replication), and to environmental factors that cause mitochondrial toxicity, depleting mitochondria and the genomes within (Borsche et al., 2021); underscoring once again, interactions between human genome, mtDNA and environment.

2.3 Microbiome

The world, as we humans perceive it, is run by microbes. Without them life would not exist. Microbes live in communities and inhabit niches within their specialized ecosystems. The terms microbiota and microbiome refer to the community of the individual microorganisms that inhabit an ecosystem, and their collective genomes, respectively. Metagenome refers to all the genetic material that are sampled from a microbiome for the purpose of studying it. Metagenomics is the study, which like genomics, takes a holistic view of all the genes. Unlike human genomics which concerns with the genome of one organism, metagenomics captures genomes of all organisms in the ecosystem—which would be tens of trillions of microorganisms if the gut microbiome is being studied.

2.3.1 The leap from single organism culture to metagenomics

For over a century, studies of the microbiota were focused on individual pathogens and entailed culturing, isolating, and characterizing the individual microorganisms. The rest of the microorganisms in the human body were not of interest because they were thought to be benign and unimportant for the host. Up until 2005, before DNA sequencing entered the field, everything we had learned about human microbiomes came from the laboratory: by growing the microbes outside their natural habitat, in isolation from their community, under optimized conditions. Ecologists pioneered the use of DNA sequencing in the 1980's to study the microbial communities in forest and oceanic ecosystems. In 2005, adopting the next generation (NG) DNA sequencing, Eckburg et al. described the human gut microflora through 13,355 prokaryotic ribosomal RNA genes that they detected in multiple colonic mucosal sites and human stool—possibly the first ever top-down view of a human microbiome (Eckburg et al., 2005). They discovered immense genetic diversity, unfathomable at the time, and recognized that the majority of the microorganisms that were detected by DNA sequencing were unknown before.

2.3.2 Methods and challenges

The human gut microbiome can now be studied, in large-scale human case-control or longitudinal studies, by obtaining fecal samples, extracting DNA, conducting NG DNA sequencing, followed by bioinformatic analyses of DNA sequences to assign taxonomy and to profile the diversity, and statistical analyses to compare groups (ex., PD vs. controls) for the global composition of their microbiomes, or differential abundance of specific microorganisms. It seems deceptively straight forward.

The most common NG sequencing approaches are 16S ribosomal RNA gene amplicon sequencing (16S rRNA for short), and whole metagenome shotgun sequencing (shotgun for short). The 16S rRNA approach targets and sequences variable regions of the 16S rRNA gene, which is present in Bacteria and Archaea. Viruses and Eukarya are not detectable by 16S rRNA sequencing. The taxonomic resolution of 16S rRNA sequencing often does not exceed the genus level, and it has limited functional resolution because the rest of the microbial genes are inferred, not directly sequenced. Shotgun sequencing, on the other hand, captures the entire metagenome, i.e., all the DNA that is present in the sample, and can therefore detect Bacteria, DNA Viruses, Eukarya and Archaea (as well as any human DNA present in stool which is removed from analysis bioinformatically). Shotgun sequencing improves taxonomic resolution to species and strain level, and functional resolution to genes and single nucleotides. Function can be directly inferred from characterized orthologous genes and the pathways they contribute to, using KEGG ORTHOLOGY and MetaCyc databases. If shotgun sequencing is performed at a sufficient depth, even the very low abundance genes and microorganisms can be detected. Shotgun metagenomics is the gold standard, but the cost can be prohibitive. Currently, commercial companies specializing in NG sequencing charge between $50 to $100 per sample for 16S rRNA sequencing, as compared to over $300 per sample for deep shotgun sequencing. The cost differential can become prohibitive for large-scale studies. The first large-scale deep shotgun study of PD recently appeared in the pre-publication repository, bioRxiv (Wallen et al., 2022).

While metagenomics is afforded the same technological advantages as human genomics, it comes with unique challenges. Foremost is the inability to discern causality. A coveted attribute of genetics is the causal inference. When a disease is discovered to be linked to a change in the human DNA, one seldom asks which is the cause and which is the consequence. The specific DNA variant that correlates with disease may not be the causal element itself, but the cause is somewhere on that stretch of DNA. In a genetic association, DNA is at the root tied to the cause, and disease is the consequence. The same cannot be said for association of disease with a dysbiotic microbiome, or certain taxa or even a microbial gene. Disease is as likely, if not more likely, to be the cause of the dysbiosis than the other way around. An unhealthy body can affect the microbiome, causing depletion of some microorganism and allowing others to bloom, which will also reflect on the differential abundances of microbial genes and pathways. Once an association is found between disease and an altered microbiome state, it will require longitudinal and clinical human studies and experiments in model systems to decipher the directionality, causality, and the mechanism of the association.

Second, is the elusive nature of microbiome as compared to the stable human genome. Mendel somehow managed to distill what became genetics into two laws: segregation of alleles (each parent contributes one copy of each gene to an offspring) and independent assortment of genes (which turned out to apply to genes on different chromosomes and not to genes that are closely linked to each other). These laws, while broken more times than held true, served as the principles that guided us into the era of molecular genetics and genomics, and to this day are used for genetic counseling. Compared to the orderly laws of the human genome, the microbiome is chaos. The human genome is stable and predictable, mutation rate is low and negligible for all practical concerns. Whereas the microbiome can be wiped out with one course of antibiotics. It changes by medication, diet, geography, and a myriad of other factors (Scepanovic et al., 2019). As if navigating this much variation were not challenging enough, the factors currently known to affect gut microbiome composition account for less than 20% of the total variability—we are 80% in the dark.

Third, is the challenges that are intrinsic to any young and emerging field. We are not operating at full potential. Much of the human microbiome is unexplored and currently unrecognizable (Pasolli et al., 2019). Reference databases used for taxonomic assignments are incomplete and are being continually updated. As unknown sequences are characterized, new taxonomic assignments are added, and old ones are re-defined. Methods are not standardized. Over 20 statistical methods have been developed for analysis of microbiome data, and they yield vastly inconsistent results (Nearing et al., 2022; Wallen, 2021) which underscores the analytic nuances that investigators must be well-versed in and able to navigate in a disease- and data-specific manner. The field is exciting and seems deceptively easy: get stool samples from your patients, send them to a company and they will give you the results to publish. Excitement is high, rigor not always. The metagenomics community is keenly aware and working towards establishing tried and proven methods and setting standards of rigor.

2.3.3 The human microbiome project

The Human Microbiome Project (HMP) was established in 2008 with funding support of the National Institutes of Health of the United States. HMP was designed as a community resource to develop and release data, reference databases for taxonomic profiling, and methods for analysis of human microbiome. In the first phase of HMP, 242 healthy individuals age 18–40 years were studied, where microbiome was sampled from five body sites, and characterized for structure and diversity using 16S sequencing and a subset were shotgun sequenced to assess function (Human Microbiome Project, 2012). The second phase of HMP investigated the changes that occur in the microbiome in three conditions: vaginal microbiome collected during 600 pregnancies, the gut microbiome in 100 individuals with inflammatory bowel disease, and host and microbial responses to dietary changes and infections in 106 individuals with pre-diabetes (Integrative, 2019). A second wave of data was generated with updated profiling and assembly methods and published in 2017 (Lloyd-Price et al., 2017). The data, software and technology are broadly available (https://www.hmpdacc.org/hmp/) and will likely be continually re-analyzed and updated with improving methodology.

2.3.4 A glimpse at the monumental importance of our microbiomes

DNA sequencing has vastly increased the fundamental knowledge of how microorganisms interact with each other in their communities and with the host. The nose is the main point of entry for inhaled toxicants and pathogens. The microbiome of the nose and upper respiratory tract (URT) is our first line of defense against them, not only as a filter for the particles we breath, but also actively neutralizing potentially harmful pathogens and toxicants via direct access to and communication with host immune system (Kumpitsch, Koskinen, Schopf, & Moissl-Eichinger, 2019). Nasal microbiome is involved in the development of olfactory epithelium and maintenance of its function, including the sense of smell and the physical connection to and communication with the brain (Koskinen et al., 2018). Loss of sense of smell occurs in 90% of person with PD early in disease and often manifests before the onset of motor signs (Doty, 2012). There is not yet sufficient data to link URT microbiome to PD (attempts have been made but studies are few and inconclusive). It is unknown if alpha-synuclein pathology can originate in the olfactory (Braak et al., 2003). That said, inarguably, URT microbiome can protect the host or make them vulnerable to air born environmental triggers of PD. The skin microbiome protects the host against the UV light, creates an environment that is not conducive to pathogens and acts as natural antibiotics to kill them, helps heal wounds, communicates with the immune system and controls inflammation (Boxberger, Cenizo, Cassir, & La Scola, 2021). The oral microbiome is the most extensively studied human microbiome because of the accessibility and the high prevalence of oral infections and inflammatory diseases. Importance of the mouth microbiome extends beyond oral health to general systemic health (Deo & Deshmukh, 2019), much like the nasal microbiome, because of open connectivity with other parts of the body, including the GI and gut microbiome. The gut microbiome helps digest food and produce vitamins. Gut microbiome is involved in metabolism of drugs and influences efficacy and toxicity of medications, including L-DOPA (van Kessel et al., 2019). Microbiota metabolize toxicants, and have the capability to activate pre-toxins as well as degrade active neurotoxins, which is a potential link to toxicant-associated PD. Most fundamental, however, is the role of the gut microbiome in the maturation and maintenance of the host immune system, and in the biochemical signaling in the gut-brain axis. The gut microbiome communicates with the brain via metabolites encoded by microbial genes. One mode of action is by acting as triggers on host signaling pathways, such as by endotoxin lipopolysaccharide which induces immune reaction and inflammation by triggering host Toll-like receptor (TLR) signaling pathway (Tang et al., 2017). Another is by epigenetic modification of human genes which can have vast and far-reaching influences, including activation and control of immune response in the brain (Erny et al., 2015; van der Hee & Wells, 2021).

3. PD and the microbiome

More than 90% of PD is idiopathic; that is, less than 10% of PD are caused by fully penetrant genetic mutations, the rest are of unknown cause (Klein & Westenberger, 2012). It is often said that idiopathic PD arises from interaction between genetic susceptibility and environmental factors, mediated by intrinsic factors such as sex and age. However, despite having discovered 90 genetic risk factors (Nalls et al., 2019) and numerous environmental factors (Bellou, Belbasis, Tzoulaki, Evangelou, & Ioannidis, 2016), we still do not have even one gene-environment combination that is proven to cause PD. It is possible that the microbiome is the missing link in the gene-environment equation.

Interestingly, in the gene-environment equation, the microbiome can be regarded as either genes or as environment. This dual nature offers the best of both worlds: defined as the collective genomes of microorganism living in a particular environment, the microbiome is measured and analyzed with the precision and rigor of molecular genetic techniques, but unlike the nuclear genome, which is inherited and almost immutable, the composition of microbiome can be modified to prevent and treat disease.

3.1 Rationale for focusing on the gut microbiome

The scientific rationale for focusing on the gut in PD is multi-fold. GI dysfunction is a common feature of PD, including constipation, which may be the first prodromal sign of PD preceding motor signs by decades (Chen et al., 2015). PD gut has signs of inflammation (Houser et al., 2018). The gut barrier is compromised early in PD, showing increased permeability with alpha-synuclein staining and proinflammatory bacterial endotoxins (Forsyth et al., 2011). Alpha-synuclein pathology has been documented in the GI tract both in early PD and in persons with REM sleep behavior disorder (RBD) which is a PD prodrome, suggesting that in a subset of PD, pathology may initiate in the gut (Berg et al., 2021). Alpha-synuclein can spread from cell to cell, like a prion, and form pathologic aggregates in the receiving cell (Henderson, Trojanowski, & Lee, 2019). The discovery of alpha-synuclein pathology in the gut and the prion like behavior was reminiscent of Braak's hypothesis that PD can start in the gut and spread to the brain (Braak, Rub, Gai, & Del Tredici, 2003). One route by which alpha-synuclein might spread from the gut to the brain is the vagus nerve. To explore this possibility, large population-based healthcare databases in Denmark and Sweden were mined to identify individuals who had undergone vagotomy decades ago and found that complete truncal vagotomy was associated with 50% reduction in incidence of PD later in life (Liu et al., 2017; Svensson et al., 2015). The hypothesis was put to test experimentally in mice by injecting alpha-synuclein fibrils into the gut and observing the formation of pathologic alpha-synuclein and its transmission to the brain, and moreover, demonstrating that if the vagus nerve were severed, the spread to the brain would halt (Kim et al., 2019). A recent study used multi-modal imaging on PD and RBD patients to trace pathological markers across body sites and found the pattern compatible with two potential subtypes for PD: one which starts in the brain and spreads down, the other starts in the gut and spreads up to the brain (Horsager et al., 2020). With the gut becoming a focal point in PD investigations, the gut microbiome began to attract interest because the microorganisms in the gut play critical roles in mechanism that are aberrant in PD, most notably, the integrity of the GI membrane, GI motility, and inflammation.

One can envision a role for gut microbiome in genetic, toxic, or even an infectious etiology for PD. Microbiome can modulate the penetrance of genetic risk factors by providing, or withholding, the second hit that would render a genetic susceptibility into causative. A striking example is genetic susceptibility to cerebral cavernous malformation (CMM). It has been shown that CMM develops in the genetically susceptible when lipopolysaccharides in the gut microbiome trigger the host TLR4 signaling pathway; and strikingly, the genetic susceptibility can be nullified and disease prevented with a course of antibiotics (Tang et al., 2017). There is also early evidence in PD that increased risk of PD conferred by regulatory elements of *SCNA* gene is modulated by the abundance of specific opportunistic pathogens in the gut microbiome (Wallen et al., 2021). Microbiome could potentially harbor pathogens that would trigger the initiation of the pathology outside the brain, as Braak first proposed. Professor Heiko Braak and colleagues originated the notion that non-familial forms of PD may be caused by an unknown pathogen that enters the body via nasal passage, initiates pathology in the gut, and spreads from the gut to the brain (Braak, Rub, et al., 2003). While an infectious etiology for PD is not yet proven, increased abundance of opportunistic pathogens has been observed in PD gut metagenome (Wallen et al., 2020). In addition, in a genetic mouse model of PD (*PINK1*$^{-/-}$) intestinal infection with Gram-negative pathogens was shown to be the triggering event of the disease, by engaging mitochondrial antigen presentation and eliciting an autoimmune reaction (Matheoud et al., 2019). This latter study exemplifies entangling of a complex interaction: they demonstrated that interaction involving all three genomes (human *PINK1*, mitochondria, gut microbiome) and an environmental trigger (infection) induced onset of neurodegeneration and motor deficits. There are likely many such complex interactions at play in PD.

One could also envision microbiome as a genomic milieu that affects the rate of disease progression and determines the inter-individual variation in efficacy of drugs and adverse outcomes. A prime example is the correlation between bacterial *tdc* gene and efficacy of L-DOPA treatment. L-DOPA is used for dopamine replacement therapy because dopamine cannot cross the blood brain barrier. To prevent metabolism of L-DOPA before it reaches the brain, it is taken with carbidopa which inhibits the human enzyme, aromatic amino acid decarboxylase, that coverts L-DOPA to dopamine. But carbidopa does not inhibit the bacterial enzyme, encoded by *tdc* genes, that can also metabolize L-DOPA to dopamine. This connection was first discovered in the 1970's (Bergmark et al., 1972). Recent studies confirmed the observation (Maini Rekdal, Bess, Bisanz, Turnbaugh, & Balskus, 2019; van Kessel et al., 2019), showed that *tdc* converts L-DOPA to dopamine even in the presence of carbidopa, and that the relative abundance of *tdc* gene in the microbiome correlates with plasma levels of L-DOPA. This suggests a vicious cycle where the higher the abundances of bacteria *tdc* in the gut microbiome, the less L-DOPA would be available to brain, limiting treatment efficacy, increasing side effects, and requiring higher drug doses to manage symptoms.

3.2 Microbiome at other body sites

The lesser emphasis here on other microbiomes does not imply that the nasal, oral, upper GI, or other sites are not involved in PD, rather the sparsity of knowledge. As speculated above, nasal microbiome is key to keeping inhaled toxicants and pathogens at bay. The upper GI microbiome is already known to play a role in efficacy of L-DOPA treatment, dating back to when microbiome studies were culture-based. *H. pylori*, which colonizes the upper intestine, interferes with absorption of L-DOPA (Pierantozzi et al., 2001), and thus limiting treatment efficacy of the drug. It is estimated that more than half the world's population is infected with *H. pylori* (Hooi et al., 2017), with prevalence in adults exceeding 80% in countries with lower socioeconomic conditions, and 20%–50% in western countries. Most infected individuals are unaware of it. Moreover, meta-analysis of ten studies with 5043 PD patients and 23,449 healthy controls, showed prevalence of *H. pylori* is even higher, by 45%, in persons with PD (Dardiotis et al., 2018). In addition to interfering with L-DOPA treatment, *H. pylori* is of concern because the chronic inflammation and the dysbiosis that it induces in the microbiome can influence systemic damage, including cardiovascular and neurological symptoms (Santos et al., 2020). *H. pylori* infection can be easily diagnosed, treated with antibiotics, and eradicated.

3.3 Analysis of gut microbiome in PD

The first question to address was whether the gut microbiome is altered in PD. The answer is a resounding yes. Every study that has investigated the global composition of gut microbiome in PD vs. non-PD controls (beta-diversity) has found it to be significantly different. The list is long, but recent meta-analyses provide excellent tabulations and summaries (Nishiwaki et al., 2020; Romano et al., 2021, 2021 #241; Toh et al., 2022).

The next step was to identify the taxa that drive the dysbiosis in PD gut. The first three studies were published in 2015 and they did not agree. Scheperjans et al. found 77% reduction in the abundance of *Prevotellaceae* family in PD compared to controls (Scheperjans et al., 2015). Keshavarzian et al. found a significant reduction in short-chain fatty acid (SCFA) producing genera and a significant increase of genes involved in biosynthesis of lipopolysaccharides in PD (Keshavarzian et al., 2015). Hasegawa et al. reported higher numbers of *Lactobacillus* in PD (Hasegawa et al., 2015). As the literature grew, so did the disparity in the results of taxa that were being associated with PD. Recently, results have started to converge. Owing to the groups who make their data publicly available, meta-analyses have begun to consolidate the data and extract robust genus and family level associations that transcend inter-study variations. In addition, emerging large-scale metagenomic studies with shotgun sequencing are clarifying some of the inconsistencies seen at genus and family level at the higher species resolution (Wallen et al., 2022). Three meta-analyses were conducted in the past 3 years (Nishiwaki et al., 2020; Romano et al., 2021; Toh et al., 2022). They all reported that PD microbiome has reduced levels of SCFA producing genera and is enriched in genus *Akkermansia*. In addition, each study reported associations that were not captured by the other two. Nishiwaki et al. reported increased abundance of *Catabacter*. Romano et al. found enrichment of the *Lactobacillus* and *Bifidobacterium*. Toh et al. found increased abundance of *Megasphaera*. Seems ironic that although they used much of the same publicly available datasets, they did not yield identical results. This is because once the raw sequences and meta data are downloaded, investigators must process and analyze the data. Since analysis protocols are not standardized, analysts have discretion at every step and their choices will determine the outcome. Hence, another source of discrepancy. More striking is the robustness of the signals for reduced SCFA genera and elevated *Akkermansia* which can be detected irrespective of where the data come from and how they are analyzed. Signals for *Catabacter*, *Lactobacillus*, *Bifidobacterium*, and *Megasphaera* are also likely true positives—as they have been reported by several studies.

Circling back to the original three papers: they were all right, despite being seemingly inconsistent. As Keshavarzian et al. reported, meta-analyses confirmed that SCFA producing taxa are reduced in PD, and the recent large-scale shotgun sequencing confirmed, by direct sequencing of all genes, that genes encoding lipopolysaccharides are elevated in PD (Wallen et al., 2022). *Lactobacillus* is also elevated in PD as Hasegawa et al. detected and confirmed by Romano et al.'s meta-analysis. And as Scheperjans et al. noted at family level and Bedarf et al. at species level (Bedarf et al., 2017), *Prevotella copri*, the most common species of the *Prevotella*, is reduced in PD (Wallen et al., 2022). Whereas *Prevotella* species with pathogenic properties are elevated, as originally observed at sub-genus level (Hill-Burns et al., 2017; Wallen et al., 2020) and recaptured at species level by shotgun sequencing (Wallen et al., 2022).

In retrospect, inconsistency across studies should not have been surprising. PD is a heterogenous disorder, and studies differed on how they selected the patients. Possibly more important is how the controls were selected. Some studies used control individuals who had other disorders and/or were taking certain medications that the PD cases they were being compared to were not. A myriad of factors affect microbiome, most notably medications, disease, and geography. The sample sizes were also small, thus none of the studies had enough power to capture every signal that others were capturing. Some of the findings were inevitably false positive, since none of the studies except one (Wallen et al., 2020) had an internal replication.

Gut microbiome may have a role in disease progression and severity. In addition, various constituents of the microbiome could have differential influences on sub-phenotypes of PD. There have been a few studies that explored association of microbial taxa with measures of disease severity, PD sub-phenotypes, and response to medication. At this time, these data are sparse and are based on limited sample sizes. We therefore refrain from presenting a summary here, as it may be premature and misleading.

Studies to date have been conducted with 16S technology, except for one published study that used shotgun on a sample size of 31 PD patients and 28 controls (Bedarf et al., 2017) and the pre-print that reports shogun metagenomics on 490 persons with PD and 234 neurologically healthy elderly controls (Wallen et al., 2022). As noted, 16S sequencing has low taxonomic and functional resolution. Whole metagenome shotgun sequencing provides high resolution at species, strain, genome, and gene level, and allows functional inference because the genes are directly sequenced. The first large-scale study of PD metagenome using shotgun was conducted recently and is posted on prepublication repositor bioRxiv (Wallen et al., 2022). We hesitate to elaborate on the new findings before peer review (other than confirmatory findings noted above). However, certain features of the study which should not change during peer review are worth mentioning: The study is the largest attempted for PD to date (490 PD and 242 neurologically healthy elderly controls); used deep NG shotgun sequencing which provides a comprehensive coverage and the ability to detect microorganisms with very low abundance; and the dataset will be public for research community to mine. More large-scale high-resolution studies will likely to emerge in near future.

4. Concluding remarks

Knowing that PD is etiologically heterogenous, but not to what extent, and having learned that each subtype will be complex with many interacting components, solving the disease seems daunting at times. We may not have all the parts, yet we strive to make the whole that would be greater than the sum. Like any other complex puzzle, as more pieces are found and assembled, the picture comes more into focus, and the rate of progress accelerates. We have made tremendous progress since our naïve beliefs of the 1980's. With an open mind we explored and built a massive repertoire of knowledge. Geneticists and epidemiologists, working on the population and organismal level, discovered the individual pieces and passed them on to basic scientists to elucidate the mechanisms at the molecular level. The next step is to complete the circle, by coalescing what we have learned, and to create the sum from the parts. There is more to be done to find all the pieces and understand the mechanisms at the basic level, but we have enough to start integrating the many lines of research that have evolved and branched out as we progressed from simplistic theories of PD to realizing and accepting the immense complexities of it. Microbiome may be a treasure trove of new information and has the promising potential of filling many gaps towards coalescing knowledge of the disease. It is quite exciting to envision and test ways in which human genome, mitochondrial genome and microbiome are interacting within the hologenome of PD, and how PD hologenome senses and responds to toxic or beneficial environmental stimuli. We will have a full bird's eye view of the entire syndrome we call PD and the details of its mechanisms and pathways at molecular level.

Funding sources

1. Endowed Chair in Parkinson's Disease Research, a gift from John T. and Juanelle D. Strain through University of Alabama at Birmingham.
2. Aligning Science Across Parkinson's [ASAP-020527] through the Michael J. Fox Foundation for Parkinson's Research (MJFF).

3. The U.S. Army Medical Research Materiel Command endorsed by the U.S. Army through the Parkinson's Research Program Investigator-Initiated Research Award under Award No. W81XWH1810508.

References

Aarsland, D., Andersen, K., Larsen, J. P., Lolk, A., Nielsen, H., & Kragh-Sorensen, P. (2001). Risk of dementia in Parkinson's disease: A community-based, prospective study. *Neurology*, *56*(6), 730–736. https://doi.org/10.1212/wnl.56.6.730.

Bedarf, J. R., Hildebrand, F., Coelho, L. P., Sunagawa, S., Bahram, M., Goeser, F., et al. (2017). Functional implications of microbial and viral gut metagenome changes in early stage L-DOPA-naive Parkinson's disease patients. *Genome Medicine*, *9*(1), 39. https://doi.org/10.1186/s13073-017-0428-y.

Bellou, V., Belbasis, L., Tzoulaki, I., Evangelou, E., & Ioannidis, J. P. (2016). Environmental risk factors and Parkinson's disease: An umbrella review of meta-analyses. *Parkinsonism & Related Disorders*, *23*, 1–9. https://doi.org/10.1016/j.parkreldis.2015.12.008.

Berg, D., Borghammer, P., Fereshtehnejad, S. M., Heinzel, S., Horsager, J., Schaeffer, E., et al. (2021). Prodromal Parkinson disease subtypes - key to understanding heterogeneity. *Nature Reviews. Neurology*, *17*(6), 349–361. https://doi.org/10.1038/s41582-021-00486-9.

Bergmark, J., Carlsson, A., Granerus, A. K., Jagenburg, R., Magnusson, T., & Svanborg, A. (1972). Decarboxylation of orally administered L-dopa in the human digestive tract. *Naunyn-Schmiedeberg's Archives of Pharmacology*, *272*(4), 437–440. https://doi.org/10.1007/BF00501249.

Borsche, M., Pereira, S. L., Klein, C., & Grunewald, A. (2021). Mitochondria and Parkinson's disease: Clinical, molecular, and translational aspects. *Journal of Parkinson's Disease*, *11*(1), 45–60. https://doi.org/10.3233/JPD-201981.

Boxberger, M., Cenizo, V., Cassir, N., & La Scola, B. (2021). Challenges in exploring and manipulating the human skin microbiome. *Microbiome*, *9*(1), 125. https://doi.org/10.1186/s40168-021-01062-5.

Braak, H., Del Tredici, K., Rub, U., de Vos, R. A., Jansen Steur, E. N., & Braak, E. (2003). Staging of brain pathology related to sporadic Parkinson's disease. *Neurobiology of Aging*, *24*(2), 197–211. https://doi.org/10.1016/s0197-4580(02)00065-9.

Braak, H., Rub, U., Gai, W. P., & Del Tredici, K. (2003). Idiopathic Parkinson's disease: Possible routes by which vulnerable neuronal types may be subject to neuroinvasion by an unknown pathogen. *Journal of Neural Transmission (Vienna)*, *110*(5), 517–536. https://doi.org/10.1007/s00702-002-0808-2.

Chen, H., Zhao, E. J., Zhang, W., Lu, Y., Liu, R., Huang, X., et al. (2015). Meta-analyses on prevalence of selected Parkinson's nonmotor symptoms before and after diagnosis. *Translational Neurodegeneration*, *4*(1), 1. https://doi.org/10.1186/2047-9158-4-1.

Dardiotis, E., Tsouris, Z., Mentis, A. A., Siokas, V., Michalopoulou, A., Sokratous, M., et al. (2018). *H. pylori* and Parkinson's disease: Meta-analyses including clinical severity. *Clinical Neurology and Neurosurgery*, *175*, 16–24. https://doi.org/10.1016/j.clineuro.2018.09.039.

Deo, P. N., & Deshmukh, R. (2019). Oral microbiome: Unveiling the fundamentals. *Journal of Oral and Maxillofacial Pathology*, *23*(1), 122–128. https://doi.org/10.4103/jomfp.JOMFP_304_18.

Doty, R. L. (2012). Olfactory dysfunction in Parkinson disease. *Nature Reviews. Neurology*, *8*(6), 329–339. https://doi.org/10.1038/nrneurol.2012.80.

Eckburg, P. B., Bik, E. M., Bernstein, C. N., Purdom, E., Dethlefsen, L., Sargent, M., et al. (2005). Diversity of the human intestinal microbial flora. *Science*, *308*(5728), 1635–1638. https://doi.org/10.1126/science.1110591.

Erny, D., Hrabe de Angelis, A. L., Jaitin, D., Wieghofer, P., Staszewski, O., David, E., et al. (2015). Host microbiota constantly control maturation and function of microglia in the CNS. *Nature Neuroscience*, *18*(7), 965–977. https://doi.org/10.1038/nn.4030.

Forsyth, C. B., Shannon, K. M., Kordower, J. H., Voigt, R. M., Shaikh, M., Jaglin, J. A., et al. (2011). Increased intestinal permeability correlates with sigmoid mucosa alpha-synuclein staining and endotoxin exposure markers in early Parkinson's disease. *PLoS One*, *6*(12), e28032. https://doi.org/10.1371/journal.pone.0028032.

Gogoi, J., Bhatnagar, A., Ann, K. J., Pottabathini, S., Singh, R., Mazeed, M., et al. (2022). Switching a conflicted bacterial DTD-tRNA code is essential for the emergence of mitochondria. *Science Advances*, *8*(2), eabj7307. https://doi.org/10.1126/sciadv.abj7307.

Hardy, J. (2010). Genetic analysis of pathways to Parkinson disease. *Neuron*, *68*(2), 201–206. https://doi.org/10.1016/j.neuron.2010.10.014.

Hasegawa, S., Goto, S., Tsuji, H., Okuno, T., Asahara, T., Nomoto, K., et al. (2015). Intestinal Dysbiosis and lowered serum lipopolysaccharide-binding protein in Parkinson's disease. *PLoS One*, *10*(11), e0142164. https://doi.org/10.1371/journal.pone.0142164.

Henderson, M. X., Trojanowski, J. Q., & Lee, V. M. (2019). Alpha-Synuclein pathology in Parkinson's disease and related alpha-synucleinopathies. *Neuroscience Letters*, *709*, 134316. https://doi.org/10.1016/j.neulet.2019.134316.

Hill-Burns, E. M., Debelius, J. W., Morton, J. T., Wissemann, W. T., Lewis, M. R., Wallen, Z. D., et al. (2017). Parkinson's disease and Parkinson's disease medications have distinct signatures of the gut microbiome. *Movement Disorders*, *32*(5), 739–749. https://doi.org/10.1002/mds.26942.

Hooi, J. K. Y., Lai, W. Y., Ng, W. K., Suen, M. M. Y., Underwood, F. E., Tanyingoh, D., et al. (2017). Global prevalence of helicobacter pylori infection: Systematic review and Meta-analysis. *Gastroenterology*, *153*(2), 420–429. https://doi.org/10.1053/j.gastro.2017.04.022.

Horsager, J., Andersen, K. B., Knudsen, K., Skjaerbaek, C., Fedorova, T. D., Okkels, N., et al. (2020). Brain-first versus body-first Parkinson's disease: A multimodal imaging case-control study. *Brain*, *143*(10), 3077–3088. https://doi.org/10.1093/brain/awaa238.

Houser, M. C., Chang, J., Factor, S. A., Molho, E. S., Zabetian, C. P., Hill-Burns, E. M., et al. (2018). Stool immune profiles evince gastrointestinal inflammation in Parkinson's disease. *Movement Disorders*, *33*(5), 793–804. https://doi.org/10.1002/mds.27326.

Human Microbiome Project Consortium. (2012). Structure, function and diversity of the healthy human microbiome. *Nature*, *486*(7402), 207–214. https://doi.org/10.1038/nature11234.

Integrative, H. M. P. R. N. C. (2019). The Integrative human microbiome project. *Nature*, *569*(7758), 641–648. https://doi.org/10.1038/s41586-019-1238-8.

Keshavarzian, A., Green, S. J., Engen, P. A., Voigt, R. M., Naqib, A., Forsyth, C. B., et al. (2015). Colonic bacterial composition in Parkinson's disease. *Movement Disorders*, *30*(10), 1351–1360. https://doi.org/10.1002/mds.26307.

Kim, S., Kwon, S. H., Kam, T. I., Panicker, N., Karuppagounder, S. S., Lee, S., et al. (2019). Transneuronal propagation of pathologic alpha-Synuclein from the gut to the brain models Parkinson's disease. *Neuron*, *103*(4), 627–641. e627 https://doi.org/10.1016/j.neuron.2019.05.035.

Klein, C., & Westenberger, A. (2012). Genetics of Parkinson's disease. *Cold Spring Harbor Perspectives in Medicine*, *2*(1), a008888. https://doi.org/10.1101/cshperspect.a008888.

Koskinen, K., Reichert, J. L., Hoier, S., Schachenreiter, J., Duller, S., Moissl-Eichinger, C., et al. (2018). The nasal microbiome mirrors and potentially shapes olfactory function. *Scientific Reports*, *8*(1), 1296. https://doi.org/10.1038/s41598-018-19438-3.

Kumpitsch, C., Koskinen, K., Schopf, V., & Moissl-Eichinger, C. (2019). The microbiome of the upper respiratory tract in health and disease. *BMC Biology*, *17*(1), 87. https://doi.org/10.1186/s12915-019-0703-z.

Langston, J. W., Ballard, P., Tetrud, J. W., & Irwin, I. (1983). Chronic parkinsonism in humans due to a product of meperidine-analog synthesis. *Science*, *219*, 979–980.

Liu, B., Fang, F., Pedersen, N. L., Tillander, A., Ludvigsson, J. F., Ekbom, A., et al. (2017). Vagotomy and Parkinson disease: A Swedish register-based matched-cohort study. *Neurology*, *88*(21), 1996–2002. https://doi.org/10.1212/WNL.0000000000003961.

Lloyd-Price, J., Mahurkar, A., Rahnavard, G., Crabtree, J., Orvis, J., Hall, A. B., et al. (2017). Strains, functions and dynamics in the expanded human microbiome project. *Nature*, *550*(7674), 61–66. https://doi.org/10.1038/nature23889.

Maini Rekdal, V., Bess, E. N., Bisanz, J. E., Turnbaugh, P. J., & Balskus, E. P. (2019). Discovery and inhibition of an interspecies gut bacterial pathway for levodopa metabolism. *Science*, *364*(6445). https://doi.org/10.1126/science.aau6323.

Matheoud, D., Cannon, T., Voisin, A., Penttinen, A. M., Ramet, L., Fahmy, A. M., et al. (2019). Intestinal infection triggers Parkinson's disease-like symptoms in Pink1(−/−) mice. *Nature*, *571*(7766), 565–569. https://doi.org/10.1038/s41586-019-1405-y.

Muller-Nedebock, A. C., Brennan, R. R., Venter, M., Pienaar, I. S., van der Westhuizen, F. H., Elson, J. L., et al. (2019). The unresolved role of mitochondrial DNA in Parkinson's disease: An overview of published studies, their limitations, and future prospects. *Neurochemistry International*, *129*, 104495. https://doi.org/10.1016/j.neuint.2019.104495.

Nalls, M. A., Blauwendraat, C., Vallerga, C. L., Heilbron, K., Bandres-Ciga, S., Chang, D., et al. (2019). Identification of novel risk loci, causal insights, and heritable risk for Parkinson's disease: A meta-analysis of genome-wide association studies. *Lancet Neurology*, *18*(12), 1091–1102. https://doi.org/10.1016/S1474-4422(19)30320-5.

Nearing, J. T., Douglas, G. M., Hayes, M. G., MacDonald, J., Desai, D. K., Allward, N., et al. (2022). Microbiome differential abundance methods produce different results across 38 datasets. *Nature Communications*, *13*(1), 342. https://doi.org/10.1038/s41467-022-28034-z.

Nishiwaki, H., Ito, M., Ishida, T., Hamaguchi, T., Maeda, T., Kashihara, K., et al. (2020). Meta-analysis of gut Dysbiosis in Parkinson's disease. *Movement Disorders*, *35*(9), 1626–1635. https://doi.org/10.1002/mds.28119.

Nunnari, J., & Suomalainen, A. (2012). Mitochondria: In sickness and in health. *Cell*, *148*(6), 1145–1159. https://doi.org/10.1016/j.cell.2012.02.035.

Nurk, S., Koren, S., Rhie, A., Rautiainen, M., Bzikadze, A. V., Mikheenko, A., et al. (2022). The complete sequence of a human genome. *Science*, *376*(6588), 44–53. https://doi.org/10.1126/science.abj6987.

Obeso, J. A., Stamelou, M., Goetz, C. G., Poewe, W., Lang, A. E., Weintraub, D., et al. (2017). Past, present, and future of Parkinson's disease: A special essay on the 200th anniversary of the shaking palsy. *Movement Disorders*, *32*(9), 1264–1310. https://doi.org/10.1002/mds.27115.

Pasolli, E., Asnicar, F., Manara, S., Zolfo, M., Karcher, N., Armanini, F., et al. (2019). Extensive unexplored human microbiome diversity revealed by over 150,000 genomes from metagenomes spanning age, geography, and lifestyle. *Cell*, *176*(3), 649–662. e620 https://doi.org/10.1016/j.cell.2019.01.001.

Payami, H., Larsen, K., Bernard, S., & Nutt, J. (1994). Increased risk of Parkinson's disease in parents and siblings of patients. *Annals of Neurology*, *36*, 659–661.

Pierantozzi, M., Pietroiusti, A., Galante, A., Sancesario, G., Lunardi, G., Fedele, E., et al. (2001). Helicobacter pylori-induced reduction of acute levodopa absorption in Parkinson's disease patients. *Annals of Neurology*, *50*(5), 686–687. https://doi.org/10.1002/ana.1267.

Postuma, R. B., & Berg, D. (2016). Advances in markers of prodromal Parkinson disease. *Nature Reviews. Neurology*, *12*(11), 622–634. https://doi.org/10.1038/nrneurol.2016.152.

Pyle, A., Anugrha, H., Kurzawa-Akanbi, M., Yarnall, A., Burn, D., & Hudson, G. (2016). Reduced mitochondrial DNA copy number is a biomarker of Parkinson's disease. *Neurobiology of Aging*, *38*. https://doi.org/10.1016/j.neurobiolaging.2015.10.033. e217-216 e210.

Roger, A. J., Munoz-Gomez, S. A., & Kamikawa, R. (2017). The origin and diversification of mitochondria. *Current Biology*, *27*(21), R1177–R1192. https://doi.org/10.1016/j.cub.2017.09.015.

Romano, S., Savva, G. M., Bedarf, J. R., Charles, I. G., Hildebrand, F., & Narbad, A. (2021). Meta-analysis of the Parkinson's disease gut microbiome suggests alterations linked to intestinal inflammation. *NPJ Parkinson's Disease*, 7(1), 27. https://doi.org/10.1038/s41531-021-00156-z.

Sagan, L. (1967). On the origin of mitosing cells. *Journal of Theoretical Biology*, *14*(3), 255–274. https://doi.org/10.1016/0022-5193(67)90079-3.

Santos, M. L. C., de Brito, B. B., da Silva, F. A. F., Sampaio, M. M., Marques, H. S., Oliveira, E. S. N., et al. (2020). Helicobacter pylori infection: Beyond gastric manifestations. *World Journal of Gastroenterology*, *26*(28), 4076–4093. https://doi.org/10.3748/wjg.v26.i28.4076.

Scepanovic, P., Hodel, F., Mondot, S., Partula, V., Byrd, A., Hammer, C., et al. (2019). A comprehensive assessment of demographic, environmental, and host genetic associations with gut microbiome diversity in healthy individuals. *Microbiome*, 7(1), 130. https://doi.org/10.1186/s40168-019-0747-x.

Scheperjans, F., Aho, V., Pereira, P. A., Koskinen, K., Paulin, L., Pekkonen, E., et al. (2015). Gut microbiota are related to Parkinson's disease and clinical phenotype. *Movement Disorders*, *30*(3), 350–358. https://doi.org/10.1002/mds.26069.

Schneider, V. A., Graves-Lindsay, T., Howe, K., Bouk, N., Chen, H. C., Kitts, P. A., et al. (2017). Evaluation of GRCh38 and de novo haploid genome assemblies demonstrates the enduring quality of the reference assembly. *Genome Research*, *27*(5), 849–864. https://doi.org/10.1101/gr.213611.116.

Schonhoff, A. M., Williams, G. P., Wallen, Z. D., Standaert, D. G., & Harms, A. S. (2020). Innate and adaptive immune responses in Parkinson's disease. *Progress in Brain Research*, *252*, 169–216. https://doi.org/10.1016/bs.pbr.2019.10.006.

Sharma, P., & Sampath, H. (2019). Mitochondrial DNA integrity: Role in health and disease. *Cell*, *8*(2). https://doi.org/10.3390/cells8020100.

Svensson, E., Horvath-Puho, E., Thomsen, R. W., Djurhuus, J. C., Pedersen, L., Borghammer, P., et al. (2015). Vagotomy and subsequent risk of Parkinson's disease. *Annals of Neurology*, *78*(4), 522–529. https://doi.org/10.1002/ana.24448.

Tang, A. T., Choi, J. P., Kotzin, J. J., Yang, Y., Hong, C. C., Hobson, N., et al. (2017). Endothelial TLR4 and the microbiome drive cerebral cavernous malformations. *Nature*, *545*(7654), 305–310. https://doi.org/10.1038/nature22075.

Toh, T. S., Chong, C. W., Lim, S. Y., Bowman, J., Cirstea, M., Lin, C. H., et al. (2022). Gut microbiome in Parkinson's disease: New insights from meta-analysis. *Parkinsonism & Related Disorders*, *94*, 1–9. https://doi.org/10.1016/j.parkreldis.2021.11.017.

van der Hee, B., & Wells, J. M. (2021). Microbial regulation of host physiology by short-chain fatty acids. *Trends in Microbiology*, *29*(8), 700–712. https://doi.org/10.1016/j.tim.2021.02.001.

van Kessel, S. P., Frye, A. K., El-Gendy, A. O., Castejon, M., Keshavarzian, A., van Dijk, G., et al. (2019). Gut bacterial tyrosine decarboxylases restrict levels of levodopa in the treatment of Parkinson's disease. *Nature Communications*, *10*(1), 310. https://doi.org/10.1038/s41467-019-08294-y.

Vazquez-Velez, G. E., & Zoghbi, H. Y. (2021). Parkinson's disease genetics and pathophysiology. *Annual Review of Neuroscience*, *44*, 87–108. https://doi.org/10.1146/annurev-neuro-100720-034518.

Wallen, Z. D. (2021). Comparison study of differential abundance testing methods using two large Parkinson disease gut microbiome datasets derived from 16S amplicon sequencing. *BMC Bioinformatics*, *22*(1), 265. https://doi.org/10.1186/s12859-021-04193-6.

Wallen, Z. D., Appah, M., Dean, M. N., Sesler, C. L., Factor, S. A., Molho, E., et al. (2020). Characterizing dysbiosis of gut microbiome in PD: Evidence for overabundance of opportunistic pathogens. *NPJ Parkinson's Disease*, *6*, 11. https://doi.org/10.1038/s41531-020-0112-6.

Wallen, Z. D., Demirkan, A., Twa, G., Cohen, G., Dean, M. N., Standaert, D. S., et al. (2022). Metagenomics of Parkinson's disease implicates the gut microbiome in multiple disease mechanisms. *bioRxiv*. https://doi.org/10.1101/2022.06.08.495316.

Wallen, Z. D., Stone, W. J., Factor, S. A., Molho, E., Zabetian, C. P., Standaert, D. G., et al. (2021). Exploring human-genome gut-microbiome interaction in Parkinson's disease. *NPJ Parkinson's Disease*, 7(1), 74. https://doi.org/10.1038/s41531-021-00218-2.

Weintraub, D., & Mamikonyan, E. (2019). The neuropsychiatry of Parkinson disease: A perfect storm. *The American Journal of Geriatric Psychiatry*, *27*(9), 998–1018. https://doi.org/10.1016/j.jagp.2019.03.002.

CHAPTER FOUR

Intestinal microbiota and neuroinflammation in Parkinson's disease: At the helm of the gut-brain axis

Richard A. Manfready[a], Christopher G. Goetz[b], and Ali Keshavarzian[a,c,*]

[a]Department of Internal Medicine, Division of Digestive Diseases and Nutrition, Rush University Medical Center, Chicago, IL, United States
[b]Department of Neurological Sciences, Rush University Medical Center, Chicago, IL, United States
[c]Rush Center for Integrated Microbiome and Chronobiology Research, Rush University Medical Center, Chicago, IL, United States
*Corresponding author: e-mail address: ali_keshavarzian@rush.edu

Contents

Abstract

Emerging data suggest that disrupted intestinal microbiota, or dysbiosis, may be responsible for multiple features of Parkinson's disease (PD), from initiation, to progression, to therapeutic response. We have progressed greatly in our understanding of microbial signatures associated with PD, and have gained important insights into how dysbiosis and intestinal permeability promote neurodegeneration through neuroinflammation and Lewy body formation. These insights underscore the potential of microbiota-directed therapies, which include dietary, pharmacologic, and lifestyle interventions.

International Review of Neurobiology, Volume 167
ISSN 0074-7742
https://doi.org/10.1016/bs.irn.2022.07.008

Abbreviations

COMT catechol-*o*-methyl transferase
DAMP damage-associated molecular pattern
FMT fecal microbiota transplantation
GLP-1 glucagon-like peptide-1
IBD inflammatory bowel disease
LPS lipopolysaccharide
MPTP 1-methyl-4-phenyl-1,2,3,6-tetrahydropyridine
PAMP pathogen-associated molecular pattern
PD Parkinson's disease
SCFA short-chain fatty acid
SIBO small-intestinal bacterial overgrowth
TLR toll-like receptor

1. Introduction: Intestinal microbiota in health and disease

Human gut microbiota comprise a vast ecosystem of bacteria, archaea, viruses, and fungi that reside in the digestive tract. Although these 100 trillion microbes represent more than 500 different species, 99% of bacterial abundance is derived from less than 50 species (Lozupone, Stombaugh, Gordon, Jansson, & Knight, 2012). The sum of genetic material from these microbes, known as the microbiome, contains more than three million unique genes, greatly outnumbering the full set of human genes (Proctor, 2011). Early human life plays a major role in shaping adult gut microbial composition, which is largely determined and stabilized by three years of life. Bacterial phyla that predominate in adults include Bacteroidetes, Firmicutes, Proteobacteria, Actinobacteria, and Verrucomicrobia, in addition to archaea, yeasts, and viruses (Keshavarzian, Engen, Bonvegna, & Cilia, 2020). Once established, the adult microbial ecosystem can be altered by environmental factors which include diet, medications (especially frequent use of antibiotics), toxicants, infections, emotional stress, biological stress including disruption of circadian rhythms and sleep, surgery, and geography.

Progress in understanding gut microbes from healthy adults in recent decades has revealed the important role of "normal" microbiota in host immune function, metabolism, and endocrine regulation. Deviation from normal composition, termed dysbiosis, has been associated with several disease states, including inflammatory bowel disease (IBD), irritable bowel syndrome, obesity, cancer, and neurological diseases (Fields et al., 2018; Lenhart et al., 2022). Taking an ecological approach, disruptions in microbial composition can be monitored by several measures. The Shannon index is a common measure of system entropy and has been adapted as a tool to study species richness and evenness in a particular community (Hill, 1973). In the field of gut microbiota, the Shannon index is often used to measure within-sample species distribution, or α-diversity, whereas diversity among samples (i.e., between PD and control subjects) is quantified by measures of β-diversity. Loss of species diversity was previously thought to signal a disease state. Although this may be generally true, we will see that in PD this simplification belies a more complex interplay of microbial signaling that cannot be captured by straightforward indices. Furthermore, microbial diversity is highly variable between human populations, across geography, and over time, which prevents us from defining a single "normal" set of microbiota or expected diversity.

It should be noted that our current view of microbial dysfunction in PD evolved significantly before it could account for phenotypic dysfunction. The earliest methods for interrogating gut microbiota relied on culture, however it became evident that the vast majority of bacteria from the gut could not be easily cultured. Introduction of amplicon and shotgun sequencing-based technologies that are not reliant on bacterial culture revolutionized our ability to take a more comprehensive account of the microbiome. These methods can identify bacteria to the species level (Matsuo et al., 2021) and can also identify viruses and fungi (Frau et al., 2019). Yet, the composition of a particular microbial community alone does not immediately imply its function. To capture functional information, it is necessary to examine mRNA, genes, proteins, and microbial byproducts, and compare them across different populations or pathological states. A disease state may be characterized by differential gene expression, protein production, or metabolite excretion that can be measured by transcriptomics, proteomics, or metabolomics (untargeted and targeted), respectively. Taxonomic and functional diversity often do not correlate, a phenomenon that may signal important between-group differences among populations. As we will see, these techniques have unraveled important differences between PD and healthy populations.

2. Microbial signatures and their significance in Parkinson's disease

The reported prevalence of PD is rising in both the developed and developing world. More than 1% of adults over the age of 60 in the United States are affected (Marras et al., 2018). As the second most common neurodegenerative disease, PD is a synucleinopathy, characterized by aggregates of misfolded α-synuclein protein called Lewy bodies that accumulate in the substantia nigra and disable the dopaminergic neurons responsible for coordinated motor activity (Tysnes & Storstein, 2017). In their misfolded state, α-synuclein monomers attach to one another in a prion-like mechanism to form the aggregates. Despite the prevalence of PD and its significant burden on patients, caregivers, and the healthcare system, available treatments focus on symptom palliation without intentional disease-modifying activity. Although the nigral Lewy body has been known as the pathologic hallmark of PD since at least 1923, the cause of α-synuclein misfolding and strategies to prevent or repair such misfolding have remained elusive (Holdorff, Silva, & Dodel, 2013). Immune dysregulation, inflammatory activity, and chaperone protein malfunction have been implicated with experimental success, but none of these factors in isolation has proven to be a root cause (Manfready, 2010). Over the last several decades, environmental factors have also been implicated. Manganese and carbon disulfide, used in pesticides and rubber manufacturing, are known to cause Parkinson-like syndromes via a mechanism that involves α-synuclein aggregation (Caudle, Guillot, Lazo, & Miller, 2012; Guilarte, 2011). Several medicinal and non-medicinal drugs also potentiate Parkinson-like syndromes, including 1-methyl-4-phenyl-1,2,3,6-tetrahydropyridine (MPTP), which is commonly used to create animal models of PD (Przedborski, Tieu, Perier, & Vila, 2004). Monogenic mutations, including several in the gene encoding α-synuclein, can cause familial PD, and mutations in other genes have been associated with heightened susceptibility (Polymeropoulos et al., 1997).

Years or even decades before the onset of motor symptoms, which typically include tremor, bradykinesia, rigidity, and postural instability, patients with PD frequently describe constipation, abdominal bloating, anosmia, sleep disorders, and autonomic dysfunction (Knudsen, Krogh, Østergaard, & Borghammer, 2017). These prodromal symptoms suggest a peripheral origin for PD, yet identifying it has proved challenging. In 2003, Braak and colleagues, already known for the PD staging system, proposed that α-synuclein aggregates form first in the gastrointestinal tract and olfactory bulb, then progress to the brain along the vagus nerve (Braak, Rüb, Gai, & Del Tredici, 2003). Evidence for this pathway includes the finding of α-synuclein aggregates in post-ganglionic enteric nervous system terminals (Shannon et al., 2012), retrograde transport of α-synuclein along the vagus nerve in animal models (Del Tredici & Braak, 2016), decreased PD frequency following truncal vagotomy (Breen, Halliday, & Lang, 2019), and detection of α-synuclein aggregates in the brain after injection into the duodenal wall of rodents and primates (Challis et al., 2020; Manfredsson et al., 2018).

Despite the presence of Lewy bodies in enteric neurons, myenteric neuron loss is an uncommon finding, and is not required for progression to the brain (Annerino et al., 2012). Although this caveat has challenged the gut-to-brain route of α-synuclein propagation, it provided an important impetus to shift attention to the GI tract as a major site of Lewy body formation. Braak proposed that pathogens or toxicants in the GI tract serve as catalysts for the initial misfolding event. However, until recently there was no epidemiologic evidence of a widespread toxicant that could account for the growing incidence of PD. Instead of a single toxicant, disrupted host intestinal microbiota may comprise a more likely enteral source. We now appreciate that the vast repository of host microbiota and their dietary determinants serve as a source of enteral inflammation across large human populations (Keshavarzian et al., 2020). From an epidemiological perspective, the relative proportion of certain "pro-inflammatory" gut bacteria is greater in developed countries and expands with economic development and industrialization—a trend that is also associated with PD incidence (Marras, Canning, & Goldman, 2019). As a generalization supported by over 30 studies to date (Aho et al., 2021; Hill-Burns et al., 2017; Keshavarzian et al., 2015; Lin et al., 2018; Unger et al., 2016), PD intestines compared to those of healthy controls have increased relative abundance of these putative "pro-inflammatory" bacteria (pathobionts) and decreased relative abundance of putative "anti-inflammatory" bacteria (Keshavarzian et al., 2020).

While there is no particular microbial signature that defines the PD gut, there have emerged several common functional patterns that provide insight into pathogenesis. Although the specific microbiota may vary between studies (Hill-Burns et al., 2017; Unger et al., 2016), the PD gut is characterized by elevated "pro-inflammatory" bacteria, or pathobionts, and diminished "anti-inflammatory" bacteria relative to the healthy gut. Within the phylum Proteobacteria which contains several pro-inflammatory pathobionts, PD patients have increased relative abundance of the family Enterobacter and the species *E. coli*. Within the phylum Firmicutes, PD patients have increased relative abundance of the family Lactobacillaceae and the genus *Lactobacillus* (Barichella et al., 2019). There is decreased relative abundance of the family Lachnospiraceae, and the genera *Blautia* and *Roseburia*; importantly, these changes were observed in *de novo* (untreated) PD patients. The PD phenotype is also characterized by increased bacteria of the phyla Bacteroidetes and Verrucomicrobia, increased abundance of the genus *Akkermansia*, and decreased abundance of the genus *Prevotella*. Notably, studies of microbial diversity in PD have been inconsistent, however as with many common diseases there is a trend toward lower species diversity (Chen, Liang et al., 2021). Instead, most insights have come from the functional characteristics of the enriched or depleted bacterial species as well as the metabolic byproducts of these bacteria.

Pathogen-associated molecular patterns (PAMPs) produced by pathobionts interact with innate immune receptors, notably toll-like receptors (TLRs) expressed by intestinal epithelial cells, intestinal mucosal and systemic immune cells, enteric and central neurons, and microglia. TLR activation triggers an inflammatory response involving NF-κB and inflammasome signaling with production of pro-inflammatory cytokines TNF-α, IL-8, IL-1, and interferon-γ (Keogh, Rude, & Gareau, 2021). The most widely-implicated PAMPs in neurodegenerative disease are the Gram-negative cell wall component lipopolysaccharide (LPS), which binds to TLR4, and the Gram-positive cell wall component lipoteichoic acid, which binds to TLR2 (Blaylock, 2017). PD stool has significantly greater abundance of bacterial genes involved in LPS synthesis (Keshavarzian et al., 2015). PAMP-TLR inflammatory signaling cascades can lead to microglial activation, which is a key feature of dysbiosis-related neuroinflammation in PD. Chronic activation of TLR4 in PD may also drive NLRP3 inflammasome activation, which is correspondingly found in postmortem brains of PD patients (Gordon et al., 2018). The inflammasome promotes caspase activation of IL-1β and IL-18, which affect both the CNS and peripheral organs, to include the intestine. Blocking this pathway notably protects against neurodegeneration in multiple animal models of PD, including those treated with α-synuclein preformed fibrils, pesticides, MPTP, and genetic mutations (Chakraborty, Kaushik, Gupta, & Basu, 2010; Gordon et al., 2018). Whether inflammasome activation directly causes misfolding and aggregation of α-synuclein remains unknown; since α-synuclein itself appears to play a role in inflammasome activation, and since neuroinflammation-microglial activation may contribute to Lewy body formation (Labzin, Heneka, & Latz, 2018), more work is needed to elucidate the relative role of each component.

Microbial community structures in PD should be interpreted with care, as some distinct features may be reactive, rather than causative. For example, increased levels of *Lactobacillus brevis* in PD mice are associated with *lower* LPS levels in feces and blood, as well as reduced levels of inflammatory markers and NF-κB activation (Jeong, Kim, Hwang, Han, & Kim, 2016). Similarly, *Lactobacillus plantarum* may protect against human intestinal epithelial cell inflammation (Noh, Kang, Yun, & Han, 2015). These data suggest that the observed increase in abundance of *Lactobacillus* may be an attempt by the host to mitigate pro-inflammatory effects of increased LPS-producing bacteria in PD. A possible explanation is the association between *Lactobacillus* and increased lipopolysaccharide binding protein (Minato et al., 2017), which mitigates LPS-induced cytokine activation.

3. Neuroprotection by short-chain fatty acids

Several enteric bacterial species can ferment complex carbohydrate fibers to produce the short-chain fatty acids (SCFAs) acetate, propionate, and butyrate. Strains of *Lactobacillus* and *Bifidobacterium* are able to produce acetate, while Firmicutes such as *Clostridium* and *Roseburia* can produce butyrate. The SCFAs, in particular butyrate, are the preferred energy substrates for colonic epithelial cells (Makki, Deehan, Walter, & Bäckhed, 2018). SCFAs are vital for maintaining a healthy epithelium, and PD patients are known to have lower SCFA content in their stool (Aho et al., 2021). A decrease in butyrate-producing bacteria is associated with increased intestinal permeability, mucosal inflammation, and enteric neuron hyperpolarization. Similarly, reduced SCFA levels in the colon may be responsible for constipation and leakiness of the intestinal barrier (Aho et al., 2021). Increased intestinal permeability in PD patients may explain higher than expected plasma SCFA content (Chen et al., 2022).

SCFAs may play several additional roles, endowed by their ability to enter the bloodstream, cross the blood-brain barrier, and inhibit histone deacetylases to enhance normal microglial development (Bourassa, Alim, Bultman, & Ratan, 2016). Locally within the colon, SCFAs interact with enteroendocrine cells to trigger the secretion of hormones, notably neuroprotective glucagon-like peptide-1 (GLP-1) by L-cells (Greiner & Bäckhed, 2016; Zhou et al., 2008), and provide nourishment for colonocytes

(Koh, De Vadder, Kovatcheva-Datchary, & Bäckhed, 2016) to maintain normal intestinal barrier integrity. Indeed, it has been demonstrated that a diminished relative proportion of SCFA-producing bacteria such as *Bifidobacterium* in the stool is associated with progression of PD motor features at over a 2-year period (Aho et al., 2019), and that *Bifidobacterium*-containing probiotic supplementation may improve central and peripheral symptoms (Sun et al., 2022). Yet, discrepancies between studies currently limit the prognostic value of fecal metabolomics in individual patients. For example, increased relative abundance of *Bifidobacterium* in PD stool reported by several groups (Keshavarzian et al., 2015; Keshavarzian et al., 2020) contradicts the proposed beneficial effects of acetate-producing *Bifidobacterium*. One possible explanation that could reconcile this apparent discrepancy is that increased abundance of *Bifidobacterium* is a compensatory attempt by the host. Further longitudinal and interventional human studies are required to clarify this important discrepancy.

4. Microbial effects on neuroinflammation

The intestinal barrier in PD is characterized by higher-than-normal permeability, which enables translocation of PAMPs and subsequent neuroinflammation (Tang et al., 2015). Mucosal bacteria, accessible by biopsy, are separated from those in the lumen exposed to feces, creating two distinct populations. In the setting of mucosal irritation or a pathologic state, these communities are more likely to mix, and mucosal contents may leak through a disrupted epithelial barrier. In PD, LPS-producing bacteria in the colonic lumen pass into the mucosal population, allowing LPS to translocate across a leaky epithelial barrier causing endotoxemia. Urine sucralose, collected over 24 hours, is a marker of total gut permeability and is increased in PD, providing evidence for a leaky gut (Shaikh, Rajan, Forsyth, Voigt, & Keshavarzian, 2015). Increased intestinal permeability in PD is likely due to microbiota-induced inflammation, as it correlates with increased TLR2, TLR4, and pro-inflammatory cytokines (Perez-Pardo et al., 2019).

Once in the circulation, LPS may directly cause neuroinflammation in the brain due to its role as a TLR agonist. Experimental treatment of microglial cells with LPS increases production of TNFα, IL-6, and IL-1b, proinflammatory cytokines that act with reactive oxygen species to potentiate neuronal destruction (Zhang et al., 2015). These LPS-induced inflammatory responses are upregulated in the presence of known genetic variants

Parkin, PINK1, and DJ-1 that increase susceptibility (Akundi et al., 2011; Mitsumoto & Nakagawa, 2001). Neuronal destruction in turn releases damage-associated molecular patterns (DAMPs) that activate nearby microglia and so propagate inflammation and neuronal destruction. α-Synuclein is a known DAMP that promotes microglial activation via a TLR4-mediated mechanism. In addition to inducing neuroinflammation by a cytokine response, microglial activation may itself promote α-synuclein aggregation by a mechanism that involves phosphorylation (Jang et al., 2009). In this way, α-synuclein may be both a cause and a result of inflammatory activation in the brain, advancing neurodegeneration in a positive feedback loop triggered by LPS.

Pro-inflammatory dysbiotic microbiota capable of producing LPS and activating microglia have been found in rodent models of PD induced by rotenone and MPTP (Perez-Pardo et al., 2018). Transplanting human PD feces or selected PD-associated bacteria into germ-free mice results in PD-like behaviors (Sampson, 2020; Sampson et al., 2016). In a rotenone-induced rodent model of PD, restraint stress worsens dysbiosis and gut leakiness, which in turn activates microglia and inflicts dopaminergic neuron loss (Dodiya et al., 2020). Thus, dysbiosis is associated with increased local pro-inflammatory signals and intestinal permeability, which are followed by endotoxemia, neuroinflammation, and α-synuclein aggregation. Next, we will explore several environmental factors that lead to dysbiosis in PD.

5. Environmental and dietary influences

Environmental factors seem to play greater roles in PD pathogenesis than genetic factors as suggested by studies of twins (Turnbaugh et al., 2009). These environmental factors include not only dietary elements, but also chemicals, herbicides, and other xenobiotics. Experimental evidence suggests that environmental factors that are associated with increased risk of PD and parkinsonism, which include xenobiotics and high fat/low fiber diets, are also associated with dysbiosis and altered intestinal barrier function (Ghaisas, Maher, & Kanthasamy, 2016). Thus, dysbiosis-mediated inflammation may be the mechanism by which these environmental factors promote PD.

Most PD cases are not caused by identifiable toxicants, but rather may result from geographical variations in diet and lifestyle as suggested earlier. In particular, the Western diet and lifestyle, characterized by high relative proportions of processed food and stress, seem to be correlated with a higher incidence of PD (Marras et al., 2019). Animal studies have demonstrated that these factors can recapitulate neuroinflammation, neurodegeneration, and Parkinson-like behavior (Dodiya et al., 2020). It is clear that certain dietary patterns affect intestinal microbiota in a manner that mimics the PD gut. Diets high in sugar and refined carbohydrates and low in fiber, typical of the Western diet, promote the growth of dysbiotic microbiota that produce LPS, while decreasing the abundance of anti-inflammatory bacteria that produce SCFAs (Cassani et al., 2017; Zinöcker & Lindseth, 2018). Similar microbial effects are triggered by stress, poor sleep patterns, and disrupted circadian rhythms, which are associated with activation of enteric glia and brain microglia, α-synuclein aggregation, and dopaminergic neuron loss (Marras et al., 2019; Paschos & FitzGerald, 2017).

Smoking and coffee are among the lifestyle habits demonstrated to have protective effects against PD. These effects may be at least partially explained by a higher abundance of Bacteroides found in the feces of smokers and SCFA-producing *Bifidobacterium* in the feces of coffee consumers (Derkinderen, Shannon, & Brundin, 2014). Yet, as stated previously, PD patients have increased relative abundance of *Bifidobacterium* in their stool at baseline, and so the "protective" effect of coffee may be attributed to further enhancing SCFA production, or might not be mediated through changes in microbiota composition.

Notably, fecal microbial dysbiosis is also observed among patients with metabolic syndrome and diabetes, which are PD risk factors, and among patients with IBD, which may also be a PD risk factor (Villumsen, Aznar, Pakkenberg, Jess, & Brudek, 2019). It is intriguing that patients with PD, IBD, and metabolic syndrome have decreased relative abundance of SCFA-producing bacteria in their stool (Keshavarzian et al., 2020; Vindigni, Zisman, Suskind, & Damman, 2016). These findings provide important evidence that intestinal microbiota are central to the gut-brain axis in PD.

6. Common medications and microbiota

Although most evidence to this point has favored dysbiosis as a trigger and/or promoter of PD rather than a consequence of PD, it should be noted that PD patients often experience worsening gastrointestinal symptoms and alter their diet and lifestyle accordingly in ways that may exacerbate dysbiosis. Constipation, a low-fiber diet, and reversal of sleep/wake cycles are features of PD that are independently associated with dysbiosis (Voigt, Forsyth, Green, Engen, & Keshavarzian, 2016). Despite the presence of dysbiosis among prodromal PD patients, those with advanced disease on medications experience significant, perhaps worse forms of dysbiosis and diminished intestinal barrier function (Heintz-Buschart et al., 2018). Emerging evidence strongly suggests that PD medications, especially levodopa and catechol-*o*-methyl transferase (COMT) inhibitors, impact microbial community structure. In patients on COMT inhibitors such as entacapone and tolcapone, diarrhea is a potential adverse effect that is negatively correlated with Firmicutes abundance and fecal butyrate concentrations (Unger et al., 2016).

Simultaneously, altered microbiota can modify the efficacy of common PD medications. A subset of gut bacteria residing in both the small intestine and colon, which include *Lactobacillus* and *Enterobacter*, contain tyrosine decarboxylase, a bacterial enzyme capable of breaking down levodopa locally within the gut. In patients requiring high doses of levodopa, the stool contains high quantities of bacterial tyrosine decarboxylase, suggesting that the high levodopa requirement is due to heightened breakdown and thereby reduced availability for absorption (Maini Rekdal, Bess, Bisanz, Turnbaugh, & Balskus, 2019; Van Kessel et al., 2019). Additionally, prolonged exposure to anti-PD medications, especially entacapone, may increase tyrosine decarboxylase over time (van Kessel, Auvinen, Scheperjans, & El Aidy, 2021).

The presence of existing gastrointestinal conditions such as *Helicobacter pylori* infection or small intestinal bacterial overgrowth (SIBO) may interfere with levodopa pharmacokinetics. Specifically, *H. pylori* interferes with the absorption of levodopa, leading to decreased frequency and duration of medication on-time (Fasano, Visanji, Liu, Lang, & Pfeiffer, 2015). The effect can be reversed by *H. pylori* eradication (Pierantozzi et al., 2006). Delayed intestinal transit in PD predisposes patients to SIBO, which may also prevent adequate levodopa absorption and thus influence motor outcomes (Fasano et al., 2015). SIBO may also promote increased intestinal permeability and thus endotoxemia and microglial activation, which mechanistically makes it a risk factor for PD onset or progression by way of neuroinflammation.

7. Therapeutic strategies

Given the mounting evidence for dysbiosis in triggering or promoting the PD phenotype, it follows that there may be a useful role for gut microbiota-directed therapeutic strategies that prevent, delay, or modify PD onset or disease course. In particular, dietary intervention may be a useful strategy for stopping or improving dysbiosis; this may be achieved by implementing diets known to be associated with putative anti-inflammatory microbiota and increased SCFA-production such as the Mediterranean diet (Cassani et al., 2017; De Filippis et al., 2016; Jackson et al., 2019), or by integrating probiotic or prebiotic supplementation into the diet. Co-administration of dopamine agonists and SCFA-producing probiotics (*Bifidobacterium*) improves motor symptoms, sleep, and gastrointestinal symptoms as demonstrated in a PD clinical trial (Sun et al., 2022). The accumulated evidence suggests that increased SCFA production may improve intestinal barrier function, which will in turn decrease endotoxemia, prevent colonic and neuronal inflammation, decrease microglial activation, and lead to decreased neuron loss.

In addition to nutritional intervention and dietary probiotics, fecal microbiota transplantation (FMT) may be an effective method to correct intestinal dysbiosis in PD. FMT in animal models may protect against neuroinflammation and LPS-mediated neurodegeneration (Sampson, 2020; Zhao et al., 2021). In a prospective study of 11 human PD patients who received FMT, there was a significant increase in the abundance of *Blautia* and *Prevotella* and a decrease in *Bacteroidetes*. These changes were accompanied by improvements in motor scores and decreases in PD-associated SIBO and constipation (Kuai et al., 2021). Improved motor symptoms were observed in a smaller trial as well (Segal, Zlotnik, Moyal-Atias, Abuhasira, & Ifergane, 2021).

Secretion of neuroprotective GLP-1 by enteroendocrine cells in the presence of intestinal SCFA has opened a new avenue of exploration in the treatment of PD (Manfready et al., 2022). As a neuroprotective agent, GLP-1 has receptors on multiple cell types, including on the vagus nerve, nigrostriatal neurons, and brain microglia. GLP-1 agonism has a protective effect against neurodegeneration in rodents and in human neuronal culture; pretreatment inhibits inflammatory cytokine production and prevents death of dopaminergic neurons induced by PD genetic mutations, MPTP, and rotenone, and can prevent or rescue PD motor phenotypes (Zhang, Li, & Hölscher, 2019, Chen, Huang et al., 2021). GLP-1 receptor agonists and dipeptidyl peptidase-4 inhibitors, such as those used in the treatment of diabetes, are being repurposed to treat PD (Athauda et al., 2017; Brauer et al., 2020; Ducastel et al., 2020). SCFA interaction with receptors on enteroendocrine L-cells throughout the gastrointestinal tract is the key source of systemic GLP-1 (Ducastel et al., 2020). Notably, PD patients have a diminished systemic GLP-1 response to a meal compared to healthy controls, suggesting that GLP-1 secretion is lacking in PD (Manfready et al., 2021). It is therefore conceivable that dietary prebiotic fibers may enhance GLP-1 secretion and produce neuroprotective effects (Manfready et al., 2022). Whether GLP-1 attenuation is purely the result of SCFA deficit or is due to compromised enteroendocrine function in PD is a focus of ongoing study.

Future well-designed randomized controlled trials are required to determine if gut microbiota-directed interventions capable of modifying dysbiosis can positively impact PD disease course. Examples of proposed interventions include dietary prebiotic supplements that promote SCFA-producing bacteria and interventions that increase production of neuroprotective gut hormones.

8. Conclusions

Recent animal and human studies suggest that a microbiota-mediated gut-brain axis may play a principal role in PD pathogenesis and progression. Microbial community structure is central to a healthy gut-brain axis; disrupted homeostasis can potentiate enteric and systemic inflammation in a susceptible host. Intestinal dysbiosis is a key feature of PD, which among other neurodegenerative diseases is characterized by alterations in gut microbiota that favor pro-inflammatory over anti-inflammatory metabolite production.

Studies suggest that pathobionts in the PD gut produce PAMPs such as LPS, which escape through an impaired intestinal barrier to cause neuroinflammatory signaling, secondary microglial activation in the brain, and eventual degeneration of dopaminergic neurons. The inflammatory signals in this process are also associated with aggregation of α-synuclein in the enteric nervous system and in the substantia nigra, the anatomic focus of Lewy body formation and neurodegeneration in PD. Altered intestinal permeability, which is a key feature of PD, may be the responsible mechanism for transmitting these pro-inflammatory signals from the gut to the brain. As such, microbial dysbiosis may be a trigger of PD disease onset. Lifestyle changes, intestinal dysmotility, and anti-PD medications act to further worsen dysbiosis, creating a vicious cycle that exacerbates and maintains neuroinflammation, leads to progressive dopamine loss, and eventually worsens the clinical course of PD.

It should be noted that the pro-inflammatory bacteria found in the PD intestine are similar to those favored by a Western diet, stress, and altered sleep/wake cycles that are common features of PD. This insight may explain the higher PD prevalence in geographical regions that have adapted to these patterns.

The PD phenotype is also characterized by a dearth of anti-inflammatory bacteria, which include those that produce SCFAs. Given their critical role in enterocyte health and in secretion of neuroprotective GLP-1 from enteroendocrine cells, SCFAs are key mediators of the gut-brain axis. Microbes and medicines that promote SCFA production and/or GLP-1 secretion have been shown to protect against neurodegeneration in multiple PD models, and may improve motor and non-motor phenotypes in existing disease. This mechanism supports the use of dietary interventions that promote SCFA production, as well as the repurposing of GLP-1 agonists for treatment of PD. In addition to justifying the use of dietary interventions, the emerging concept of microbial dysbiosis also provides rationale for the effects of xenobiotics and preexisting disease states (e.g., metabolic syndrome and diabetes) on PD pathogenesis, and offers mechanistic insight into the lack of medication efficacy observed in advanced disease states.

Since Braak's initial proposal that hypothesized a central role for the gut in PD pathogenesis, we have made tremendous progress toward understanding the gut-brain axis in PD. Microbial dysbiosis is a key feature of PD and its risk factors, and is likely a major determinant of neuroinflammation and Lewy body formation that are the underpinnings of PD pathology. Understanding gut-brain communication will enable improved PD therapeutic options, which include dietary, lifestyle-related, and pharmacologic interventions to prevent, slow, or reverse disease.

References

Aho, V. T. E., Houser, M. C., Pereira, P. A. B., Chang, J., Rudi, K., Paulin, L., et al. (2021). Relationships of gut microbiota, short-chain fatty acids, inflammation, and the gut barrier in Parkinson's disease. *Molecular Neurodegeneration*, *16*, 1–14.

Aho, V. T. E., Pereira, P. A. B., Voutilainen, S., Paulin, L., Pekkonen, E., Auvinen, P., et al. (2019). Gut microbiota in Parkinson's disease: Temporal stability and relations to disease progression. *eBioMedicine*, *44*, 691–707.

Akundi, R. S., Huang, Z., Eason, J., Pandya, J. D., Zhi, L., Cass, W. A., et al. (2011). Increased mitochondrial calcium sensitivity and abnormal expression of innate immunity genes precede dopaminergic defects in Pink1-deficient mice. *PLoS One*, *6*, e16038.

Annerino, D. M., Arshad, S., Taylor, G. M., Adler, C. H., Beach, T. G., & Greene, J. G. (2012). Parkinson's disease is not associated with gastrointestinal myenteric ganglion neuron loss. *Acta Neuropathologica*, *124*, 665–680.

Athauda, D., Maclagan, K., Skene, S. S., Bajwa-Joseph, M., Letchford, D., Chowdhury, K., et al. (2017). Exenatide once weekly versus placebo in Parkinson's disease: A randomised, double-blind, placebo-controlled trial. *The Lancet.*, *390*(10103), 1664–1675.

Barichella, M., Severgnini, M., Cilia, R., Cassani, E., Bolliri, C., Caronni, S., et al. (2019). Unraveling gut microbiota in Parkinson's disease and atypical parkinsonism. *Movement Disorders*, *34*, 396–405.

Blaylock, R. L. (2017). Parkinson's disease: Microglial/macrophage-induced immuno-excitotoxicity as a central mechanism of neurodegeneration. *Surgical Neurology International*, *8*, 65.

Bourassa, M. W., Alim, I., Bultman, S. J., & Ratan, R. R. (2016). Butyrate, neuroepigenetics and the gut microbiome: Can a high fiber diet improve brain health? *Neuroscience Letters*, *625*, 56–63.

Braak, H., Rüb, U., Gai, W. P., & Del Tredici, K. (2003). Idiopathic Parkinson's disease: Possible routes by which vulnerable neuronal types may be subject to neuroinvasion by an unknown pathogen. *Journal of Neural Transmission*, *110*, 517–536.

Brauer, R., Wei, L., Ma, T., Athauda, D., Girges, C., Vijiaratnam, N., et al. (2020). Diabetes medications and risk of Parkinson's disease: A cohort study of patients with diabetes. *Brain*, *143*, 3067–3076.

Breen, D. P., Halliday, G. M., & Lang, A. E. (2019). Gut-brain axis and the spread of α-synuclein pathology: Vagal highway or dead end? *Movement Disorders*, *34*, 307–316.

Cassani, E., Barichella, M., Ferri, V., Pinelli, G., Iorio, L., Bolliri, C., et al. (2017). Dietary habits in Parkinson's disease: Adherence to Mediterranean diet. *Parkinsonism & Related Disorders*, *42*, 40–46.

Caudle, W. M., Guillot, T. S., Lazo, C. R., & Miller, G. W. (2012). Industrial toxicants and Parkinson's disease. *Neurotoxicology*, *33*, 178–188.

Chakraborty, S., Kaushik, D. K., Gupta, M., & Basu, A. (2010). Inflammasome signaling at the heart of central nervous system pathology. *Journal of Neuroscience Research*, *88*, 1615–1631.

Challis, C., Hori, A., Sampson, T. R., Yoo, B. B., Challis, R. C., Hamilton, A. M., et al. (2020). Gut-seeded α-synuclein fibrils promote gut dysfunction and brain pathology specifically in aged mice. *Nature Neuroscience*, *23*, 327–336.

Chen, S.-J., Chen, C.-C., Liao, H.-Y., Lin, Y.-T., Wu, Y.-W., Liou, J.-M., et al. (2022). Association of Fecal and Plasma Levels of short-chain fatty acids with gut microbiota and clinical severity in patients with Parkinson disease. *Neurology*, *98*, e848–e858.

Chen, X., Huang, Q., Feng, J., Xiao, Z., Zhang, X., & Zhao, L. (2021). GLP-1 alleviates NLRP3 inflammasome-dependent inflammation in perivascular adipose tissue by inhibiting the NF-κB signalling pathway. *The Journal of International Medical Research*, *49*. 300060521992981.

Chen, Z. J., Liang, C. Y., Yang, L. Q., Ren, S. M., Xia, Y. M., Cui, L., et al. (2021). Association of Parkinson's disease with microbes and microbiological therapy. *Frontiers in Cellular and Infection Microbiology*, *11*, 619354.

De Filippis, F., Pellegrini, N., Vannini, L., Jeffery, I. B., La Storia, A., Laghi, L., et al. (2016). High-level adherence to a Mediterranean diet beneficially impacts the gut microbiota and associated metabolome. *Gut*, *65*, 1812–1821.

Del Tredici, K., & Braak, H. (2016). Review: Sporadic Parkinson's disease: Development and distribution of α-synuclein pathology. *Neuropathology and Applied Neurobiology*, *42*, 33–50.

Derkinderen, P., Shannon, K. M., & Brundin, P. (2014). Gut feelings about smoking and coffee in Parkinson's disease. *Movement Disorders*, *29*, 976–979.

Dodiya, H. B., Forsyth, C. B., Voigt, R. M., Engen, P. A., Patel, J., Shaikh, M., et al. (2020). *Chronic stress-induced gut dysfunction exacerbates Parkinson's disease phenotype and pathology in a rotenone-induced mouse model of Parkinson's disease*. Academic Press Inc.

Ducastel, S., Touche, V., Trabelsi, M. S., Boulinguiez, A., Butruille, L., Nawrot, M., et al. (2020). The nuclear receptor FXR inhibits glucagon-like Peptide-1 secretion in response to microbiota-derived short-chain fatty acids. *Scientific Reports*, *10*, 1–10.
Fasano, A., Visanji, N. P., Liu, L. W., Lang, A. E., & Pfeiffer, R. F. (2015). Gastrointestinal dysfunction in Parkinson's disease. *Lancet Neurology*, *14*, 625–639.
Fields, C. T., Sampson, T. R., Bruce-Keller, A. J., Kiraly, D. D., Hsiao, E. Y., & De Vries, G. J. (2018). Defining Dysbiosis in disorders of movement and motivation. *The Journal of Neuroscience*, *38*, 9414–9422.
Frau, A., Kenny, J. G., Lenzi, L., Campbell, B. J., Ijaz, U. Z., Duckworth, C. A., et al. (2019). DNA extraction and amplicon production strategies deeply influence the outcome of gut mycobiome studies. *Scientific Reports*, *9*(1), 1–17.
Ghaisas, S., Maher, J., & Kanthasamy, A. (2016). Gut microbiome in health and disease: Linking the microbiome–gut–brain axis and environmental factors in the pathogenesis of systemic and neurodegenerative diseases. *Pharmacology & Therapeutics*, *158*, 52–62.
Gordon, R., Albornoz, E. A., Christie, D. C., Langley, M. R., Kumar, V., Mantovani, S., et al. (2018). Inflammasome inhibition prevents α-synuclein pathology and dopaminergic neurodegeneration in mice. *Science Translational Medicine*, *10*, eaah4066.
Greiner, T. U., & Bäckhed, F. (2016). Microbial regulation of GLP-1 and L-cell biology. *Molecular Metabolism*, *5*, 753–758.
Guilarte, T. R. (2011). Manganese and Parkinson's disease: A critical review and new findings. *Ciência & Saúde Coletiva*, *16*, 4549–4566.
Heintz-Buschart, A., Pandey, U., Wicke, T., Sixel-Döring, F., Janzen, A., Sittig-Wiegand, E., et al. (2018). The nasal and gut microbiome in Parkinson's disease and idiopathic rapid eye movement sleep behavior disorder. *Movement Disorders*, *33*, 88–98.
Hill, M. O. (1973). Diversity and evenness: A unifying notation and its consequences. *Ecology*, *54*, 427–432.
Hill-Burns, E. M., Debelius, J. W., Morton, J. T., Wissemann, W. T., Lewis, M. R., Wallen, Z. D., et al. (2017). Parkinson's disease and Parkinson's disease medications have distinct signatures of the gut microbiome. *Movement Disorders*, *32*, 739–749.
Holdorff, B., Silva, A. M. R. E., & Dodel, R. (2013). Centenary of Lewy bodies (1912–2012). *Journal of Neural Transmission*, *120*, 509–516.
Jackson, A., Forsyth, C. B., Shaikh, M., Voigt, R. M., Engen, P. A., Ramirez, V., et al. (2019). Diet in Parkinson's disease: Critical role for the microbiome. *Frontiers in Neurology*, *10*, 1–21.
Jang, H., Boltz, D., Sturm-Ramirez, K., Shepherd, K. R., Jiao, Y., Webster, R., et al. (2009). Highly pathogenic H5N1 influenza virus can enter the central nervous system and induce neuroinflammation and neurodegeneration. *Proceedings of the National Academy of Sciences of the United States of America*, *106*, 14063–14068.
Jeong, J. J., Kim, K. A., Hwang, Y. J., Han, M. J., & Kim, D. H. (2016). Anti-inflammaging effects of Lactobacillus brevis OW38 in aged mice. *Beneficial Microbes*, *7*, 707–718.
Keogh, C. E., Rude, K. M., & Gareau, M. G. (2021). Role of pattern recognition receptors and the microbiota in neurological disorders. *The Journal of Physiology*, *599*, 1379–1389.
Keshavarzian, A., Engen, P., Bonvegna, S., & Cilia, R. (2020). The gut microbiome in Parkinson's disease: A culprit or a bystander? *Progress in Brain Research*, *252*, 357–450.
Keshavarzian, A., Green, S. J., Engen, P. A., Voigt, R. M., Naqib, A., Forsyth, C. B., et al. (2015). Colonic bacterial composition in Parkinson's disease. *Movement Disorders*, *30*(10), 1351–1360.
Knudsen, K., Krogh, K., Østergaard, K., & Borghammer, P. (2017). Constipation in parkinson's disease: Subjective symptoms, objective markers, and new perspectives. *Movement Disorders*, *32*, 94–105.
Koh, A., De Vadder, F., Kovatcheva-Datchary, P., & Bäckhed, F. (2016). From dietary Fiber to host physiology: Short-chain fatty acids as key bacterial metabolites. *Cell*, *165*, 1332–1345.

Kuai, X.-Y., Yao, X.-H., Xu, L.-J., Zhou, Y.-Q., Zhang, L.-P., Liu, Y., et al. (2021). Evaluation of fecal microbiota transplantation in Parkinson's disease patients with constipation. *Microbial Cell Factories*, *20*(1), 1–9.

Labzin, L. I., Heneka, M. T., & Latz, E. (2018). Innate immunity and Neurodegeneration. *Annual Review of Medicine*, *69*, 437–449.

Lenhart, A., Dong, T., Joshi, S., Jaffe, N., Choo, C., Liu, C., et al. (2022). Effect of exclusion diets on symptom severity and the gut microbiota in patients with irritable bowel syndrome. *Clinical Gastroenterology and Hepatology*, *20*, e465–e483.

Lin, A., Zheng, W., He, Y., Tang, W., Wei, X., He, R., et al. (2018). Gut microbiota in patients with Parkinson's disease in southern China. *Parkinsonism & Related Disorders*, *53*, 82–88.

Lozupone, C. A., Stombaugh, J. I., Gordon, J. I., Jansson, J. K., & Knight, R. (2012). Diversity, stability and resilience of the human gut microbiota. *Nature*, *489*, 220–230.

Maini Rekdal, V., Bess, E. N., Bisanz, J. E., Turnbaugh, P. J., & Balskus, E. P. (2019). Discovery and inhibition of an interspecies gut bacterial pathway for levodopa metabolism. *Science*, *364*.

Makki, K., Deehan, E. C., Walter, J., & Bäckhed, F. (2018). The impact of dietary Fiber on gut microbiota in host health and disease. *Cell Host & Microbe*, *23*, 705–715.

Manfready, R. A. (2010). Probing yeast for insights into neurodegenerative disease: ORFeome-wide screens for genetic modifiers of [alpha]-synuclein cytotoxicity. *Biotechnology and Molecular Biology Reviews*, *5*, 67–70.

Manfready, R. A., Engen, P. A., Verhagen Metman, L., Sanzo, G., Goetz, C. G., Hall, D. A., et al. (2021). Attenuated postprandial GLP-1 response in Parkinson's disease. *Frontiers in Neuroscience*, *15*, 660942.

Manfready, R. A., Forsyth, C. B., Voigt, R. M., Hall, D. A., Goetz, C. G., & Keshavarzian, A. (2022). Gut-brain communication in Parkinson's disease: Enteroendocrine regulation by GLP-1. *Current Neurology and Neuroscience Reports*, *22*(7), 335–342.

Manfredsson, F. P., Luk, K. C., Benskey, M. J., Gezer, A., Garcia, J., Kuhn, N. C., et al. (2018). Induction of alpha-synuclein pathology in the enteric nervous system of the rat and non-human primate results in gastrointestinal dysmotility and transient CNS pathology. *Neurobiology of Disease*, *112*, 106–118.

Marras, C., Beck, J. C., Bower, J. H., Roberts, E., Ritz, B., Ross, G. W., et al. (2018). Prevalence of Parkinson's disease across North America. *NPJ Parkinson's Disease*, *4*, 21.

Marras, C., Canning, C. G., & Goldman, S. M. (2019). Environment, lifestyle, and Parkinson's disease: Implications for prevention in the next decade. *Movement Disorders*, *34*, 801–811.

Matsuo, Y., Komiya, S., Yasumizu, Y., Yasuoka, Y., Mizushima, K., Takagi, T., et al. (2021). Full-length 16S rRNA gene amplicon analysis of human gut microbiota using MinIONTM nanopore sequencing confers species-level resolution. *BMC Microbiology*, *21*(1), 1–13.

Minato, T., Maeda, T., Fujisawa, Y., Tsuji, H., Nomoto, K., Ohno, K., et al. (2017). Progression of Parkinson's disease is associated with gut dysbiosis: Two-year follow-up study. *PLoS One*, *12*, e0187307.

Mitsumoto, A., & Nakagawa, Y. (2001). DJ-1 is an indicator for endogenous reactive oxygen species elicited by endotoxin. *Free Radical Research*, *35*, 885–893.

Noh, S. Y., Kang, S. S., Yun, C. H., & Han, S. H. (2015). Lipoteichoic acid from Lactobacillus plantarum inhibits Pam2CSK4-induced IL-8 production in human intestinal epithelial cells. *Molecular Immunology*, *64*, 183–189.

Paschos, G. K., & FitzGerald, G. A. (2017). Circadian clocks and metabolism: Implications for microbiome and aging. *Trends in Genetics*, *33*, 760–769.

Perez-Pardo, P., Dodiya, H. B., Engen, P. A., Forsyth, C. B., Huschens, A. M., Shaikh, M., et al. (2019). Role of TLR4 in the gut-brain axis in Parkinson's disease: A translational study from men to mice. *Gut*, *68*, 829–843.

Perez-Pardo, P., Dodiya, H. B., Engen, P. A., Naqib, A., Forsyth, C. B., Green, S. J., et al. (2018). Gut bacterial composition in a mouse model of Parkinson's disease. *Beneficial Microbes*, *9*, 799–814.

Pierantozzi, M., Pietroiusti, A., Brusa, L., Galati, S., Stefani, A., Lunardi, G., et al. (2006). Helicobacter pylori eradication and l-dopa absorption in patients with PD and motor fluctuations. *Neurology*, *66*, 1824–1829.

Polymeropoulos, M. H., Lavedan, C., Leroy, E., Ide, S. E., Dehejia, A., Dutra, A., et al. (1997). Mutation in the alpha-synuclein gene identified in families with Parkinson's disease. *Science*, *276*, 2045–2047.

Proctor, M. L. (2011). The human microbiome project in 2011 and beyond. *Cell Host & Microbe*, *10*, 287–291.

Przedborski, S., Tieu, K., Perier, C., & Vila, M. (2004). MPTP as a mitochondrial neurotoxic model of Parkinson's disease. *Journal of Bioenergetics and Biomembranes*, *36*, 375–379.

Sampson, T. (2020). The impact of indigenous microbes on Parkinson's disease. *Neurobiology of Disease*, *135*, 104426.

Sampson, T. R., Debelius, J. W., Thron, T., Janssen, S., Shastri, G. G., Ilhan, Z. E., et al. (2016). Gut microbiota regulate motor deficits and Neuroinflammation in a model of Parkinson's disease. *Cell*, *167*(6), 1469–1480.

Segal, A., Zlotnik, Y., Moyal-Atias, K., Abuhasira, R., & Ifergane, G. (2021). Fecal microbiota transplant as a potential treatment for Parkinson's disease - a case series. *Clinical Neurology and Neurosurgery*, *207*, 106791.

Shaikh, M., Rajan, K., Forsyth, C. B., Voigt, R. M., & Keshavarzian, A. (2015). Simultaneous gas-chromatographic urinary measurement of sugar probes to assess intestinal permeability: Use of time course analysis to optimize its use to assess regional gut permeability. *Clinica Chimica Acta*, *442*.

Shannon, K. M., Keshavarzian, A., Mutlu, E., Dodiya, H. B., Daian, D., Jaglin, J. A., et al. (2012). Alpha-synuclein in colonic submucosa in early untreated Parkinson's disease. *Movement Disorders*, *27*, 709–715.

Sun, H., Zhao, F., Liu, Y., Ma, T., Jin, H., Quan, K., et al. (2022). Probiotics synergized with conventional regimen in managing Parkinson's disease. *NPJ Parkinson's Disease*, *8*.

Tang, M. S., Poles, J., Leung, J. M., Wolff, M. J., Davenport, M., Lee, S. C., et al. (2015). Inferred metagenomic comparison of mucosal and fecal microbiota from individuals undergoing routine screening colonoscopy reveals similar differences observed during active inflammation. *Gut Microbes*, *6*, 48–56.

Turnbaugh, P. J., Hamady, M., Yatsunenko, T., Cantarel, B. L., Duncan, A., Ley, R. E., et al. (2009). A core gut microbiome in obese and lean twins. *Nature*, *457*, 480–484.

Tysnes, O.-B., & Storstein, A. (2017). Epidemiology of Parkinson's disease. *Journal of Neural Transmission*, *124*, 901–905.

Unger, M. M., Spiegel, J., Dillmann, K. U., Grundmann, D., Philippeit, H., Bürmann, J., et al. (2016). Short chain fatty acids and gut microbiota differ between patients with Parkinson's disease and age-matched controls. *Parkinsonism & Related Disorders*, *32*, 66–72.

van Kessel, S. P., Auvinen, P., Scheperjans, F., & El Aidy, S. (2021). Gut bacterial tyrosine decarboxylase associates with clinical variables in a longitudinal cohort study of Parkinsons disease. *NPJ Parkinson's Disease*, 7, 115.

Van Kessel, S. P., Frye, A. K., El-Gendy, A. O., Castejon, M., Keshavarzian, A., Van Dijk, G., et al. (2019). Gut bacterial tyrosine decarboxylases restrict levels of levodopa in the treatment of Parkinson's disease. *Nature Communications*, *10*.

Villumsen, M., Aznar, S., Pakkenberg, B., Jess, T., & Brudek, T. (2019). Inflammatory bowel disease increases the risk of Parkinson's disease: A Danish nationwide cohort study 1977–2014. *Gut*, *68*, 18–24.

Vindigni, S. M., Zisman, T. L., Suskind, D. L., & Damman, C. J. (2016). The intestinal microbiome, barrier function, and immune system in inflammatory bowel disease: A tripartite pathophysiological circuit with implications for new therapeutic directions. *Therapeutic Advances in Gastroenterology*, *9*, 606–625.

Voigt, R. M., Forsyth, C. B., Green, S. J., Engen, P. A., & Keshavarzian, A. (2016). Circadian rhythm and the gut microbiome. *International Review of Neurobiology*, *131*, 193–205.

Zhang, X., Dong, H., Zhang, S., Lu, S., Sun, J., & Qian, Y. (2015). Enhancement of LPS-induced microglial inflammation response via TLR4 under high glucose conditions. *Cellular Physiology and Biochemistry*, *35*, 1571–1581.

Zhang, L., Li, L., & Hölscher, C. (2019). Semaglutide is Neuroprotective and reduces α-Synuclein levels in the chronic MPTP mouse model of Parkinson's disease. *Journal of Parkinson's Disease*, *9*, 157–171.

Zhao, Z., Ning, J., Bao, X. Q., Shang, M., Ma, J., Li, G., et al. (2021). Fecal microbiota transplantation protects rotenone-induced Parkinson's disease mice via suppressing inflammation mediated by the lipopolysaccharide-TLR4 signaling pathway through the microbiota-gut-brain axis. *Microbiome*, *9*, 226.

Zhou, J., Martin, R. J., Tulley, R. T., Raggio, A. M., McCutcheon, K. L., Shen, L., et al. (2008). Dietary resistant starch upregulates total GLP-1 and PYY in a sustained day-long manner through fermentation in rodents. *American Journal of Physiology - Endocrinology and Metabolism*, *295*, 1160–1166.

Zinöcker, M., & Lindseth, I. (2018). The Western diet–microbiome-host interaction and its role in metabolic disease. *Nutrients*, *10*, 365.

CHAPTER FIVE

The emerging role of the microbiome in Alzheimer's disease

Caroline Wasén[a,b], Ella Simonsen[a,c], Millicent N. Ekwudo[a], Martin R. Profant[d], and Laura M. Cox[a,*]

[a]Ann Romney Center for Neurologic Diseases, Brigham & Women's Hospital, Harvard Medical School, Boston, MA, United States

[b]Department of Biology and Biological Engineering, Chalmers University of Technology, Gothenburg, Sweden

[c]University of Massachusetts, Amherst, MA, United States

[d]Department of Immunology, University of Toronto, Toronto, ON, Canada

*Corresponding author: e-mail address: lcox@bwh.harvard.edu

Contents

International Review of Neurobiology, Volume 167
ISSN 0074-7742
https://doi.org/10.1016/bs.irn.2022.09.001

Abstract

Alzheimer's disease (AD) is the most prevalent form of dementia and can be influenced by genetic and environmental factors. Recent studies suggest that the intestinal microbiota is altered in AD patients when compared to healthy individuals and may play a role in disease onset and progression. Aging is the greatest risk factor for AD, and age-related changes in the microbiota can affect processes that contribute to cognitive decline. The microbiota may affect AD by modulating peripheral and central immunity or by secreting factors that influence neurogenesis or neuronal cell death. Finally, probiotic and dietary interventions that target the microbiome may have therapeutic potential to prevent or treat AD.

1. Alzheimer's disease

Alzheimer's disease (AD) is a neurodegenerative disease characterized by the accumulation of misfolded proteins, neuronal cell death, and cognitive impairment (Hebert, Weuve, Scherr, & Evans, 2013; Alzheimer's's, 2013; Masters et al., 2015). Mutations associated with familial AD provide evidence that Aβ and tau contribute to the disease, however, 90% of AD is sporadic (Alzheimer's's, 2013), suggesting that there may be important environmental drivers of the disease. While altered genetics can increase the risk of developing AD (Tanzi, 2012), the largest risk factor is advanced age (Alzheimer's Association, 2013). Recent studies suggest that age-related changes in the gut microbiota may contribute to immunologic and neurologic decline, raising the question of whether the aging microbiome contributes to AD (Fig. 1). Furthermore, studies have found that carrying genetic variants that increase the risk of developing AD may also lead to an altered gut microbiota, for example, the presence of the apolipoprotein E4 (Apoe4) allele (Tran et al., 2019), suggesting that there may be important environmental and genetic interactions between the gut microbiota and AD.

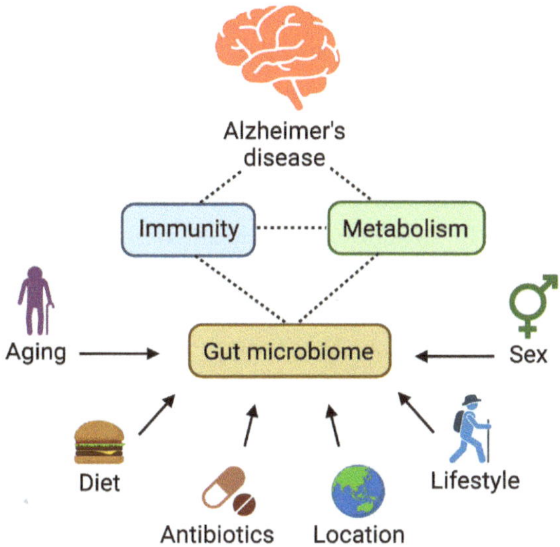

Fig. 1 Many factors influence the gut microbiota that interact with host immunologic and metabolic processes and play a role in Alzheimer's disease. These factors include aging, diet, use of antibiotics, geographical location, lifestyle, and sex.

2. The gut microbiota influence health and disease in aging

The intestinal microbiota encodes a metabolic capacity that can influence the host physiology (O'Hara & Shanahan, 2006). It is environmentally acquired, making it a modifiable therapeutic target. Its diverse functions may benefit the host, such as maintaining gastrointestinal health, producing vitamins, and promoting neuronal health (O'Hara & Shanahan, 2006), but disruption can result in disease (Cho et al., 2012). Altered microbiota, whether induced by diet (Turnbaugh, Backhed, Fulton, & Gordon, 2008), medication (Cox et al., 2014), hormone status (Markle et al., 2013), or host immune and metabolic status (Vijay-Kumar et al., 2010) can play an active role in disease. The microbiota influence the brain (Ochoa-Reparaz, Mielcarz, Begum- Haque, & Kasper, 2011) and interact with AD (Bekkering, Jafri, van Overveld, & Rijkers, 2013; Bhattacharjee & Lukiw, 2013), although precise mechanisms are still being actively investigated. Identification of key microbial changes could potentially have diagnostic or therapeutic value if successfully translated to humans.

In the elderly, age-related changes in host physiology and immunosenescence can destabilize the microbiota (Tiihonen, Ouwehand, & Rautonen, 2010). Because the incidence of AD increases with increasing age, these changes to the gut microbiota are believed to contribute to the risk of developing AD. Notably, there is an increase in intestinal permeability in aging (Tran & Greenwood-Van Meerveld, 2013) that leads to a higher exposure of the host to microbial products. This exposure may disrupt immune signaling pathways that protect against AD pathology. Certain lifestyle changes may prevent age-related changes to the gut. Long-term calorie-restriction substantially alters the gut microbiota and decreases gut permeability leading to lower circulating lipopolysaccharide-binding protein that indicates less translocation of pro-inflammatory bacterial products from the gut to the circulation (Zhang et al., 2013). This suggests that lifestyle modifications that improve AD may be partially mediated by controlling microbe-induced systemic inflammation.

3. The gut microbiota of patients with Alzheimer's disease

Several studies have investigated the gut microbiome in patients with AD. It is important to note that the AD risk and the gut microbiota composition have a high degree of geographic variation, which may be shaped by the host genetics, lifestyle factors, diet, and indigenous microbes to that location (Gupta, Paul, & Dutta, 2017). Furthermore, lifestyle factors within a population, including mobility and living independently vs. living in a long-term care facility can affect microbiota composition. Thus, for human studies of the microbiome in AD, it is critical to obtain well-matched controls for both geography and lifestyle.

One of the earliest studies of the microbiome in AD investigated changes in 6 microbial groups by qPCR in an Italian cohort of 40 amyloid positive subjects with cognitive impairments, 33 amyloid negative subjects with cognitive impairment, and 10 healthy controls. They found that amyloid positive subjects had higher *Escherichia/Shigella*, and a lower abundance of *Bacteroides fragilis* and *Eubacterium rectale* compared to amyloid negative subjects (Cattaneo et al., 2017). They also found elevated levels of proinflammatory cytokines, including elevated IL-6, IL-1β, CXCL-2, and NLRP3 compared to both amyloid negative subjects and healthy controls, suggesting that microbial alterations were associated with increased markers of inflammation. In 2017, Vogt et al. investigated differences in the gut microbiome of 25 patients with AD and 25 age- and sex-matched healthy controls. The study participants were recruited in Wisconsin, USA, and were all home-dwelling with no significant differences in diet. According to 16S rRNA sequencing of stool samples, AD patients had reduced *Firmicutes* and *Actinobacteria*, and increased *Bacteroidetes* at the phylum level compared to the healthy controls. On the family level, bacteria that were increased in AD patient included *Bacteroidaceae* and *Rikenellaceae* and families less abundant included *Ruminococcaceae* and *Bifidobacteriaceae*. Furthermore, the levels of *Bacteroides* and *Blautia* correlated with a lower Aβ-42/Aβ-40 ratio, and higher pTau in cerebrospinal fluid of AD patients and controls (Vogt et al., 2017).

Haran et al. investigated the gut microbiome of elderly individuals residing in nursing homes in Massachusetts, USA. The study included a total of 108 participants diagnosed with AD, other dementia or no dementia and aimed to use a statistical model to predict AD diagnosis based on the gut microbial composition of the participants. Increased levels of *Bacteroides* ssp., together with *Alistipes* ssp., *Odoribacter* ssp., *Barnesiella* ssp. and lower *Lachnoclostridium* ssp. were found in AD patients compared to controls. A random forest classification algorithm was used to classify study participants into AD patients and healthy controls based on clinical variables and the abundance of bacterial species in stool. The 30 most important predictors for AD included the abundance of *B. fragilis* and other bacterial species measured in the stool samples, many of them identified as butyrate producers, in addition to frailty and malnutrition (Haran et al., 2019).

4. Antibiotics and Alzheimer's disease

Antibiotics can have a profound effect on the gut microbiota and may have beneficial effects when targeting pathogens in a setting of infectious diseases or may have detrimental off-target effects. Disruption of the gut microbiota with antibiotics can reduce colonization resistance and increase the risk of infection with gastrointestinal pathogens (Kim, Covington, & Pamer, 2017) and has been linked to several chronic conditions, including obesity and metabolic syndrome (Cox & Blaser, 2015). The Nurses' Health Study II examined changes in cognition with previous antibiotic use in a population of 14,542 women and found that long-term (> 2 months) antibiotic use in midlife (mean age of 55) was associated with decreases in cognition 7 years later even while controlling for diet, socioeconomic status, body mass index, smoking, exercise, use of anti-inflammatory medications and comorbidities (Mehta et al., 2022). This data suggests that microbiota disruptions in midlife, when AD pathogenesis is estimated to start, could be linked to later cognitive decline. While the mechanisms are unknown, long-term antibiotic use may increase the risk of cognitive decline by permanently depleting beneficial microbiota, decreasing neuroprotective functions of the gut microbiota including the production of neurotransmitters and modulating immune responses, or by favoring colonization by detrimental microbiota post antibiotics. One limitation that was acknowledged by the Nurses' Health Study was that the data on cognitive decline could be stratified by antibiotic class, which can have markedly different effects on the gut microbiota.

An important consideration in the influence of antibiotics in cognitive decline is the reason for antibiotic use and the timing in relation to diagnosis. In the Nurses' Health Study, all antibiotics were given to cognitively normal individuals the most common reasons were respiratory infection, urinary tract infection, acne/rosacea, dental, and chronic bronchitis. It is possible that an acute treatment for a specific infection may have a beneficial effect. A study of 6628 individuals in Taiwan found that individuals hospitalized for infections with *Chlamydia pneumoniae* had an increased risk of developing AD, while those with *C. pneumoniae* infections receiving a macrolide or fluoroquinolone for more than 15 days had a decreased risk of developing AD (Ou et al., 2021). Interestingly, in a pilot study of 10 individuals with mild to moderate probable AD dementia (mini-mental state examination (MMSE) score 17 ±3), 3-month antibiotic treatment with rifaximin reduced serum levels of phosphorylated tau and neurofilament light chain and increased levels of *Erysipeloclostridium* (Suhocki, Ronald, Diehl, Murdoch, & Doraiswamy, 2022). Rifaximin is a broad-spectrum antibiotic that is poorly absorbed from the gastrointestinal tract; thus, effects are likely to be limited to modulating the gut microbiota. On the other hand, a multicenter trial of 406 patients with mild to moderate AD (MMSE score 14–26) found that 12 months of treatment with rifampin or doxycycline was associated with significant decreases in cognition, as measured by the Standardized Alzheimer's Disease Assessment Scale-Cognitive Subscale. Both rifampin and doxycycline are well absorbed from the GI tract, thus may influence microbial colonization at multiple sites, as well as directly affect host physiology.

The use of antibiotics for the treatment of infections in patients with severe AD and dementia may vary by country and clinical practice. For example, in the Netherlands, antibiotic treatment is commonly withheld from end-stage dementia patients (van der Steen, Ooms, Adèr, Ribbe, & van der Wal, 2002). In Canada, a survey of attitudes toward withholding antibiotics from individuals with advanced dementia from stakeholders, including senior citizens, family caregivers, nurses, and physicians, found that 75% of individuals supported withholding antibiotics from patients with advanced stage dementia and 98% of physicians supported withholding antibiotics from terminal stage dementia patients (Bravo, Van den Block, Downie, Arcand, & Trottier, 2021). Altogether, these studies highlight the need for a better understanding of antibiotics in several stages of AD, including assessing the degree to which chronic midlife antibiotics may be involved in the initiation of processes that lead to AD dementia, whether specific classes of antibiotics may have benefit for early- and mid-stage AD, and whether antibiotics can or should be used for end-stage AD.

5. The gut microbiota of murine models of Alzheimer's disease

Microbiota differences have been reported in several animal models of AD. Many of these animal models carry transgenes coding for proteins involved in amyloid processing which lead to the overproduction of amyloid peptides and formation of Aβ plaques. Other models carry transgenes with mutated forms of tau leading to the formation of neurofibrillary tangles. Depending on the mutations and expression levels of the transgenes, the animal models may exhibit pathology at different time points. The experimental timeline of different models may be a critical consideration for modeling age-dependent changes in the gut microbiota, for that reason each model is discussed separately in the following description of changes to the gut microbiota to mice with experimental AD.

The **Tg2576** model overexpresses a mutant form of APP with the Swedish mutation KM670/671NL and exhibits an age-dependent accumulation of Aβ plaques by 12 months of age (Hsiao et al., 1996). A study by Cox et al., investigated longitudinal changes in Tg2576 mice which typically develops plaques between 11 and 13 months of age (Cox et al., 2019). Because mice were singly housed, it was possible to measure microbiota drift over time in individual animals. The microbiota in aging Tg2576 females showed a substantial change in β-diversity from 5 to 15 months of age, which was not observed in littermate controls or in male mice. This suggests that the microbiota showed pronounced age-related changes in female AD mice. The largest age-related changes in female Tg2576 mice were an age-related decrease in *Faecalibaculum rodentium*, a member of the *Erysipleotrichaceae* family, and an age-related increase of *Bacteroides*, which correlated with the Aβ40 and AB42 levels found in the brain at 15 months of age, male mice did not have this association (Cox et al., 2019). However, another study by Honarpisheh et al. found lower levels of *Bacteroidetes* and higher levels of *Firmicutes* at the phylum level in Tg2576 mice at 15 months compared to 6 months of age and compared to wildtype controls (Honarpisheh et al., 2020). Differences in age-related microbiota changes in the same model may be due to sex-specific differences, as both sexes were used in a combined analysis in Honarpisheh et al., or may be due to animal facility dependent effects.

The **APP/PS1-ΔE9** model carries a transgene inserted into chromosome 9, which has a chimeric mouse/human APP Mo/HuAPP695swe and PSEN1 with a deletion of exon 9 (Goodwin et al., 2019) and shows the first signs of Aβ deposition at 6 months of age (Jankowsky et al., 2004). **The APP/PS1-21**, also known as the APP/PS1-Jucker mouse model of Alzheimer's disease expresses human transgenes for both APP with the Swedish mutation and PSEN1 with an L166P mutation under the control of the Thy1 promoter. The expression of APP is approximately three times higher than the endogenous APP expression in these mice leading to the formation of amyloid plaques starting around 1.5 months of age (Radde et al., 2006). Harach et al. analyzed the microbial composition of fecal pellets from APP/PS1 mice at 1 month (before amyloid pathology), 3.5 months (early pathology) and 8 months (late pathology) of age by sequencing the 16S rRNA gene. They reported that while the 1-month-old APP/PS1 mice had a microbial composition of their gut that was similar to wildtype C57BL/6 mice, aging APP/PS1 mice had an increasing fraction of *Bacteroidetes* and decreasing levels of *Firmicutes* in the stool (Harach et al., 2017). Furthermore, Cox et al. found that administering *Bacteroides* to APP/PS1-21 female mice by weekly oral gavages between 2.5 and 5 months of age, increased plaque burden in the cortex compared to controls (Cox et al., 2019). Chen et al., on the other hand, found an age-related decrease *Bacteroidetes* in APP/PS1 mice vs. non-transgenic controls 1 and 9 months of age. APP/PS1 mice also had higher levels of *Verrucomicrobia, Actinobacteria, and Proteobacteria.* Of note, several of the changes in the gut microbiota were detected at early time points before changes in AD pathology of changes in microglia were observed (Chen et al., 2020). Bäuerl et al. analyzed the gut microbiome of APP/PS1 mice at 3, 6 and 24 months of age and reported that *Turicibacteriaceae* and *Rikenellaceae* were decreased in aging wildtype. In addition, APP/PS1 mice had increased *Sutterella* and *Erysipelotrichaceae* in aging and also compared to wildtype mice (Bauerl, Collado, Diaz Cuevas, Vina, & Perez Martinez, 2018). *Sutterella* is a bacteria associated with autism disorder and Down Syndrome (Biagi et al., 2014; De Angelis et al., 2013) and the *Erysipelotrichaceae* family is highly immunogenic and linked to inflammatory bowel disease (Schaubeck et al., 2016), colorectal cancer (Chen, Liu, Ling, Tong, & Xiang, 2012), and Th17 responses in an animal model of multiple sclerosis (Miyauchi et al., 2020). Shen et al. reported higher levels of *Prevotella* in APP/PS1 mice compared to wildtype mice, and increasing levels of *Prevotellaceae* in 6- and 8-month-old mice compared to APP/PS1 mice at 3 months of age (Shen, Liu, & Ji, 2017). Zhang et al. found no difference in *Bacteroides* between APP/PS1 mice that were 3 and 9 months of age, differences on phylum level were not analyzed in this study (Zhang et al., 2021). Zhang et al. found that aging APP/PS1-ΔE9 mice had an increased ratio of *Firmicutes/Bacteroidetes* between 3 and 12 months of age (Zhang et al., 2017). Furthermore, the level of *Parabacteroides distasonis* was lower in aging wildtype and APP/PS1-ΔE9 mice. Cuervo-Zanatta et al. found that in APP/PS1-ΔE9 mice, female mice had higher *Bacteroidetes* than male mice (Cuervo-Zanatta, Garcia-Mena, & Perez-Cruz, 2021). Another study investigating the APP/PS1-ΔE9 strain concluded that the abundance of *Bacteroides* and *Lactobacillus* correlated with the age of the mice and were mostly found in mice that were 12 months of age (Li, Zhu, Guo, Du, & Qin, 2020). Furthermore, Harach et al. also demonstrated that APP/PS1 mice raised in germ-free conditions had decreased amyloid peptides levels in their brains and fewer amyloid plaques.

The **5xFAD** mice express human APP with the Swedish, Florida and London mutations and human PS1 with the M146L and L286V mutations, also under the Thy1 promoter, and develop plaques around 2 months of age (Oakley et al., 2006). Brandscheid et al. reported higher levels of *Bacteroidetes* between 1.5 and 4 months of age (Brandscheid et al., 2017), but Shukla et al. reported lower levels between 5 and 15 months of age although levels of *Bacteroidetes* was higher than wildtype controls at both timepoints (Shukla et al., 2021). Shukla et al. also reported increasing levels of *Firmicutes* and *Bifidobacterium bifidum* in older mice, although levels were lower in

5xFAD mice compared to wildtype mice. Furthermore, Tran et al. compared the gut microbiome of 5xFAD mice that has been crossed with APOE target replacement mice to express human APOE3 and APOE4, called E3FAD and E4FAD, respectively. EFAD mice develop cognitive impairment at 1–2 months and amyloid plaques around 4 months of age. At 4 months of age, E4FAD mice had lower levels of *Bacteroides* than E3FAD mice and a similar trend was seen in mice at 18 months. The E4FAD mice also had lower levels of *Alistipes* and *Johnsonella* at 4 months and higher levels of six other genera at either 4 or 18 months including *Desulfovibrio* and *Butyricicoccus* (Tran et al., 2019). Maldonado et al. reported lower levels of *Prevotella Sutterella* and *Ruminococcus,* higher levels of *Anaeroplasma* and no difference in *Bacteroides* between 4-month-old E3FAD and E4FAD mice (Maldonado Weng et al., 2019). Parikh et al. found differences in the levels of *Erysipelotrichaceae*, *Lactobacillaceae*, *Ruminococcaceae* and *Rikencellaceae* between mice with APOE2, −3 or −4 (Parikh et al., 2020).

The **3xTg-AD** mouse model harbor human transgenes of APP with the Swedish mutation and PS1 with the M146V mutation but also tau with the P301L mutation, meaning that these mice develop both amyloid and tau pathology (Oddo et al., 2003). This model develops amyloid plaques at 6 months of age and tau neurofibrillary tangles at 12 months. Bello-Medina et al. investigated changes in the microbiota of wild type B6129SF1/J mice at 3- vs. 5-months of age and found that both male and female transgenic mice showed increased levels of minor gut microbial populations, including phyla *Fusobacteria* and *Cyanobacteria* in comparison to non-transgenic mice (Bello-Medina et al., 2021). Transgenic female mice showed a decrease in the phylum *Actinobacteria* in aging which was not observed in wild type mice, and this bacterium has been linked to anti-inflammatory properties as well as a decreased gut permeability. Non-transgenic male mice showed an increase in *Firmicutes* and *Actinobacteria*, while also showing a decrease in *Proteobacteria*, *Fusobacteria*, *Cyanobacteria*, and *Bacteroidetes*. Overall, this study suggested that changes in the gut microbiota could be observed in models of AD that precede the development of cognitive symptoms (Bello-Medina et al., 2021).

Taken together, no specific species or group of bacteria has been consistently reported to have a higher or lower abundance in the gut of AD mouse models. As described above, a number of studies has reported higher levels of *Bacteroidetes* or *Bacteroides* in aging AD mice, but there are also reports indicating the opposite direction. However, a detrimental role of *Bacteroides* was supported by the report of higher levels of amyloid plaques found in mice administered with *B. fragilis*, supporting a direct relationship between these bacteria and amyloid pathology (Cox et al., 2019). There can be many reasons why studies come to different conclusions with regard to which bacteria are associated with amyloid and tau pathology in mice. We have discussed that the timing for microbiome collection could be an important factor as these mice develop pathology at different ages. In addition, the gut microbiota is sensitive to the environment, and may be affected by the facility that the mice were bred and raised in. Another important factor is that most studies are relying on sequencing of specific regions of the 16S rRNA gene, and there might be important genetic differences in these bacteria on species and strain levels that this methodology is not designed to find. To support this notion, Cox et al. reported that while the abundance of *Akkermansia* is generally associated with multiple sclerosis (MS), administration of MS-derived strains of *Akkermansia* to mice with experimental MS led to ameliorated disease in a strain-specific manner (Cox et al., 2021). Therefore, it is important to focus future research on functional aspects of the gut microbiota that may be linked to AD.

5.1 The role of the microbiota in modifying Aβ pathology in animal models

While several studies provide evidence that the gut microbiota shows age- and AD-dependent alterations in several animal models, microbiota changes may be a consequence of disease-specific changes in host physiology. To test whether there is a causal relationship between the microbiota and AD, several groups have investigated the effect of microbiota interventions, including depletion of the microbiota, addition of whole communities or specific bacteria, or dietary interventions that may shape the gut microbiota.

5.2 Depleting the gut microbiome alleviates amyloid and tau pathology

To test whether the gut microbiota was necessary for the development of Aβ plaques, researchers have utilized germ-free mice and antibiotics to deplete the gut microbiota. Germ-free APP/PS1 mice showed reduced amyloid plaque deposition, less neuroinflammation, and increased levels of Aβ-degrading enzymes compared to conventionally raised mice (Harach et al., 2017). Furthermore, the colonization of 4-month-old germ-free APP/PS1 mice with gut microbiota from 12-week-old conventionally raised wildtype or APP/PS1 mice for 8 weeks resulted in increased Aβ pathology (Harach et al., 2017), suggesting that the gut microbiota is both necessary and sufficient for the promotion of amyloid deposition. However, germ-free mice have substantially altered development and physiology, including immune and neurologic functions.

To address whether depleting the microbiota after development could affect AD pathology, several studies have investigated the effects of antibiotics in AD animal models. Minter et al. administered a combination of eight antibiotics given at a moderate to low dose to modulate the microbiome without inducing substantial host toxicity from long-term use of antibiotics. Strikingly, only male mice showed reduced Aβ plaque accumulation, which was linked to a reduction in proinflammatory cytokines in the serum (Minter et al., 2016). This low-dose antibiotic treatment did not entirely deplete the microbiota but was associated with a shift in microbiota composition over time. Further work from the same group found that antibiotics limited to early-life was sufficient to decrease plaque development in adult mice (Minter et al., 2017), and that the full combination of eight antibiotics was necessary for the effect, rather than single antibiotics (Dodiya et al., 2020). Altogether, these studies suggest that the microbiome can contribute to amyloid deposition, which may be influenced by antibiotic choice, the timing of administration, and sex-specific effects.

5.3 Fecal microbiota transfers reveal an important link between the gut microbial composition and memory function

There is a bidirectional connection between the gut microbiota and AD-like pathology and the influence of the gut microbiota on the brain has been directly investigated by transferring fecal matter between wildtype healthy mice and AD mouse models. A transfer of wildtype fecal microbiome to 6-month-old APPswe/PS1-ΔE9 mice decreased amyloid levels in the brain and improved spatial learning and recognition memory (Sun et al., 2019). The reduced plaque levels were associated with an increased abundance of *Bacteroidetes*, reduced abundance of *Verrucomicrobia* and increased butyrate production (Sun et al., 2019). Transfer of wildtype gut microbiome to $ADLP^{APT}$ also led to reduced levels of amyloid plaques in addition to phosphorylated tau and improved spatial memory (Kim et al., 2020). Interestingly, when the microbiota of 9-month-old 5xFAD mice were transferred to wildtype mice, the 5XFAD microbiota recipient wildtype mice demonstrated reduced spatial memory and reduced neurogenesis in the hippocampus (Kim et al., 2021). Furthermore, colonizing germ-free

APP/PS1 with the microbiome of aged conventionally raised wildtype and APP/PS1 mice increased the accumulation of amyloid peptides in the brains of the mice regardless of genotype, although mice colonized with the microbiome from APP/PS1 mice had higher levels than those colonized with the microbiome of wildtype mice (Harach et al., 2017), suggesting that the overproduction of Aβ could select for a microbiota that promotes plaque deposition.

6. The gut microbiota and the immune system in Alzheimer's disease

6.1 The gut microbiome modulates peripheral immunity

There is substantial evidence that both peripheral immune activation and immunosenescence can play a role in AD (Fig. 2). Studies have found that patients with AD have increased levels of circulating proinflammatory cytokines (Licastro et al., 2000). These proinflammatory cytokines may be produced by peripheral immune cells or microglia and may lead to disruptions in the blood-brain barrier, leading to reduced expression of tight junction proteins and increased permeability to serum proteins, reviewed in (Desai, Monahan, Carvey, & Hendey, 2007).

Peripheral immune cells may also have protective functions in AD. Monocytes can be recruited from the periphery to the central nervous system and play an important role in phagocytosing Aβ plaques. Peripheral monocytes from AD subjects have been reported to have impaired Aβ phagocytosis, suggesting a loss of innate immune function involved in Aβ clearance (Fiala et al., 2005). Nasally administering a proteosome adjuvant, derived from Gram-negative bacterial cell walls, leads to the activation of the peripheral innate immune system and improved clearance of Aβ in mice (Weiner & Frenkel, 2006). This is also associated with upregulating CCL2 and CCR2, critical factors that aid in monocyte trafficking to the brain (El Khoury et al., 2007). Thus, immunostimulatory products from the microbiota may play an important role in maintaining innate immune function that can prevent processes associated with AD. Furthermore, regulatory T cells have been shown to support microglia clearance of amyloid plaques (Dansokho et al., 2016).

Age- and AD-related immune changes may be affected by intestinal bacteria. Several studies demonstrate that the gut microbiota can modulate immune function during development and throughout the lifespan. Furthermore, gastrointestinal infections have been associated with an increased risk for dementia (Fink, Doblhammer, & Tamgüney, 2021).

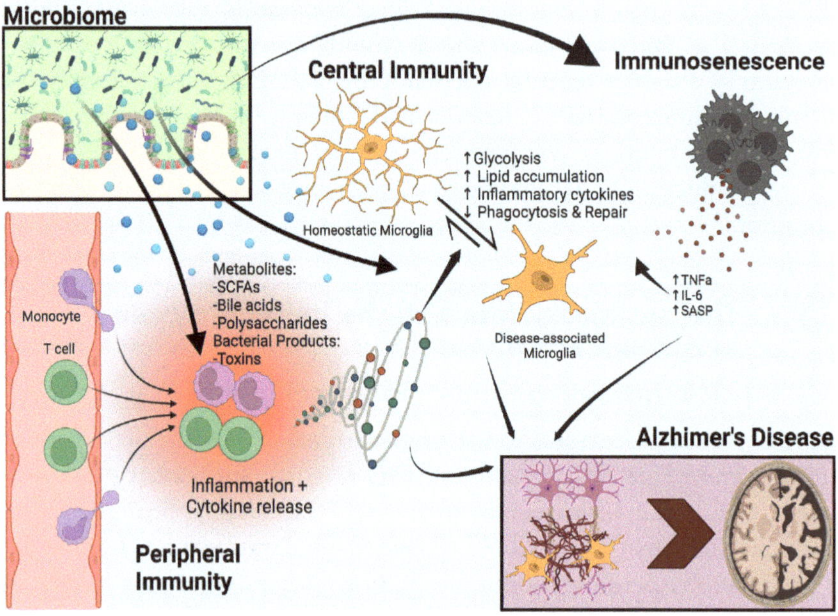

Fig. 2 The microbiome modulates peripheral and central immunity and immunosenescence to exacerbate AD pathogenesis. Three potential mechanisms by which the microbiome is able to modulate AD is through peripheral immunity, central immunity, and immunosenescence. The microbiome interacts with the peripheral and central immune systems through the secretion of metabolites including short-chain fatty acids (SCFAs), bile acids, polysaccharides, and bacterial components such as toxins. In the periphery, these microbial products can act to enhance or hinder the recruitment of monocytes to the brain to participate in phagocytosis of Aβ plaques. These microbial products can also promote systemic pro-inflammatory cytokine release, including tumor necrosis factor a (TNFa) and interleukins (IL), and promote the senescence-associated secretory phenotype (SASP) in immune cells. Microbial derived products, monocytes, and cytokines can then interact with brain immunity. The overall effect is the immune system's inability to clear plaques and resolve neuroinflammatory insults leading to exacerbated AD pathology.

6.2 The gut microbiome is important for microglia function

Microglia are the primary immune cell of the brain and play an important role in AD. In both mice and humans, microglia function diminishes with age due to changes in phagocytotic ability, morphology, activity, and even the production of toxic chemicals. In AD, microglia undergo functional change that contribute to neuroinflammation, which results in increased Aβ plaque deposition and neural atrophy.

Studies using germ-free or antibiotic-treated mice demonstrate that the gut microbiota affect microglia in both wildtype animals and models of AD. Germ-free mice have fewer microglia and fewer stromal choroid plexus macrophages than specific pathogen-free (SPF) mice (Erny et al., 2015; Sankowski et al., 2021). The depletion of the gut microbiota by administration of antibiotics to young male APP/PS1 mice did not reduce microglia numbers, but caused morphological and transcriptional changes associated with an inactive and immature state of microglia that could be reversed by fecal microbiota transfers from either wildtype or APP/PS1 mice (Dodiya et al., 2022). The mechanism by which the fecal matter transfer restores amyloidosis in mice with a depleted microbiota is unclear, but administration of short-chain fatty acids (SCFAs) resulted in higher microglia numbers in germ-free mice (Colombo et al., 2021), suggesting that microbial metabolites play an important role in supporting the microglia population in the brain. Furthermore, it was determined that the reduction in Aβ-plaques observed in antibiotics-treated APP/PS1 mice was diminished when microglia was depleted by an CSF-1R inhibitor (Dodiya et al., 2022), demonstrating that microglia play a critical role in mediating the effect of gut microbiota interventions on AD pathology. Similar observations were made in antibiotic-treated or germ-free 5xFAD mice. Germ-free mice, but not antibiotics-treated mice, had reduced microglia numbers (Mezö et al., 2020). Furthermore, microglia isolated from germ-free 5xFAD mice had higher expression of genes involved in phagosome maturation and were more efficient at phagocytosis ex vivo, while no difference was observed in microglia from antibiotics-treated mice. This observation suggests that lower plaque burden in germ-free 5xFAD mice may be the result of improved Aβ phagocytosis by microglia. Interestingly, these differences were only observed in microglia from relatively young animals (4 months old). At an older age (10 months old), no differences were observed in total microglia numbers or numbers of microglia that phagocyte Aβ, suggesting that long-term exposure to Aβ makes the microglia less responsive to gut microbiota modulations.

6.3 The gut microbiome modulates metabolic pathways in microglia

Cellular metabolic pathways regulate immune cell function (O'Neill, Kishton, & Rathmell, 2016). The gut microbiome and immune system are inextricably linked and one source of their connection is through microbial metabolites altering the metabolism and function of immune cells (Michaudel & Sokol, 2020).

Microglia, which have been widely implicated in their role in AD, are metabolically plastic and dependent on metabolic reprogramming for their function (Bernier, York, & MacVicar, 2020). A switch from mitochondrial oxidative phosphorylation to glycolysis in microglia contributes to AD pathology by impairing phagocytosis of Aβ (Pan, Ma, & Kong, 2019). Studies have shown that inflammatory stimuli from bacterial product may alter metabolic function in microglia. For example, stimulation of microglia with Aβ in combination with LPS or IFNγ led to increased glycolysis and a reduced ability to phagocytize Aβ in vitro (McIntosh et al., 2019; Rubio-Araiz, Finucane, Keogh, & Lynch, 2018). Another study found that Aβ alone was sufficient to induce reprogramming of microglia from oxidative phosphorylation to glycolysis in an mTOR-HIF1α dependent pathway (Baik et al., 2019). Increased glycolysis was also found in microglia isolated from APP/PS1 mice compared to wildtype mice (Holland et al., 2018).

The gut microbiome derived SCFA butyrate polarizes macrophages to an M2 phenotype by promoting a shift of the metabolism toward oxidative phosphorylation (Scott et al., 2018). According to a network-based ranking algorithm that were developed to investigate interactions between gut microbiome metabolites, microglia and AD, SCFAs was identified as the top microbial metabolites modulating microglia in the context of AD (Wang et al., 2021). Indeed, microbially derived SCFAs supplementation is sufficient to reverse the immature phenotype of microglia observed in germ-free mice (Erny et al., 2015). The mechanism by which SCFAs modulate microglia was determined to be through metabolic reprogramming of microglia metabolism. In the absence of the gut microbiome-derived SCFAs, there is modulations of histone acetylation and methylation in the proximity of promoter regions of genes that regulate metabolic functions in microglia. Bacteria-derived acetate was also shown to participate in the tricarboxylic acid cycle of microglia, and acetate-treated and SPF mice had lower levels of oxaloacetate leading to higher activity of complex II of the electron transport chain (Erny et al., 2021). It was further shown that acetate, but not butyrate or propionate, promoted microglia maturation in germ-free mice, and acetate-treated mice had impaired microglia phagocytosis of Aβ leading to higher plaque burden in germ-free 5xFAD mice.

Immunometabolism and its connection with the gut microbiome offer insight into potential explanations for sex dimorphism in AD. It is known that the gut microbiome and immune axis is different between males and females, especially in the context of AD (Cox et al., 2019). Immune cells and their metabolism are sex dimorphic in numerous metabolic pathways (Lee, Profant, & Wang, 2022). In APP/PS1 mice female and male microglia are distinct in their transcriptome and these differences can be mapped to metabolic genes. In functional assays, female microglia showed increased glycolytic flux and reduced phagocytosis suggesting that host sex is determinant in microglial metabolism and subsequent function (Guillot-Sestier et al., 2021). Several studies have begun to elucidate parts of this pathway. Thion et al. found that the microbiome contributes to the sex dimorphic development of microglia beginning in the prenatal stage (Thion et al., 2018). This establishes that microglial development is sex-specific and regulated by the gut microbiome. Rothhammer et al. found that type I interferons along with microbial metabolites modulate neuroinflammation (Rothhammer et al., 2016). Collectively, these findings suggest that sexual dimorphism in AD is present in the gut microbiome and contributes to altered immune cell metabolism.

7. Interactions between the diet, the microbiota, and Alzheimer's disease

The intestinal microbiota is shaped by diet among other factors and in turn, these microbes and their metabolites significantly influence gut health. Dietary patterns have been shown to influence AD pathology. There is a plethora of evidence of the neurocognitive improving actions of various diets including ketogenic and Mediterranean, as well as the consumption of prebiotics such as polyphenols in individuals and mouse models. However, there is a need to investigate the effects of these interventions on neurodegeneration biomarkers, especially in AD and dementia.

Studies have found that **calorie restriction** may have beneficial effects in aging and Alzheimer's disease. In mice, a 30% reduction in calorie intake in Tg2576 prior to plaque development led to improved glucose tolerance, reduced Aβ peptides and reduced amyloid plaque burden (Cox et al., 2019; Wang et al., 2005). As the mice aged, calorie restriction counteracted the increase in *Bacteroides* and the reduction in *Faecalibaculum* that were observed in Tg2576 mice fed ad libitum. Furthermore, the abundance of *Bacteroides* in feces correlated with higher levels of Aβ peptides in the brain and *Faecalibaculum* correlated with lower levels (Cox et al., 2019). In APPswe/PS1ΔE9 mice, calorie restriction led to improved glucose metabolism, improved spatial learning and lower amyloid plaque levels, which was associated with improved autophagy (Müller et al., 2021).

The Mediterranean diet plan supports the high intake of meals rich in vegetables, legumes, fruits, cereals, fish, and unsaturated fatty acids with low consumption of dairy products, and processed red meat. Most of the supported foods are rich in dietary fiber and polyphenols which are known for their numerous health benefits, including anti-inflammatory, anti-apoptotic and antioxidant action. Interestingly, adherence to the Mediterranean diet attenuated disease phenotype, characterized by decreased amyloidosis, decreased mediotemporal atrophy, and improved memory in participants of the German Longitudinal Cognitive Impairment and Dementia Study (Ballarini et al., 2021). A modified Mediterranean-ketogenic diet has been shown to modulate the gut microbiome and mycobiome in human subjects with MCI (Nagpal et al., 2020; Nagpal, Neth, Wang, Craft, & Yadav, 2019).

AD is often characterized by impaired brain metabolism and cognitive function. A positive correlation between physiological ketosis and improved cognitive function in the AD population has been suggested. The consumption of high-fat and low-carbohydrate food, called the **ketogenic diet (KD)**, promotes the formation of ketone bodies such as acetone, beta-hydroxybutyrate, and acetoacetate which are excellent fuel sources (Altayyar, Nasser, Thomopoulos, & Bruneau Jr., 2022). The use of these substrates can alleviate impaired glucose metabolism often associated with AD (Phillips et al., 2021). Although the mechanisms through which KD may improve AD are not fully understood, the anti-inflammatory and antioxidant potential of ketones have been reported (McDonald & Cervenka, 2018).

The ketogenic diet has been shown to modify the gut microbiota in humans and mice (Park, Zhang, Wu, & Yi Qiu, 2020; Ang et al., 2020; Olson et al., 2018). The shift in microbial profiles in KD-fed overweight/class I obese men was attributed to ketone bodies produced by the host. FMT from KD-fed human donors decreased the population of Th17 cells in the gut and b-hydroxybutyrate selectively repressed the growth of *Bifidobacteria* in germ-free mice (Ang et al., 2020). The neuroprotective role and cognitive benefits of KD in AD including clinical trials have been extensively reviewed here (Hersant & Grossberg, 2022; Rusek, Pluta, Ulamek-Koziol, & Czuczwar, 2019). However, these studies did not assess the microbial profiles of the intervened population. In mice, KD led to worsened disease phenotype characterized by gut dysbiosis with an elevated abundance of the phylum *Proteobacteria*, increased neuroinflammation, and glucose intolerance (Park et al., 2020). Other studies have demonstrated that the antiseizure effect of KD in mice is gut microbiota-dependent and increased the abundance of *Akkermansia muciniphila* and *Parabacteroides spp* (Olson et al., *2018)*.

Polyphenols are bioactive compounds produced by various plants such as fruits, legumes, herbs, and teas. Phenolic compounds are known for their evident health benefits including autoinflammatory, antioxidant, antimicrobial, and radical scavenging action, and have been considered potential therapeutics for diverse health conditions. Phenolic compounds often possess low bioavailability and are transformed into metabolites by the intestinal microbiota. These metabolites are released into the systemic circulation. In patients with mild cognitive decline, grape juice formulation, rich in phenols offered protection against cerebral metabolic decline and improved cognitive features such as attention and working memory (Lee, Torosyan, & Silverman, 2017) Similar improvements were reported for chlorogenic acids richly found in coffee beans administered to MCI patients in a randomized controlled crossover trial (Ochiai, Saitou, Suzukamo, Osaki, & Asada, 2019). Further, in a placebo-controlled, double-blind, randomized clinical trial, a homogenized mixture of fruits and vegetables improved neurocognitive patterns in a healthy non-elderly population (Carrillo, Arcusa, Zafrilla, & Marhuenda, 2021).

Trimethylamine (TMA) is a gut microbiota metabolite derived from dietary choline and carnitine and can be oxidized to Trimethylamine-N-oxide (TMAO) in the liver. The latter has been implicated in the pathophysiology of AD. In a recent study to investigate the connection between AD pathology and TMAO levels in the cerebrospinal fluid (CSF) of individuals with Alzheimer's clinical syndrome and those with mild cognitive impairment, elevated CSF TMAO levels were reported in both groups, compared to healthy control (Vogt et al., 2018). Additionally, elevated TMAO levels correlated with disease phenotype as characterized by AD biomarkers(Vogt et al., 2018). Similarly, elevated levels of TMAO were observed in urine samples from human AD and patients with mild cognitive impairment compared to healthy controls (Yilmaz et al., 2020). 3xTg-AD mice exhibited elevated levels of TMAO in an age-dependent manner and these mice showed less synaptic strength (Govindarajulu et al., 2020). Importantly, TMAO impaired synaptic transmission in hippocampal tissues retrieved from healthy mice and treated with TMAO and upregulated PKR-like endoplasmic reticulum kinase (PERK) pathway and consequently endoplasmic reticulum stress signaling (Govindarajulu et al., 2020). Overall, the mechanisms through which TMAO may influence AD pathology include protein aggregation (Aβ, tau), endoplasmic reticulum stress as well as microglial activation (Arrona Cardoza, Spillane, & Morales Marroquin, 2022). Intriguingly, attempts to lower TMAO production have shown promising therapeutic effects in mice. Coadministration of *Lactobacillus plantarum* and memantine decreased Aβ plaques in AD mice and cognitive decline, this was associated with decreased plasma TMAO levels (Wang et al., 2020). Similar interventions have been demonstrated here (Gao et al., 2019; Li et al., 2019; Liu et al., 2020).

8. Prebiotics

Prebiotics are dietary compounds that are not fully digested by the host and serve as growth substrates for beneficial microbiota. These can be a relatively low-risk intervention to modulate the gut microbiota, but the efficacy depends on having beneficial microbes already present in the microbiota, albeit at low levels.

8.1 Sodium oligomannate

Oligosaccharides, indigestible carbohydrates, are classical examples of prebiotics and are fermented by the gut microbiota. Sodium oligomannate, (GV-971) a derivative of marine algae has shown great therapeutic potential (Xiao et al., 2021). In a recently concluded phase III trial for mild for moderate AD/dementia, with 818 participants (placebo $n = 410$, GV-971 $n = 408$), GV-971 improved cognitive performance (Xiao et al., 2021). Elevated levels of amino acids such as phenylalanine and isoleucine have been associated with gut dysbiosis in AD patients and consequently proinflammatory responses in the brain (Wang et al., 2019). In AD mice, GV-971 countered cognitive impairment and neuroinflammation, by reconstituting the gut microbiota composition and decreasing Th1 cells in the brain. They also reported a reduction in the levels of phenylalanine and isoleucine in the blood and stool post-treatment (Wang et al., 2019).

8.2 Dietary inulin

Dietary inulin is another prebiotic with great health benefits. In APOE4 mouse models, E3FAD and E4FAD mice, inulin decreased hippocampal inflammatory gene expression, and increased metabolism characterized by increased abundance of SCFAs, tryptophan-derived metabolites and bile acids (Hoffman et al., 2019; Yanckello et al., 2021). This enhanced metabolism could be attributed to gut microbial alterations as inulin increased the abundance of taxa such as *Prevotella* and *Lactobacillus but decreased the relative abundance of Escherichia and Turicibacter spp (*Hoffman et al., 2019; Yanckello et al., 2021*)*.

9. Microbial mediators associated with Alzheimer's disease

The gut microbiota is estimated to contain as many cells as the human body and is estimated to carry 100 times the number of genes as the human genome. Along with this, the species within the gut microbiota can secrete a wide variety of neuromodulatory and immunomodulatory substances, may influence AD by direct or indirect mechanisms (Fig. 2).

9.1 Short-chain fatty acids (SCFAs)

We previously stated that germ-free mice have reduced plaque burden and it has been suggested that reduced plasma levels of SCFAs are responsible for this effect (Colombo et al., 2021). Supplementing germ-free mice with SCFA through drinking water significantly increased Aβ pathology in APP/PS1 mice. This difference was determined not to be due to a change in Aβ production, but instead to the deposition and clearance of the plaques, which is linked to the function of microglia. The introduction of SCFAs to germ-free and SPF mice was noted to induct a more active shape of microglia; however, this was not associated with an increased microglial phagocytic capacity toward Aβ, but rather promoted microglia activation marked by increased expression of TREM2. Other studies have reported an immunosuppressive role of SCFAs. Administration of acetate to APP/PS1 mice led to improved spatial learning by binding to the Free Fatty Acid Receptor 3 (FFAR3), inhibiting the ERK/JNK/NF-κB pathway and suppressing the production of pro-inflammatory cytokines (Liu et al., 2020). This anti-inflammatory effect of SCFA was also demonstrated in a cell culture model of human microglia-like cells, in which it also suppressed phagocytic activity (Wenzel, Gates, Ranger, & Klegeris, 2020). It was further demonstrated that propionate can protect blood-brain barrier integrity from LPS insults using an in vitro system to simulate the blood-brain barrier (Hoyles et al., 2018).

9.2 Bile acids

Bile acids (BAs) are the main metabolic end products of cholesterol metabolism. Primary bile acids such as cholic acid and chenodeoxycholic acid are produced in the liver while secondary bile acids such as deoxycholic acid and lithocholic acid are synthesized by colonic gut resident communities (Zeng, Umar, Rust, Lazarova, & Bordonaro, 2019). Both primary and secondary bile acids have been implicated as important contributors to AD pathophysiology. Importantly, altered BAs synthesis and metabolism in AD patients have been highlighted. Transcriptomic analysis of 2114 postmortem brains of AD and HC subjects revealed prevalent expression of BA synthesis pathways genes (Baloni et al., 2020). Targeted metabolomic profiling of these brain samples showed higher ratios of primary and secondary BAs in AD subjects compared to controls, suggesting a correlation between BAs and cognitive deficits in AD (Baloni et al., 2020). Additionally, the detection of secondary BAs in the brain suggests a potential influence of the gut microbiota(Baloni et al., 2020). Similar profiling of sera from subjects enrolled in the AD Neuroimaging Initiative revealed a significant association between altered serum-based BA with the Aβ, tau and neurodegeneration biomarkers for AD (Nho et al., 2019). Also, impaired glucose metabolism and larger atrophy of the brain were associated with the altered BA profiles (Nho et al., 2019). In a separate study, the bile acid profiles of AD patients indicated that increased bacterial 7α-dehydroxylation of citric acid into deoxycholic acid was associated with cognitive decline and immune-related AD-risk gene variants (MahmoudianDehkordi et al., 2019). Mass spectrometry profiling of the livers of APP/PS1 and App^{NL-G-F} male and female mice showed decreased cholesterol metabolism and primary bile acid synthesis compared to controls, but no difference in secondary bile acid production was observed (Kaur, Seeger, Golovko, Golovko, & Combs, 2021). Sex-specific differences were also reported, with females showing higher levels of cholesterol (Kaur et al., 2021). Importantly, genotype-associated bile acid alteration was reported as male APP/PS1 unlike App^{NL-G-F}, showed a selective reduction of the secondary bile acid taurourosdeoxycholic compared to controls (Kaur et al., 2021).

9.3 Polysaccharides

Some *B. fragilis* strains produce factors that are beneficial for non-AD neurologic diseases, including polysaccharide A (PSA), a cell wall glycan that induces T-regulatory cells (Tregs) (Mazmanian, Round, & Kasper, 2008).

B. fragilis packages PSA and other immunomodulatory cargo in outer membrane vesicles (OMVs) that enter the circulation, travel to the brain and decrease inflammation in the EAE animal model of multiple sclerosis (Ochoa-Reparaz et al., 2010). Although Tregs are beneficial in MS, they may play a detrimental role in AD by inhibiting Aβ phagocytosis by infiltrating monocytes and microglia (Baruch et al., 2015), highlighting the importance of understanding disease-specific interactions.

9.4 Toxins

Some *Bacteroides* strains carry the pathogenicity factor *Bacteroides* fragilysin toxin (BFT), a matrix metalloprotease that increases gut barrier permeability and leads to colitis (Wu, Dreyfus, Tzianabos, Hayashi, & Sears, 2002). BFT interacts with Psen1 (Wu, Rhee, Zhang, Franco, & Sears, 2007), a γ-secretase that cleaves amyloid-precursor protein (APP). Further studies are needed to determine if BFT affects γ-secretase activity in the brain. In addition to the gut microbiota, the oral microbiota may also contribute to AD. One study has found the Gingipains, toxins from the oral microbe *Porphyromonas gingivalis* in the brain of AD patients (Dominy et al., 2019), suggesting that there may be multiple important routes for microbial toxin infiltration.

10. Potential use of probiotics for the treatment of AD

10.1 Probiotic interventions can improve memory in models of Alzheimer's disease

A formulation of lactic acid bacteria and *Bifidobacterium species* (SLAB51) was given to triple-transgenic mice (3xTg-AD) for 4 months starting at 2 months of age. Administration of SLAB51 increased the abundance of *Bifidobacterium spp* in the gut of the mice and a reduction in *Campylobacterales*, which led to reduced accumulation of Aβ aggregates, improved recognition memory, and reduced cortical degradation. The reduced plaques were associated with higher plasma concentrations of gut hormones ghrelin, leptin, GLP1 and GIP, which have been shown to aid in learning, memory, and neuroprotection against Aβ plaques. Furthermore, mice treated with SLAB51 had reduced plasma levels of several pro-inflammatory cytokines including IL-1β and IFNγ, but higher levels of G-CSF, GM-CSF and IL-4. Within the brain, SLAB51 led to improved proteasome functionality and autophagy (Bonfili et al., 2017).

The probiotic *Bifidobacterium breve* strain A1 (*B. breve* A1) was given to 10-week-old male wildtype mouse (ddY strain) that were intracerebroventricularly given a 2mM solution of Aβ protein 25–35 to induce the effects of AD. ddY (Deutschland[Germany], Denken[Tokyo], Yoken [Japan]) mice are an outbred mouse strain derived from "dd" stock mice in the pre 1920s and then fully raised in the National Institute of Infectious Diseases in Yoken, Japan (NIBIOHN JLARB, 2022). The ddY strain has been described as a general-use mouse model and are extensively employed in diverse biomedical research to investigate therapeutic interventions in various pathologies, including Alzheimer's disease (The Jackson Laboratory, 2022; Farid, Yang, Kuboyama, & Tohda, 2020; Min et al., 2017; Tohda et al., 2021; Tomino, 2010). *B. breve* A1 was administered daily starting 2 days before Aβ injection. 6 days after Aβ injection, *B. breve* A1 led to improved spatial memory. Hippocampal gene expression was also profiled and indicated a reduced immune response following *B. breve* A1 treatment in AD mice only. Mice treated with *B. breve* A1 had higher levels of SCFAs in serum, and administration of acetate also improved memory deficits, but not to the same extent as mice treated with live bacteria. Transcriptional analysis of brain tissue indicated that differentially expressed genes were involved in immune responses (Kobayashi et al., 2017).

However, not all probiotic formulation has demonstrated beneficial effects. The probiotic VSL#3, which is a mixture of lactic-acid producing bacteria was administered to 6-month-old female AppNL-G-F mice for 8 weeks, which was not observed to reduce amyloid plaques nor improve memory as measured by a cross-maze rodent behavior test (Kaur et al., 2020).

10.2 Human trials of probiotics for Alzheimer's disease

As the medical field has gained knowledge about the role of the gut microbiome in AD, there have been several attempts to leverage the gut microbiome to limit the disease. Most notably is the use of probiotics which involve directly colonizing the gut microbiome with beneficial microbes (Naomi et al., 2021). In humans, there have been clinical trials to assess the effects of probiotics on cognitive function (Table 1).

A double-blind randomized controlled trial in 2016 treated a cohort of 60 AD patients with either probiotics containing *Lactobacillus acidophilus, Lactobacillus casei, Bifidobacterium bifidum*, and *Lactobacillus fermentum*, or milk.

Table 1 Probiotic trials for Alzheimer's disease.

Author and study design	Intervention	Probiotic species	Study population	Results
Akbari et al. (2016): • Double blind randomized control clinical trial • Two study groups: control (milk) and treatment (milk + probiotic mix)	200 mL/day of probiotic milk containing 2×10^9 CFU/g of each bacterial species for 12 weeks	*Lactobacillus acidophilus*, *Lactobacillus casei*, *Bifidobacterium bifidum*, and *Lactobacillus fermentum*	• $n = 30$ per group • Diagnosed AD patients • Mean Age: 82.0 (ctrl) and 77.67 (treatment) • Mean BMI: 22.73 (ctrl) and 23.77 (treatment) • Sex: 6 males and 24 females per group	• Improved cognition in probiotic treated group (measured by mini mental state examination, MMSE) • Decreased CRP levels, serum triglycerides, insulin resistance in the probiotic treated group • Increased beta cell function and insulin sensitivity in the probiotic treated group
Agahi et al. (2018): • Double blind randomized control clinical trial • Two study groups: placebo and probiotics • Participants were matched for disease severity, age, sex, and BMI	500 mg capsules of either placebo (maltodextrin) or 3×10^9 CFU/g probiotics every other day for 12 weeks	*Lactobacillus fermentum*, *Lactobacillus plantarum*, and *Bifidobacterium lactis* Or *Lactobacillus acidophilus*, *Bifidobacterium bifidum*, and *Bifidobacterium longum*	• $n = 23$ (ctrl) and 25 (probiotic) • Participants were tested using the TYM cognitive test and patients with scores indicative of severe AD were included in the study • Mean Age: 80.57 (ctrl) and 79.7 (treatment) • Mean BMI: 24.44 (ctrl) and 24.05 (treatment) • Sex: 10 M and 13F (ctrl) and 7 M and 18F (treatment)	• No significant change in cognitive scoring between probiotic treated and control groups • No significant change in serum cytokine levels or markers of oxidative stress

Continued

Table 1 Probiotic trials for Alzheimer's disease.—cont'd

Author and study design	Intervention	Probiotic species	Study population	Results
Leblhuber, Steiner, Schuetz, Fuchs, & Gostner, (2018): • Explorative longitudinal study in which a cohort of moderate AD patients were assessed for cognitive abilities and inflammatory biomarkers before and after prebiotic treatment • One study group, two time points (before treatment and after)	Probiotic treatment was given daily for 4 weeks	*Lactobacillus casei* W56, *Lactococcus lactis* W19, *Lactobacillus acidophilus* W22, *Bifidobacterium lactis* W52, *Lactobacillus paracasei* W20, *Lactobacillus plantarum* W62, *Bifidobacterium lactis* W51, *Bifidobacterium bifidum* W23 and *Lactobacillus salivarius* W24	• $n=19$ • Participants had moderate AD based on MMSE scores • Mean Age: 76.7 • Sex: 9 males and 11 females	• Increased fecal zonulin levels and Faecalibacterium prausnitzii concentrations • Increased serum kynurenine concentrations • Increased nitrite and neopterin levels • Increased RNA content in fecal bacteria
Tamtaji et al. (2019): • Double blind randomized control clinical trial • Three study groups: selenium alone, selenium + probiotic, or placebo	200 μg/day of selenium and 2×10^9 CFU/day of probiotic for 12 weeks	*Lactobacillus acidophilus*, *Bifidobacterium bifidum*, and *Bifidobacterium longum*	• $n=26$ (Se alone), $n=27$ probiotic + Se, n = 26 (placebo) • Patients diagnosed with AD • Mean Age: ~78 • Mean BMI: ~21	• Significant reduction in CRP, insulin, and insulin resistance, LDL-cholesterol, and total cholesterol:HDL ratio • Significant increase in glutathione, insulin sensitivity, and anti-oxidant capacity • Significant increase in MMSE

(Ton et al., 2020): • Uncontrolled clinical investigation to assess effects of probiotics on AD patients • One study group, two time points (before treatment and after)	4% kefir including the probiotic species was added to pasteurized milk and incubated for 24 h. Then the fermented product was blended with 500 g of strawberries and 2 L of milk. Treatment duration 3 months	*Acetobacter aceti*, *Acetobacter* sp., *Lactobacillus delbrueckii delbrueckii*, *Lactobacillus fermentum*, *Lactobacillus fructivorans*, *Enterococcus faecium*, *Leuconostoc* spp., *Lactobacillus kefiranofaciens*, *Candida famata*, and *Candida krusei*	• $n = 13$ • AD patients displaying some kind of cognitive deficit and no comorbidities • Mean age: ~78 • Mean BMI: ~26 • Sex: 2 males and 11 females	• Significant improvement in cognitive abilities such as memory, abstraction, and language as well as higher MMSE scores with treatment • Significant reduction in TNFa, IL-8 (CXCL8), and IL-12p70 levels in the serum, as well as reduced pro- to anti-inflammatory cytokine levels with treatment • Significant reduction in ROS (superoxide, hydrogen peroxide, and hydroxyl radical) in the serum with treatment • Significant increase in mitochondrial membrane potential, DNA content and reduction in DNA fragmentation and apoptosis

These patients were treated for 12 weeks and after the trial, it was found that patients treated with probiotics showed improvement on the MMSE. These probiotics also had an improved effect on the metabolic health as determined by insulin sensitivity test, serum triglycerides, and beta cell function. Although no other biomarkers for inflammation or oxidative stress were altered with treatment, there was a significant decrease in serum C reactive protein in the probiotic-treated patients (Akbari et al., 2016). The same group followed up this study with one in which severe AD patients were given a similar cocktail of bacteria. In this trial, the AD patients showed some elevation in serum glutathione levels, but no improvement in memory function and no inflammatory or oxidative stress markers were altered between the treated and untreated groups including pro-inflammatory cytokines (Agahi et al., 2018).

In another trial, moderate AD patients (average MMSE of 18.5) were treated with either a probiotic mix of *Lactobacillus* strains and *Bifidobacterium* strains for 28 days. The major finding from the trial showed a significant increase the serum kynurenine levels and increased kynurenine/tryptophan ratio as well as neopterin. The authors claim that this change in kynurenine levels is suggestive of myeloid cell activation. However, the authors did not assess the effect of probiotic treatment on cognitive impairment (Leblhuber, et al., 2018).

In a relatively large study, 79 patients were randomly assigned to three groups: placebo or selenium alone, or selenium plus probiotics. The probiotics contained *Lactobacillus acidophilus, Bifidobacterium bifidum,* and *Bifidobacterium longum.* The treatment course was 12 weeks, after which the results showed that the selenium plus probiotic group significantly outperformed the other groups on the MMSE test. The probiotic group also showed significant reductions in C reactive protein, increased glutathione and total antioxidant capacity, as well as improved metabolic health such as lowered serum insulin, cholesterol, and triglycerides (Tamtaji et al., 2019).

Another approach to manipulate the microbiome for other acute and chronic diseases is the use of a **fecal microbiota transplant (FMT)** from a healthy donor to an affected individual via an enema, a nasal gastric infusion, or from encapsulated pills, and may have therapeutic potential to combat infectious and non-infectious diseases (Xu et al., 2021). FMT has been most widely used in the treatment of *Clostridiodes difficile* diarrhea, which is an opportunistic infection that can occur when the normal microbiota is diminished by antibiotic treatment. In cases of recurrent *C. difficile* infection, FMT efficacy is over 90% (van Nood et al., 2013). This high degree of efficacy with multiple microbiota donor sources may be mechanistically linked to restoring colonization from the microbiome, which then inhibits the growth of *C. difficile* in the setting of an acute infection. Whether these same principals can be applied to chronic diseases is an active area of investigation and may have relevance to Alzheimer's disease.

Because elderly individuals are at a greater risk of *C. difficile* infection (CDI), there have been a few case reports in which AD subject have received FMT from a cognitively normal subject. In one case, an 82-year old man with AD and an MMSE of 20 (mild/moderate dementia) was initially hospitalized for a case of methicillin resistant *Staphylococcus aureus*, then acquired a GI infection with *C. difficile* (Hazan, *2020)*. Following antibiotic treatment with vancomycin, metronidazole, fidaxomicin, and bezlotoxumab and recurrence of CDI, the patient received an FMT infusion of a stool sample from his 85-year-old intellectually acute wife, which cured the *C. difficile* infection. Over the next 2 and 6 months, the subject's MMSE improved to 26 and 29, respectively, indicating normal cognition.

In a second case report (Park et al., 2021), a 90-year-old woman had a diagnosis of AD and an MMSE of 15 (moderate dementia) recorded 30 days prior to a hospitalization event. Later, she acquired severe and antibiotic-refractory CDI. She was treated with an FMT from 27-year-old male donor who underwent detailed health screening, which initially increased her MMSE to 18 at 1-week post FMT and 20 at 3-months post FMT. In addition, at 3-months post FMT, her CDI symptoms returned, including fever, diarrhea, and continuous abdominal pain. A second FMT was performed and 1 week later, her MMSE was stable at 20. Her microbiota was sequenced at pre-FMT and post first FMT. FMT increased the levels of *Tannerellaceae* and decreased the levels of *Negativicutes* and *Lachnospira*.

The same group then evaluated cognitive changes in 10 subjects with dementia and severe CDI that received FMT therapy vs. 10 subjects with dementia and severe CDI that received antibiotic therapy (Park et al., 2022). The group receiving FMT had a baseline mean MMSE of 10.0 (range of 7.0–14.0) that increased to a mean of 16 (range of 13–18) following treatment, whereas the group receiving antibiotics had a baseline MMSE of 14.0 (range 12.8–16.0) that decreased to a mean of 10.0 (range of 9.8–15.3) after treatment. In the pre-FMT microbiota, the phylum *Proteobacteria* was approximately 44% of the relative abundance. This phylum is typically a minor component of the microbiota compared to the phyla *Bacteroidetes* and *Firmicutes,* and high levels detected pre-FMT may be linked to antibiotic-mediated disruption in the gut. FMT lowered *Proteobacteria* levels (including *Enterobacteriaceae*, *Sutterella*, and *Klebsiella*) to 22%, increased microbiota diversity, and increased prevalence of bacteria more typical of a normal gut microbiota including *Bacteroides, Alistipes, Blautia, and Bifidobacterium.*

These limited FMT case studies in the context of CDI and AD raise the question of whether this could be a promising new therapy for AD not suffering from CDI. One advantage of an FMT vs. delivery of a probiotic is that it delivers of a complex community of bacteria that may be more stable and may retain beneficial functions supported by microbial cross-feeding. In addition, this approach does not rely on the identification of a specific bacteria for treatment efficacy. However, there are considerable disadvantages including variability of microbes delivered from different donors and the potential for rare serious adverse effects, including infections with antibiotic resistance bacteria (Gupta, Mullish, & Allegretti, 2021). While regulatory measures have been established by the FDA and are continuing to be improved (Gupta et al., 2021), AD subjects represent an elderly population that may be more vulnerable to infection than younger subjects. Nonetheless, more work may be warranted in this area to evaluate potential safety and efficacy. Because of the considerable health and ethical risk of performing clinical trials in elderly and cognitively impaired individuals, there may be substantial barriers for large scale clinical trials. Thus, small case reports in which AD patients receive FMT used for non-AD clinical indications can provide valuable information for the field.

The collective data from both probiotic and FMT trials suggest that while manipulating the microbiome has shown some promise and should undergo further investigation, the studies conducted to this day are relatively small and show some variation. Future trials should aim to improve the assays performed on the AD patients and include the measurement of biomarkers in cerebrospinal fluid to detect improvements in early AD pathology. Additionally, future trials should attempt to corroborate cognitive function data by using multiple types of tests. Basic research should continue to explore new strains of gut bacteria that could potentially mediate a stronger and more specific effect. The precise mechanisms of prebiotics in AD are still unknown, but clinical trials suggest mechanisms involved in improving body metabolism, immune activation, and serum metabolites and more studies are needed in both preclinical research and in AD patients to determine new targets and refine existing therapies.

References

Agahi, A., Hamidi, G. A., Daneshvar, R., Hamdieh, M., Soheili, M., Alinaghipour, A., et al. (2018). Does severity of Alzheimer's disease contribute to its responsiveness to modifying gut microbiota? A double blind clinical trial. *Frontiers in Neurology*, *9*, 662.

Akbari, E., Asemi, Z., Daneshvar Kakhaki, R., Bahmani, F., Kouchaki, E., Tamtaji, O. R., et al. (2016). Effect of probiotic supplementation on cognitive function and metabolic status in Alzheimer's disease: A randomized, double-blind and controlled trial. *Frontiers in Aging Neuroscience*, *8*, 256.

Altayyar, M., Nasser, J. A., Thomopoulos, D., & Bruneau, M., Jr. (2022). The implication of physiological ketosis on the cognitive brain: A narrative review. *Nutrients*, *14*, 513.

Alzheimer's Association. (2013). Alzheimer's disease facts and figures. *Alzheimer's & Dementia*, *9*(2013), 208–245.

Alzheimer's, A. (2013). Alzheimer's disease facts and figures. *Alzheimers Dement*, *9*(2013), 208–245.

Ang, Q. Y., Alexander, M., Newman, J. C., Tian, Y., Cai, J., Upadhyay, V., et al. (2020). Ketogenic diets Alter the gut microbiome resulting in decreased intestinal Th17 cells. *Cell*, *181*, 1263–1275.e16.

Arrona Cardoza, P., Spillane, M. B., & Morales Marroquin, E. (2022). Alzheimer's disease and gut microbiota: Does trimethylamine N-oxide (TMAO) play a role? *Nutrition Reviews*, *80*, 271–281.

Baik, S. H., Kang, S., Lee, W., Choi, H., Chung, S., Kim, J. I., et al. (2019). A breakdown in metabolic reprogramming causes microglia dysfunction in Alzheimer's disease. *Cell Metabolism*, *30*, 493–507.e6.

Ballarini, T., Melo van Lent, D., Brunner, J., Schroder, A., Wolfsgruber, S., Altenstein, S., et al. (2021). Mediterranean diet, Alzheimer disease biomarkers and brain atrophy in old age. *Neurology*, *96*, 2920–2932.

Baloni, P., Funk, C. C., Yan, J., Yurkovich, J. T., Kueider-Paisley, A., Nho, K., et al. (2020). Metabolic network analysis reveals altered bile acid synthesis and metabolism in Alzheimer's disease. *Cell Reports. Medicine*, *1*, 100138.

Baruch, K., Rosenzweig, N., Kertser, A., Deczkowska, A., Sharif, A. M., Spinrad, A., et al. (2015). Breaking immune tolerance by targeting Foxp3(+) regulatory T cells mitigates Alzheimer's disease pathology. *Nature Communications*, *6*, 7967.

Bauerl, C., Collado, M. C., Diaz Cuevas, A., Vina, J., & Perez Martinez, G. (2018). Shifts in gut microbiota composition in an APP/PSS1 transgenic mouse model of Alzheimer's disease during lifespan. *Letters in Applied Microbiology*, *66*, 464–471.

Bekkering, P., Jafri, I., van Overveld, F. J., & Rijkers, G. T. (2013). The intricate association between gut microbiota and development of type 1, type 2 and type 3 diabetes. *Expert Review of Clinical Immunology*, *9*, 1031–1041.

Bello-Medina, P. C., Hernández-Quiroz, F., Pérez-Morales, M., González-Franco, D. A., Cruz-Pauseno, G., García-Mena, J., et al. (2021). Spatial memory and gut microbiota alterations are already present in early adulthood in a pre-clinical transgenic model of Alzheimer's disease. *Frontiers in Neuroscience*, *15*, 595583.

Bernier, L. P., York, E. M., & MacVicar, B. A. (2020). Immunometabolism in the brain: How metabolism shapes microglial function. *Trends in Neurosciences*, *43*, 854–869.

Bhattacharjee, S., & Lukiw, W. J. (2013). Alzheimer's disease and the microbiome. *Frontiers in Cellular Neuroscience*, 7, 153.

Biagi, E., Candela, M., Centanni, M., Consolandi, C., Rampelli, S., Turroni, S., et al. (2014). Gut microbiome in Down syndrome. *PLoS One*, *9*, e112023.

Bonfili, L., Cecarini, V., Berardi, S., Scarpona, S., Suchodolski, J. S., Nasuti, C., et al. (2017). Microbiota modulation counteracts Alzheimer's disease progression influencing neuronal proteolysis and gut hormones plasma levels. *Scientific Reports*, 7, 2426.

Brandscheid, C., Schuck, F., Reinhardt, S., Schafer, K. H., Pietrzik, C. U., Grimm, M., et al. (2017). Altered gut microbiome composition and tryptic activity of the 5xFAD Alzheimer's mouse model. *Journal of Alzheimer's Disease*, *56*, 775–788.

Bravo, G., Van den Block, L., Downie, J., Arcand, M., & Trottier, L. (2021). Attitudes toward withholding antibiotics from people with dementia lacking decisional capacity: Findings from a survey of Canadian stakeholders. *BMC Medical Ethics*, *22*, 119.

Carrillo, J. A., Arcusa, R., Zafrilla, M. P., & Marhuenda, J. (2021). Effects of fruit and vegetable-based nutraceutical on cognitive function in a healthy population: Placebo-controlled, double-blind, and randomized clinical trial. *Antioxidants (Basel)*, *10*, 116.

Cattaneo, A., Cattane, N., Galluzzi, S., Provasi, S., Lopizzo, N., Festari, C., et al. (2017). Association of brain amyloidosis with pro-inflammatory gut bacterial taxa and peripheral inflammation markers in cognitively impaired elderly. *Neurobiology of Aging*, *49*, 60–68.

Chen, W., Liu, F., Ling, Z., Tong, X., & Xiang, C. (2012). Human intestinal lumen and mucosa-associated microbiota in patients with colorectal cancer. *PLoS One*, *7*, e39743.

Chen, Y., Fang, L., Chen, S., Zhou, H., Fan, Y., Lin, L., et al. (2020). Gut microbiome alterations precede cerebral amyloidosis and microglial pathology in a mouse model of Alzheimer's disease. *BioMed Research International*, *2020*, 8456596.

Cho, I., Yamanishi, S., Cox, L., Methe, B. A., Zavadil, J., Li, K., et al. (2012). Antibiotics in early life alter the murine colonic microbiome and adiposity. *Nature*, *488*, 621–626.

Colombo, A. V., Sadler, R. K., Llovera, G., Singh, V., Roth, S., Heindl, S., et al. (2021). Microbiota-derived short chain fatty acids modulate microglia and promote Aβ plaque deposition. *eLife*, *10*, e59826.

Cox, L. M., & Blaser, M. J. (2015). Antibiotics in early life and obesity. *Nature Reviews. Endocrinology*, *11*, 182–190.

Cox, L. M., Yamanishi, S., Sohn, J., Alekseyenko, A. V., Leung, J. M., Cho, I., et al. (2014). Altering the intestinal microbiota during a critical developmental window has lasting metabolic consequences. *Cell*, *158*, 705–721.

Cox, L. M., Schafer, M. J., Sohn, J., Vincentini, J., Weiner, H. L., Ginsberg, S. D., et al. (2019). Calorie restriction slows age-related microbiota changes in an Alzheimer's disease model in female mice. *Scientific Reports*, *9*, 17904.

Cox, L. M., Maghzi, A. H., Liu, S., Tankou, S. K., Dhang, F. H., Willocq, V., et al. (2021). Gut microbiome in progressive multiple sclerosis. *Annals of Neurology*, *89*, 1195–1211.

Cuervo-Zanatta, D., Garcia-Mena, J., & Perez-Cruz, C. (2021). Gut microbiota alterations and cognitive impairment are sexually dissociated in a transgenic mice model of Alzheimer's disease. *Journal of Alzheimer's Disease*, *82*, S195–s214.

Dansokho, C., Ahmed, D. A., Aid, S., Toly-Ndour, C., Chaigneau, T., Calle, V., et al. (2016). Regulatory T cells delay disease progression in Alzheimer-like pathology. *Brain*, *139*, 1237–1251.

De Angelis, M., Piccolo, M., Vannini, L., Siragusa, S., De Giacomo, A., Serrazzanetti, D. I., et al. (2013). Fecal microbiota and metabolome of children with autism and pervasive developmental disorder not otherwise specified. *PLoS One*, *8*, e76993.

Desai, B. S., Monahan, A. J., Carvey, P. M., & Hendey, B. (2007). Blood-brain barrier pathology in Alzheimer's and Parkinson's disease: Implications for drug therapy. *Cell Transplantation*, *16*, 285–299.

Dodiya, H. B., Frith, M., Sidebottom, A., Cao, Y., Koval, J., Chang, E., et al. (2020). Synergistic depletion of gut microbial consortia, but not individual antibiotics, reduces amyloidosis in APPPS1-21 Alzheimer's transgenic mice. *Scientific Reports*, *10*, 8183.

Dodiya, H. B., Lutz, H. L., Weigle, I. Q., Patel, P., Michalkiewicz, J., Roman-Santiago, C. J., et al. (2022). Gut microbiota-driven brain Aβ amyloidosis in mice requires microglia. *The Journal of Experimental Medicine*, *219*, e20200895.

Dominy, S. S., Lynch, C., Ermini, F., Benedyk, M., Marczyk, A., Konradi, A., et al. (2019). Porphyromonas gingivalis in Alzheimer's disease brains: Evidence for disease causation and treatment with small-molecule inhibitors. *Science Advances*, *5*, eaau3333.

El Khoury, J., Toft, M., Hickman, S. E., Means, T. K., Terada, K., Geula, C., et al. (2007). Ccr2 deficiency impairs microglial accumulation and accelerates progression of Alzheimer-like disease. *Nature Medicine*, *13*, 432–438.

Erny, D., Hrabě de Angelis, A. L., Jaitin, D., Wieghofer, P., Staszewski, O., David, E., et al. (2015). Host microbiota constantly control maturation and function of microglia in the CNS. *Nature Neuroscience*, *18*, 965–977.

Erny, D., Dokalis, N., Mezö, C., Castoldi, A., Mossad, O., Staszewski, O., et al. (2021). Microbiota-derived acetate enables the metabolic fitness of the brain innate immune system during health and disease. *Cell Metabolism*, *33*, 2260–2276.e7.

Farid, M. M., Yang, X., Kuboyama, T., & Tohda, C. (2020). Trigonelline recovers memory function in Alzheimer's disease model mice: Evidence of brain penetration and target molecule. *Scientific Reports*, *10*, 16424.

Fiala, M., Lin, J., Ringman, J., Kermani-Arab, V., Tsao, G., Patel, A., et al. (2005). Ineffective phagocytosis of amyloid-beta by macrophages of Alzheimer's disease patients. *Journal of Alzheimer's Disease*, 7, 221–232. discussion 255-62.

Fink, A., Doblhammer, G., & Tamgüney, G. (2021). Recurring gastrointestinal infections increase the risk of dementia. *Journal of Alzheimer's Disease*, *84*, 797–806.

Gao, Q., Wang, Y., Wang, X., Fu, S., Zhang, X., Wang, R. T., et al. (2019). Decreased levels of circulating trimethylamine N-oxide alleviate cognitive and pathological deterioration in transgenic mice- a potential therapeutic approach for Alzheimer's disease. *Aging (Albany NY)*, *11*(19), 8642–8663. Qiang Gao.pdf.

Goodwin, L. O., Splinter, E., Davis, T. L., Urban, R., He, H., Braun, R. E., et al. (2019). Large-scale discovery of mouse transgenic integration sites reveals frequent structural variation and insertional mutagenesis. *Genome Research*, *29*, 494–505.

Govindarajulu, M., Pinky, P. D., Steinke, I., Bloemer, J., Ramesh, S., Kariharan, T., et al. (2020). Gut metabolite TMAO induces synaptic plasticity deficits by promoting endoplasmic reticulum stress. *Frontiers in Molecular Neuroscience*, *13*, 138.

Guillot-Sestier, M. V., Araiz, A. R., Mela, V., Gaban, A. S., O'Neill, E., Joshi, L., et al. (2021). Microglial metabolism is a pivotal factor in sexual dimorphism in Alzheimer's disease. *Communications Biology*, *4*, 711.

Gupta, V. K., Paul, S., & Dutta, C. (2017). Geography, ethnicity or subsistence-specific variations in human microbiome composition and diversity. *Frontiers in Microbiology*, *8*, 1162.

Gupta, S., Mullish, B. H., & Allegretti, J. R. (2021). Fecal microbiota transplantation: The evolving risk landscape. *The American Journal of Gastroenterology*, *116*, 647–656.

Harach, T., Marungruang, N., Duthilleul, N., Cheatham, V., Mc Coy, K. D., Frisoni, G., et al. (2017). Reduction of Abeta amyloid pathology in APPPS1 transgenic mice in the absence of gut microbiota. *Scientific Reports*, 7, 41802.

Haran, J. P., Bhattarai, S. K., Foley, S. E., Dutta, P., Ward, D. V., Bucci, V., et al. (2019). Alzheimer's disease microbiome is associated with dysregulation of the anti-inflammatory P-glycoprotein pathway. *MBio*, *10*, e00632-19.

Hazan, S. (2020). Rapid improvement in Alzheimer's disease symptoms following fecal microbiota transplantation: A case report. *The Journal of International Medical Research*, *48*. https://doi.org/10.1177/0300060520925930. 300060520925930.

Hebert, L. E., Weuve, J., Scherr, P. A., & Evans, D. A. (2013). Alzheimer disease in the United States (2010–2050) estimated using the 2010 census. *Neurology*, *80*, 1778–1783.

Hersant, H., & Grossberg, G. (2022). The Ketogenic Diet and Alzheimer's Disease. *The Journal of Nutrition, Health & Aging*, *26*, 606–614.

Hoffman, J. D., Yanckello, L. M., Chlipala, G., Hammond, T. C., McCulloch, S. D., Parikh, I., et al. (2019). Dietary inulin alters the gut microbiome, enhances systemic metabolism and reduces neuroinflammation in an APOE4 mouse model. *PLoS One*, *14*, e0221828.
Holland, R., McIntosh, A. L., Finucane, O. M., Mela, V., Rubio-Araiz, A., Timmons, G., et al. (2018). Inflammatory microglia are glycolytic and iron retentive and typify the microglia in APP/PS1 mice. *Brain, Behavior, and Immunity*, *68*, 183–196.
Honarpisheh, P., Reynolds, C. R., Blasco Conesa, M. P., Moruno Manchon, J. F., Putluri, N., Bhattacharjee, M. B., et al. (2020). Dysregulated gut homeostasis observed prior to the accumulation of the brain amyloid-beta in Tg2576 mice. *International Journal of Molecular Sciences*, *21*, 1711.
Hoyles, L., Snelling, T., Umlai, U. K., Nicholson, J. K., Carding, S. R., Glen, R. C., et al. (2018). Microbiome-host systems interactions: Protective effects of propionate upon the blood-brain barrier. *Microbiome*, *6*, 55.
Hsiao, K., Chapman, P., Nilsen, S., Eckman, C., Harigaya, Y., Younkin, S., et al. (1996). Correlative memory deficits, Abeta elevation, and amyloid plaques in transgenic mice. *Science*, *274*, 99–102.
Jankowsky, J. L., Fadale, D. J., Anderson, J., Xu, G. M., Gonzales, V., Jenkins, N. A., et al. (2004). Mutant presenilins specifically elevate the levels of the 42 residue beta-amyloid peptide in vivo: Evidence for augmentation of a 42-specific gamma secretase. *Human Molecular Genetics*, *13*, 159–170.
Kaur, H., Nagamoto-Combs, K., Golovko, S., Golovko, M. Y., Klug, M. G., & Combs, C. K. (2020). Probiotics ameliorate intestinal pathophysiology in a mouse model of Alzheimer's disease. *Neurobiology of Aging*, *92*, 114–134.
Kaur, H., Seeger, D., Golovko, S., Golovko, M., & Combs, C. K. (2021). Liver bile acid changes in mouse models of Alzheimer's disease. *International Journal of Molecular Sciences*, *22*, 7451.
Kim, S., Covington, A., & Pamer, E. G. (2017). The intestinal microbiota: Antibiotics, colonization resistance, and enteric pathogens. *Immunological Reviews*, *279*, 90–105.
Kim, M. S., Kim, Y., Choi, H., Kim, W., Park, S., Lee, D., et al. (2020). Transfer of a healthy microbiota reduces amyloid and tau pathology in an Alzheimer's disease animal model. *Gut*, *69*, 283–294.
Kim, N., Jeon, S. H., Ju, I. G., Gee, M. S., Do, J., Oh, M. S., et al. (2021). Transplantation of gut microbiota derived from Alzheimer's disease mouse model impairs memory function and neurogenesis in C57BL/6 mice. *Brain, Behavior, and Immunity*, *98*, 357–365.
Kobayashi, Y., Sugahara, H., Shimada, K., Mitsuyama, E., Kuhara, T., Yasuoka, A., et al. (2017). Therapeutic potential of Bifidobacterium breve strain A1 for preventing cognitive impairment in Alzheimer's disease. *Scientific Reports*, 7, 13510.
Leblhuber, F., Steiner, K., Schuetz, B., Fuchs, D., & Gostner, J. M. (2018). Probiotic supplementation in patients with Alzheimer's dementia—An explorative intervention study. *Current Alzheimer Research*, *15*, 1106–1113.
Lee, J., Torosyan, N., & Silverman, D. H. (2017). Examining the impact of grape consumption on brain metabolism and cognitive function in patients with mild decline in cognition: A double-blinded placebo controlled pilot study. *Experimental Gerontology*, *87*, 121–128.
Lee, J. W., Profant, M., & Wang, C. (2022). Metabolic sex dimorphism of the brain at the gene, cell, and tissue level. *Journal of Immunology*, *208*, 212–220.
Li, L., Chen, B., Zhu, R., Li, R., Tian, Y., Liu, C., et al. (2019). Fructus Ligustri Lucidi preserves bone quality through the regulation of gut microbiota diversity, oxidative stress, TMAO and Sirt6 levels in aging mice. *Aging (Albany NY)*, *11*(21), 9348–9368. Lin Li.pdf.

Li, Z., Zhu, H., Guo, Y., Du, X., & Qin, C. (2020). Gut microbiota regulate cognitive deficits and amyloid deposition in a model of Alzheimer's disease. *Journal of Neurochemistry, 155*, 448–461.

Licastro, F., Pedrini, S., Caputo, L., Annoni, G., Davis, L. J., Ferri, C., et al. (2000). Increased plasma levels of interleukin-1, interleukin-6 and alpha-1-antichymotrypsin in patients with Alzheimer's disease: Peripheral inflammation or signals from the brain? *Journal of Neuroimmunology, 103*, 97–102.

Liu, J. M., Li, H. J., Gong, T. Y., Chen, W. Y., Mao, S. Y., Kong, Y., et al. (2020). Anti-neuroinflammatory effect of short-chain fatty acid acetate against Alzheimer's disease via upregulating GPR41 and inhibiting ERK/JNK/NF-kappa B. *Journal of Agricultural and Food Chemistry, 68*, 7152–7161.

Liu, J., Zhang, T., Wang, Y., Si, C., Wang, X., Wang, R. T., et al. (2020). Baicalin ameliorates neuropathology in repeated cerebral ischemiareperfusion injury model mice by remodeling the gut microbiota. *Aging (Albany NY), 12*(4), 3791–3806. Jianfeng Liu.pdf.

MahmoudianDehkordi, S., Arnold, M., Nho, K., Ahmad, S., Jia, W., Xie, G., et al. (2019). Altered bile acid profile associates with cognitive impairment in Alzheimer's disease-an emerging role for gut microbiome. *Alzheimers Dement, 15*, 76–92.

Maldonado Weng, J., Parikh, I., Naqib, A., York, J., Green, S. J., Estus, S., et al. (2019). Synergistic effects of APOE and sex on the gut microbiome of young EFAD transgenic mice. *Molecular Neurodegeneration, 14*, 47.

Markle, J. G. M., Frank, D. N., Mortin-Toth, S., Robertson, C. E., Feazel, L. M., Rolle-Kampczyk, U., et al. (2013). Sex differences in the gut microbiome drive hormone-dependent regulation of autoimmunity. *Science, 339*, 1084–1088.

Masters, C. L., Bateman, R., Blennow, K., Rowe, C. C., Sperling, R. A., & Cummings, J. L. (2015). Alzheimer's disease. *Nature Reviews. Disease Primers, 1*, 15056.

Mazmanian, S. K., Round, J. L., & Kasper, D. L. (2008). A microbial symbiosis factor prevents intestinal inflammatory disease. *Nature, 453*, 620–625.

McDonald, T. J. W., & Cervenka, M. C. (2018). The expanding role of ketogenic diets in adult neurological disorders. *Brain Sciences, 8*, 148.

McIntosh, A., Mela, V., Harty, C., Minogue, A. M., Costello, D. A., Kerskens, C., et al. (2019). Iron accumulation in microglia triggers a cascade of events that leads to altered metabolism and compromised function in APP/PS1 mice. *Brain Pathology, 29*, 606–621.

Mehta, R. S., Lochhead, P., Wang, Y., Ma, W., Nguyen, L. H., Kochar, B., et al. (2022). Association of midlife antibiotic use with subsequent cognitive function in women. *PLoS One, 17*, e0264649.

Mezö, C., Dokalis, N., Mossad, O., Staszewski, O., Neuber, J., Yilmaz, B., et al. (2020). Different effects of constitutive and induced microbiota modulation on microglia in a mouse model of Alzheimer's disease. *Acta Neuropathologica Communications, 8*, 119.

Michaudel, C., & Sokol, H. (2020). The gut microbiota at the Service of Immunometabolism. *Cell Metabolism, 32*, 514–523.

Min, L. J., Kobayashi, Y., Mogi, M., Tsukuda, K., Yamada, A., Yamauchi, K., et al. (2017). Administration of bovine casein-derived peptide prevents cognitive decline in Alzheimer disease model mice. *PLoS One, 12*, e0171515.

Minter, M. R., Zhang, C., Leone, V., Ringus, D. L., Zhang, X., Oyler-Castrillo, P., et al. (2016). Antibiotic-induced perturbations in gut microbial diversity influences neuro-inflammation and amyloidosis in a murine model of Alzheimer's disease. *Scientific Reports, 6*, 30028.

Minter, M. R., Hinterleitner, R., Meisel, M., Zhang, C., Leone, V., Zhang, X., et al. (2017). Antibiotic-induced perturbations in microbial diversity during post-natal development alters amyloid pathology in an aged APPSWE/PS1DeltaE9 murine model of Alzheimer's disease. *Scientific Reports, 7*, 10411.

Miyauchi, E., Kim, S. W., Suda, W., Kawasumi, M., Onawa, S., Taguchi-Atarashi, N., et al. (2020). Gut microorganisms act together to exacerbate inflammation in spinal cords. *Nature*, *585*, 102–106.

Müller, L., Power Guerra, N., Stenzel, J., Rühlmann, C., Lindner, T., Krause, B. J., et al. (2021). Long-term caloric restriction attenuates β-amyloid neuropathology and is accompanied by autophagy in APPswe/PS1delta9 mice. *Nutrients*, *13*, 985.

Nagpal, R., Neth, B. J., Wang, S., Craft, S., & Yadav, H. (2019). Modified Mediterranean-ketogenic diet modulates gut microbiome and short-chain fatty acids in association with Alzheimer's disease markers in subjects with mild cognitive impairment. *eBioMedicine*, *47*, 529–542.

Nagpal, R., Neth, B. J., Wang, S., Mishra, S. P., Craft, S., & Yadav, H. (2020). Gut mycobiome and its interaction with diet, gut bacteria and alzheimer's disease markers in subjects with mild cognitive impairment: A pilot study. *eBioMedicine*, *59*, 102950.

Naomi, R., Embong, H., Othman, F., Ghazi, H. F., Maruthey, N., & Bahari, H. (2021). Probiotics for Alzheimer's disease: A systematic review. *Nutrients*, *14*, 20.

Nho, K., Kueider-Paisley, A., MahmoudianDehkordi, S., Arnold, M., Risacher, S. L., Louie, G., et al. (2019). Alzheimer's disease neuroimaging, and C. the Alzheimer disease metabolomics, altered bile acid profile in mild cognitive impairment and Alzheimer's disease: Relationship to neuroimaging and CSF biomarkers. *Alzheimers Dement*, *15*, 232–244.

NIBIOHN JLARB (2022). JCRB Laboatory Animal Resource Bank at NIBIOHN, ddY and DDY mice Description. Available from: https://animal.nibiohn.go.jp/e_ddys.html.

O'Hara, A., & Shanahan, F. (2006). The gut flora as a forgotten organ. *EMBO Reports*, 7, 688.

O'Neill, L. A., Kishton, R. J., & Rathmell, J. (2016). A guide to immunometabolism for immunologists. *Nature Reviews. Immunology*, *16*, 553–565.

Oakley, H., Cole, S. L., Logan, S., Maus, E., Shao, P., Craft, J., et al. (2006). Intraneuronal beta-amyloid aggregates, neurodegeneration, and neuron loss in transgenic mice with five familial Alzheimer's disease mutations: Potential factors in amyloid plaque formation. *The Journal of Neuroscience*, *26*, 10129–10140.

Ochiai, R., Saitou, K., Suzukamo, C., Osaki, N., & Asada, T. (2019). Effect of Chlorogenic acids on cognitive function in mild cognitive impairment: A randomized controlled crossover trial. *Journal of Alzheimer's Disease*, *72*, 1209–1216.

Ochoa-Reparaz, J., Mielcarz, D. W., Wang, Y., Begum-Haque, S., Dasgupta, S., Kasper, D. L., et al. (2010). A polysaccharide from the human commensal Bacteroides fragilis protects against CNS demyelinating disease. *Mucosal Immunology*, *3*, 487–495.

Ochoa-Reparaz, J., Mielcarz, D. W., Begum- Haque, S., & Kasper, L. H. (2011). Gut, bugs, and brain: Role of commensal bacteria in the control of central nervous system disease. *Annals of Neurology*, *69*, 240–247.

Oddo, S., Caccamo, A., Shepherd, J. D., Murphy, M. P., Golde, T. E., Kayed, R., et al. (2003). Triple-transgenic model of Alzheimer's disease with plaques and tangles: Intracellular Aβ and synaptic dysfunction. *Neuron*, *39*, 409–421.

Olson, C. A., Vuong, H. E., Yano, J. M., Liang, Q. Y., Nusbaum, D. J., & Hsiao, E. Y. (2018). The gut microbiota mediates the anti-seizure effects of the ketogenic diet. *Cell*, *173*, 1728–1741.e13.

Ou, H., Chien, W.-C., Chung, C.-H., Chang, H.-A., Kao, Y.-C., Wu, P.-C., et al. (2021). Association between antibiotic treatment of chlamydia pneumoniae and reduced risk of Alzheimer dementia: A Nationwide cohort study in Taiwan. *Frontiers in Aging Neuroscience*, *13*, 701899.

Pan, R., Ma, J., & Kong, X. (2019). Sodium rutin ameliorates Alzheimer's disease-like pathology by enhancing microglia amyloid-B clearance. *Science Advances*, *5*, eaau6328.

Parikh, I. J., Estus, J. L., Zajac, D. J., Malik, M., Maldonado Weng, J., Tai, L. M., et al. (2020). Murine gut microbiome association with APOE alleles. *Frontiers in Immunology*, *11*, 200.

Park, S., Zhang, T., Wu, X., Yi Qiu, J. (2020). Ketone production by ketogenic diet and by intermittent fasting has different effects on the gut microbiota and disease progression in an Alzheimer's disease rat model. Journal of Clinical Biochemistry and Nutrition, 67 (2), 188-198. Epub 20200320. https://doi.org/10.3164/jcbn.19-87. PubMed PMID: 33041517; PMCID: PMC7533860.

Park, S. H., Lee, J. H., Shin, J., Kim, J. S., Cha, B., Lee, S., et al. (2021). Cognitive function improvement after fecal microbiota transplantation in Alzheimer's dementia patient: A case report. *Current Medical Research and Opinion*, *37*, 1739–1744.

Park, S. H., Lee, J. H., Kim, J. S., Kim, T. J., Shin, J., Im, J. H., et al. (2022). Fecal microbiota transplantation can improve cognition in patients with cognitive decline and Clostridioides difficile infection. *Aging (Albany NY)*, *14*, 6449–6466.

Phillips, M. C. L., Deprez, L. M., Mortimer, G. M. N., Murtagh, D. K. J., McCoy, S., Mylchreest, R., et al. (2021). Randomized crossover trial of a modified ketogenic diet in Alzheimer's disease. *Alzheimer's Research & Therapy*, *13*, 51.

Radde, R., Bolmont, T., Kaeser, S. A., Coomaraswamy, J., Lindau, D., Stoltze, L., et al. (2006). Abeta42-driven cerebral amyloidosis in transgenic mice reveals early and robust pathology. *EMBO Reports*, 7, 940–946.

Rothhammer, V., Mascanfroni, I. D., Bunse, L., Takenaka, M. C., Kenison, J. E., Mayo, L., et al. (2016). Type I interferons and microbial metabolites of tryptophan modulate astrocyte activity and central nervous system inflammation via the aryl hydrocarbon receptor. *Nature Medicine*, *22*, 586–597.

Rubio-Araiz, A., Finucane, O. M., Keogh, S., & Lynch, M. A. (2018). Anti-TLR2 antibody triggers oxidative phosphorylation in microglia and increases phagocytosis of β-amyloid. *Journal of Neuroinflammation*, *15*, 247.

Rusek, M., Pluta, R., Ulamek-Koziol, M., & Czuczwar, S. J. (2019). Ketogenic Diet in Alzheimer's Disease. *International Journal of Molecular Sciences*, *20*, 3892.

Sankowski, R., Ahmari, J., Mezö, C., Hrabě de Angelis, A. L., Fuchs, V., Utermöhlen, O., et al. (2021). Commensal microbiota divergently affect myeloid subsets in the mammalian central nervous system during homeostasis and disease. *The EMBO Journal*, *40*, e108605.

Schaubeck, M., Clavel, T., Calasan, J., Lagkouvardos, I., Haange, S. B., Jehmlich, N., et al. (2016). Dysbiotic gut microbiota causes transmissible Crohn's disease-like ileitis independent of failure in antimicrobial defence. *Gut*, *65*, 225–237.

Scott, N. A., Andrusaite, A., Andersen, P., Lawson, M., Alcon-Giner, C., Leclaire, C., et al. (2018). Antibiotics induce sustained dysregulation of intestinal T cell immunity by perturbing macrophage homeostasis. *Science Translational Medicine*, *10*, eaao4755.

Shen, L., Liu, L., & Ji, H. F. (2017). Alzheimer's disease histological and behavioral manifestations in transgenic mice correlate with specific gut microbiome state. *Journal of Alzheimer's Disease*, *56*, 385–390.

Shukla, P. K., Delotterie, D. F., Xiao, J., Pierre, J. F., Rao, R., McDonald, M. P., et al. (2021). Alterations in the gut-microbial-inflammasome-brain axis in a mouse model of Alzheimer's disease. *Cell*, *10*, 779.

Suhocki, P. V., Ronald, J. S., Diehl, A. M. E., Murdoch, D. M., & Doraiswamy, P. M. (2022). Probing gut-brain links in Alzheimer's disease with rifaximin. *Alzheimers Dement (N Y)*, *8*, e12225.

Sun, J., Xu, J., Ling, Y., Wang, F., Gong, T., Yang, C., et al. (2019). Fecal microbiota transplantation alleviated Alzheimer's disease-like pathogenesis in APP/PS1 transgenic mice. *Translational Psychiatry*, *9*, 189.

Tamtaji, O. R., Heidari-Soureshjani, R., Mirhosseini, N., Kouchaki, E., Bahmani, F., Aghadavod, E., et al. (2019). Probiotic and selenium co-supplementation, and the effects on clinical, metabolic and genetic status in Alzheimer's disease: A randomized, double-blind, controlled trial. *Clinical Nutrition*, *38*, 2569–2575.

Tanzi, R. E. (2012). The genetics of Alzheimer disease. *Cold Spring Harbor Perspectives in Medicine*, *2*, a006296.
The Jackson Laboratory (2022). Strain details. 002243. DDY/JclSidSeyFrkJ mice. Available from: https://www.jax.org/strain/002243.
Thion, M. S., Low, D., Silvin, A., Chen, J., Grisel, P., Schulte-Schrepping, J., et al. (2018). Microbiome influences prenatal and adult microglia in a sex-specific manner. *Cell*, *172*, 500–516.e16.
Tiihonen, K., Ouwehand, A. C., & Rautonen, N. (2010). Human intestinal microbiota and healthy ageing. *Ageing Research Reviews*, *9*, 107–116.
Tohda, C., Kogure, C., Nomoto, K., de Toledo, A., Yang, X., & Hirano, E. (2021). A novel Heptapeptide, GPPGPAG transfers to the brain, and ameliorates memory dysfunction and dendritic atrophy in Alzheimer's disease model mice. *Frontiers in Pharmacology*, *12*, 680652.
Tomino, Y. (2010). Spontaneous animal model, ddY mouse, for studying the pathogenesis and treatment in patients with immunoglobulin a nephropathy. *Nephrology (Carlton)*, *15*, 1–6.
Ton, A. M. M., et al. (2020). Oxidative stress and dementia in Alzheimer's patients: Effects of synbiotic supplementation. *Oxidative Medicine and Cellular Longevity*, *2020*. https://doi.org/10.1155/2020/2638703.
Tran, L., & Greenwood-Van Meerveld, B. (2013). Age-associated remodeling of the intestinal epithelial barrier. *Journals of Gerontology Series A: Biological Sciences and Medical Sciences*, *68*, 1045–1056.
Tran, T. T. T., Corsini, S., Kellingray, L., Hegarty, C., Le Gall, G., Narbad, A., et al. (2019). APOE genotype influences the gut microbiome structure and function in humans and mice: Relevance for Alzheimer's disease pathophysiology. *The FASEB Journal*, *33*, 8221–8231.
Turnbaugh, P., Backhed, F., Fulton, L., & Gordon, J. (2008). Diet-induced obesity is linked to marked but reversible alterations in the mouse distal gut microbiome. *Cell Host & Microbe*, *3*, 213–223.
van der Steen, J. T., Ooms, M. E., Adèr, H. J., Ribbe, M. W., & van der Wal, G. (2002). Withholding antibiotic treatment in pneumonia patients with dementia: A quantitative observational study. *Archives of Internal Medicine*, *162*, 1753–1760.
van Nood, E., Vrieze, A., Nieuwdorp, M., Fuentes, S., Zoetendal, E. G., de Vos, W. M., et al. (2013). Duodenal infusion of donor feces for recurrent Clostridium difficile. *The New England Journal of Medicine*, *368*, 407–415.
Vijay-Kumar, M., Aitken, J. D., Carvalho, F. A., Cullender, T. C., Mwangi, S., Srinivasan, S., et al. (2010). Metabolic syndrome and altered gut microbiota in mice lacking toll-like receptor 5. *Science*, *328*, 228–231.
Vogt, N. M., Kerby, R. L., Dill-McFarland, K. A., Harding, S. J., Merluzzi, A. P., Johnson, S. C., et al. (2017). Gut microbiome alterations in Alzheimer's disease. *Scientific Reports*, 7, 13537.
Vogt, N. M., Romano, K. A., Darst, B. F., Engelman, C. D., Johnson, S. C., Carlsson, C. M., et al. (2018). The gut microbiota-derived metabolite trimethylamine N-oxide is elevated in Alzheimer's disease. *Alzheimer's Research & Therapy*, *10*, 124.
Wang, J., Ho, L., Qin, W., Rocher, A. B., Seror, I., Humala, N., et al. (2005). Caloric restriction attenuates β-amyloid neuropathology in a mouse model of Alzheimer's disease. *The FASEB Journal*, *19*, 1–18.
Wang, X., Sun, G., Feng, T., Zhang, J., Huang, X., Wang, T., et al. (2019). Sodium oligomannate therapeutically remodels gut microbiota and suppresses gut bacterial amino acids-shaped neuroinflammation to inhibit Alzheimer's disease progression. *Cell Research*, *29*, 787–803.

Wang, Q. J., Shen, Y. E., Wang, X., Fu, S., Zhang, X., Zhang, Y. N., et al. (2020). Concomitant memantine and *Lactobacillus plantarum* treatment attenuates cognitive impairments in APP:PS1 mice. *Aging (Albany NY)*, *12*, 628–649. Qiu-Jun Wang.pdf>.

Wang, Q., Davis, P. B., Qi, X., Chen, S. G., Gurney, M. E., Perry, G., et al. (2021). Gut-microbiota-microglia-brain interactions in Alzheimer's disease: Knowledge-based, multi-dimensional characterization. *Alzheimer's Research & Therapy*, *13*, 177.

Weiner, H. L., & Frenkel, D. (2006). Immunology and immunotherapy of Alzheimer's disease. *Nature Reviews. Immunology*, *6*, 404–416.

Wenzel, T. J., Gates, E. J., Ranger, A. L., & Klegeris, A. (2020). Short-chain fatty acids (SCFAs) alone or in combination regulate select immune functions of microglia-like cells. *Molecular and Cellular Neurosciences*, *105*, 103493.

Wu, S., Dreyfus, L. A., Tzianabos, A. O., Hayashi, C., & Sears, C. L. (2002). Diversity of the metalloprotease toxin produced by enterotoxigenic Bacteroides fragilis. *Infection and Immunity*, *70*, 2463–2471.

Wu, S., Rhee, K. J., Zhang, M., Franco, A., & Sears, C. L. (2007). Bacteroides fragilis toxin stimulates intestinal epithelial cell shedding and gamma-secretase-dependent E-cadherin cleavage. *Journal of Cell Science*, *120*, 1944–1952.

Xiao, S., Chan, P., Wang, T., Hong, Z., Wang, S., Kuang, W., et al. (2021). A 36-week multicenter, randomized, double-blind, placebo-controlled, parallel-group, phase 3 clinical trial of sodium oligomannate for mild-to-moderate Alzheimer's dementia. *Alzheimer's Research & Therapy*, *13*, 62.

Xu, H. M., Huang, H. L., Zhou, Y. L., Zhao, H. L., Xu, J., Shou, D. W., et al. (2021). Fecal microbiota transplantation: A new therapeutic attempt from the gut to the brain. *Gastroenterology Research and Practice*, *2021*, 6699268.

Yanckello, L. M., Hoffman, J. D., Chang, Y. H., Lin, P., Nehra, G., Chlipala, G., et al. (2021). Apolipoprotein E genotype-dependent nutrigenetic effects to prebiotic inulin for modulating systemic metabolism and neuroprotection in mice via gut-brain axis. *Nutritional Neuroscience*, *25*(8), 1669–1679.

Yilmaz, A., Ugur, Z., Bisgin, H., Akyol, S., Bahado-Singh, R., Wilson, G., et al. (2020). Targeted metabolic profiling of urine highlights a potential biomarker panel for the diagnosis of Alzheimer's disease and mild cognitive impairment: A pilot study. *Metabolites*, *10*, 357.

Zeng, H., Umar, S., Rust, B., Lazarova, D., & Bordonaro, M. (2019). Secondary bile acids and short chain fatty acids in the Colon: A focus on colonic microbiome, cell proliferation, inflammation, and Cancer. *International Journal of Molecular Sciences*, *20*, 1214.

Zhang, C., Li, S., Yang, L., Huang, P., Li, W., Wang, S., et al. (2013). Structural modulation of gut microbiota in life-long calorie-restricted mice. *Nature Communications*, *4*, 1–10.

Zhang, L., Wang, Y., Xiayu, X., Shi, C., Chen, W., Song, N., et al. (2017). Altered gut microbiota in a mouse model of Alzheimer's disease. *Journal of Alzheimer's Disease*, *60*, 1241–1257.

Zhang, L., Yang, C., Li, Y., Niu, S., Liang, X., Zhang, Z., et al. (2021). Dynamic changes in the levels of amyloid-β(42) species in the brain and periphery of APP/PS1 mice and their significance for Alzheimer's disease. *Frontiers in Molecular Neuroscience*, *14*, 723317.

CHAPTER SIX

The microbiota-gut-brain axis in Huntington's disease

Chloe J. Love[a,†], Bethany A. Masson[a,†], Carolina Gubert[a], and Anthony J. Hannan[a,b,*]
[a]Florey Institute of Neuroscience and Mental Health, Melbourne Brain Centre, University of Melbourne, Parkville, Victoria, Australia
[b]Department of Anatomy and Neuroscience, University of Melbourne, Parkville, Victoria, Australia
*Corresponding author: e-mail address: anthony.hannan@florey.edu.au

Contents

† These authors contributed equally to this work.

International Review of Neurobiology, Volume 167
ISSN 0074-7742
https://doi.org/10.1016/bs.irn.2022.06.005

Abstract

Huntington's disease (HD) is a fatal neurodegenerative disorder caused by an autosomal dominant trinucleotide (CAG) tandem repeat, resulting in complex motor, psychiatric and cognitive symptoms as well as gastrointestinal disturbances and other peripheral symptoms. There are currently no disease-modifying treatments, and the peripheral pathology of the disorder is not well understood. Emerging evidence suggests that the bi-directional communication pathways between the gut and the brain, including the microbiota-gut-brain axis, can affect motor, psychiatric and cognitive symptoms as well as weight loss and sexual dimorphism seen in HD. Furthermore, both HD and the microbiota-gut-brain axis can be influenced by environmental factors, opening potential new avenues to explore therapeutic options for this devastating disorder.

1. Microbiota-gut-brain axis

The human body hosts a wide range of microorganisms that populate the oral and vaginal cavities, as well as the skin and gut, with distinct microbial communities within each area (Brown, Tanner, & Goldman, 2018). The term gut microbiota describes all commensal, symbiotic and pathogenic microorganisms that densely populate the gastrointestinal (GI) tract, consisting primarily of bacteria as well as fungi, archaea and viruses (Cryan et al., 2019).

1.1 Gut microbiota development

During early childhood, the gut microbiota forms a symbiotic relationship with the body and allows for a diverse colony of microorganisms to colonize the GI tract (Cryan et al., 2019). The route of delivery at birth is the first known modulator of the acquisition of microorganisms that populate the body, with caesarean or vaginal procedures resulting in different levels of exposure to the maternal microbiota, which then play an essential role in the development and growth of the offspring's gut microbiota (Dominguez-Bello, Godoy-Vitorino, Knight, & Blaser, 2018). After the initial exposure to the maternal microbiota, environmental factors such as antibiotics, breast-feeding or formula consumption can all shape the development of an individual's microbiota, which resembles adult composition after 2–3 years (Dominguez-Bello et al., 2018; Koenig et al., 2011; Pannaraj et al., 2017;

Schwiertz et al., 2003). Although well established at a young age, the gut microbiota population is dynamic throughout the lifespan and is continually influenced by external factors. Sexual dimorphism is present in the human gut microbiota, with species diversity and composition differing between sexes (Ma & Li, 2019). Dimorphism is driven in part by sex hormones and can affect the way the gut microbiota responds to disruption in homeostasis, environmental changes, and disease states. The main roles of the gut microbiota (acquired during its coevolution with hosts, including humans and other animals) include preserving homeostasis within the gut through the maintenance of the GI tract, nutrient and drug metabolism, aiding the mucous layer to protect from pathogenic microbes and assisting in immune maturation and modulation (Jandhyala et al., 2015).

1.2 Studying the gut microbiome

The gut microbiome refers to all the microorganisms and their collective genetic material that reside in the GI tract (Cresci & Izzo, 2019). Powerful DNA sequencing technology uses the microbiome to analyze the composition, diversity and richness of the gut microbiota and their subsequent functional pathways. 16S ribosomal RNA (rRNA) sequencing utilizes polymerase chain reactions to amplify a conserved region of the bacterial genome to understand the genera and phyla of bacteria in the gut microbiota (Ranjan, Rani, Metwally, McGee, & Perkins, 2016). 16S rRNA analysis can provide insight into both alpha and beta diversity of the gut microbiota. Alpha diversity looks at diversity within the sample, assessing measures of richness (number of species within a sample) and evenness (species abundance within a sample) and is commonly measured using observed, inverse Simpson and Shannon diversity measures (Liu et al., 2021). Beta diversity explores similarities or dissimilarities between samples, with common measures including the Bray Curtis index and Unifrac distances (Liu et al., 2021). Whole-genome shotgun sequencing alternatively allows for analysis of the bacterial species of the gut microbiome and their metabolic capacity; however, is more resource-intensive (Ranjan et al., 2016).

Studies investigating the role of the gut microbiota using in vivo models such as germ-free (GF) mice, which lack microbial colonization, have also been fundamental to investigating rudimentary relationships between the gut microbiota and neurological deficits through the introduction of specific microbial species or gut microbiota profiles to colonize the gut (Cryan et al., 2019). GF mice exhibit physiological changes such as decreased body weight

(Cryan et al., 2019) as well as changes in their GI tract, including an enlarged caecum, smaller intestinal surface area and slower peristalsis leading to compromised nutrient utilization (Al-Asmakh & Zadjali, 2015; Shanahan, 2002). These changes reflect the importance of the microbiota in establishing and maintaining the structure and function of the GI tract.

Together, these tools have developed our understanding of the gut microbiota, including which bacteria phyla dominate, the role of bacterial diversity in gut health, the metabolic abilities of different species and inter-individual differences in microbiota profiles. Due to the complexity of the gut microbiota, these current analytical tools do provide limited insight into microbe–microbe interactions (Brüssow, 2020) and how the gut microbiota can influence functional and signaling pathways. The development of metagenomic and metabolomic techniques has provided insight into the relationship between the gut microbiota, circulating gut-derived metabolites and how they communicate with the brain (Cryan et al., 2019).

1.3 Bi-directional communication

Bi-directional communication between the gut and the brain consists of a variety of pathways including the nervous system, endocrine signaling and the immune system, collectively known as the microbiota-gut-brain axis (Cryan et al., 2019; Gubert, Renoir, & Hannan, 2020). Studies in GF mice suggest that lack of microbial colonization impacts cognition, anxiety-like behavior, and sociability, highlighting the importance of the gut microbiota in these paths of communication (reviewed in Luczynski et al., 2016) (Fig. 1).

The enteric nervous system (ENS) widely innervates the GI tract from the esophagus to the anus and is the primary neural communication pathway between the gut and the brain, relaying information regarding blood flow, motility, mucus and gut barrier permeability to the central nervous system (CNS) directly via the vagus nerve (Furness, 2006; Gershon, 2004; Herath, Hosie, Bornstein, & Franks, 2020). The microbiota can signal afferent fibers of the vagal nerve through the diffusion of metabolites such as neurotransmitters and short-chain fatty acids (SCFAs), to chemical and mechanical sensitive endings that can detect changes in gut motility, epithelial permeability and gut hormones (Bonaz, Bazin, & Pellissier, 2018). The vagal nerve contains 80% afferent fibers and 20% efferent fibers, resulting in bi-directional communication (Agostoni, Chinnock, De Burgh Daly, & Murray, 1957). The mucosal layer of the GI tract is also highly innervated by the ENS,

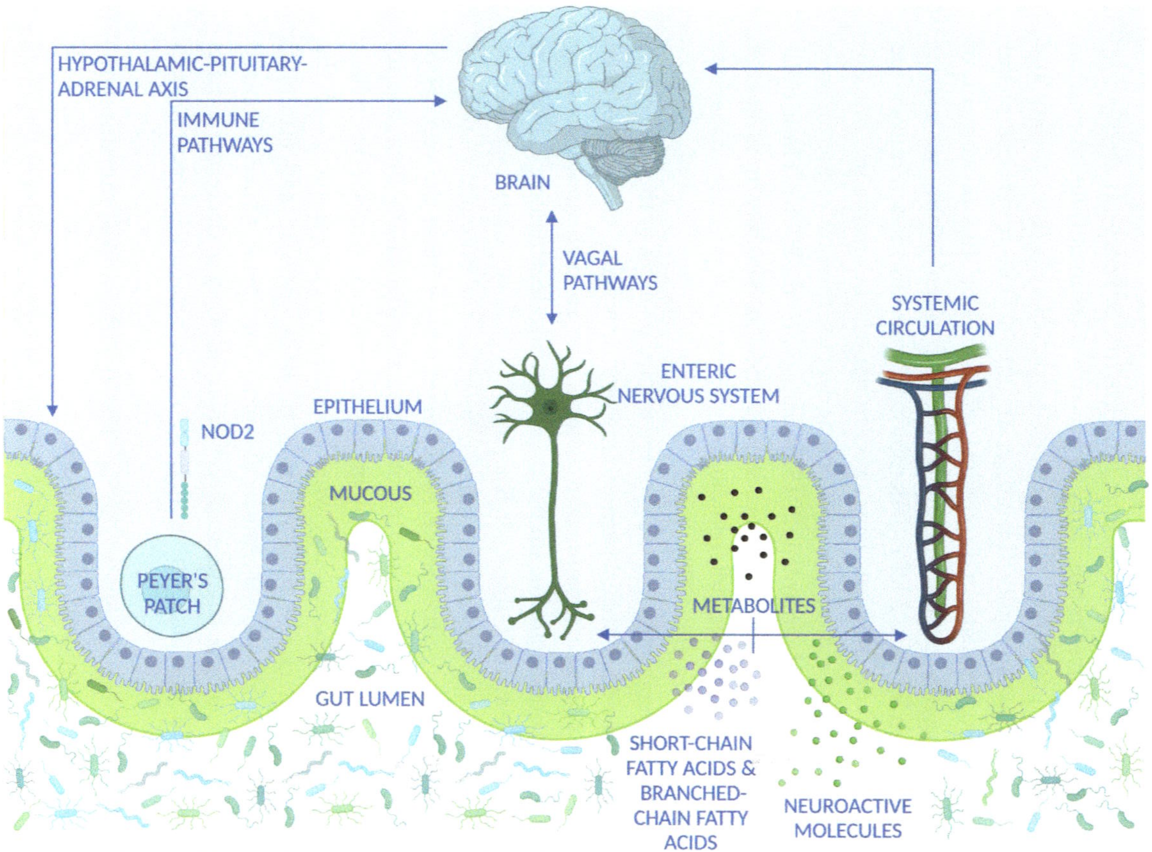

Fig. 1 The microbiota-gut-brain axis. Diagrammatic representation of the communication pathways between the microbiota, its metabolic by-products, the gut and the brain. Created with BioRender.com

which differentially regulates mucous secretion (Birchenough, Johansson, Gustafsson, Bergström, & Hansson, 2015); therefore, factors influencing the ENS such as widespread neurodegeneration could in turn influence mucous production.

The mucous layer of the GI tract is a network of polymers secreted by Goblet cells of the epithelium, with its primary roles including lubrication during digestion, protecting the intestinal epithelial cells from acidity and maintaining gut permeability (Herath et al., 2020). The myenteric plexus of the ENS also regulates GI motility, coordinating the movement of unwanted undigested materials, microbiota and mucus along the gut for removal (Herath et al., 2020). Microbiota composition determines the thickness of the outer mucosal layer, as commensal microbiota use mucous as a source of energy secreting enzymes that degrade the outer mucous layer (Donaldson, Lee, & Mazmanian, 2015; Johansson et al., 2008). These findings suggest that changes in mucous would affect its viability as an energy source, which could affect microbiota populations and possibly exacerbate disruption (Herath et al., 2020).

Growing evidence suggests that compositional changes in the microbiota may lead to increased epithelial permeability or a "leaky gut" that is associated with increased systemic inflammation which in turn then could promote and/or exacerbate neuroinflammation in neurodegenerative disorders (reviewed in Spielman, Gibson, & Klegeris, 2018). Understanding the role of the microbiota and subsequent inflammation has the potential to become a promising therapeutic tool to delay the onset and decrease the severity of these disorders. It has also been hypothesized that the blood–brain barrier (BBB) integrity can be influenced by the gut microbiota, with GF mice having an increased BBB permeability when compared to control mice, with the reduced expression of the tight junction proteins occludin and claudin-5 (Braniste, Al-Asmakh, Kowal, Anuar, & Abbaspour, 2014). Once colonized with pathogen-free gut microbiota, GF mice showed a decrease in BBB permeability and upregulation of tight junction proteins, supporting this hypothesis.

The hypothalamic-pituitary-adrenal (HPA) axis is a predominant regulator of the stress response (Frankiensztajn, Elliott, & Koren, 2020), with HPA-axis dysfunction seen in mental health disorders, metabolic disorders and inflammatory bowel disease (IBD) (Abautret-Daly, Dempsey, Parra-Blanco, Medina, & Harkin, 2018; Mawdsley & Rampton, 2005). Male GF mice were found to have a hyperresponsive HPA axis in response to stress, which was modulated upon the introduction of specific pathogen-free (SPF) feces from SPF mice which are raised with a normal gut microbiota excluding specific pathogens (Sudo et al., 2004). The HPA axis response can be attenuated by the gut microbiota-derived metabolites SCFAs, with one study finding that the treatment of mice with oral SCFA supplementation resulted in a downregulation in genes that encode proteins associated with the HPA axis and HPA-axis responsiveness (van de Wouw et al., 2018). The HPA axis interacts with other communication pathways within the microbiota-gut-brain axis, with afferent vagus nerve stimulation found to activate the HPA axis in rats (Cryan et al., 2019; Hosoi, Okuma, & Nomura, 2000). The HPA axis is also implicated in the interaction of the gut microbiota and immune system, as pro-inflammatory cytokines, interleukin (IL)-6 and tumor necrosis factor-alpha, are both able to cross the BBB and are thought to activate the HPA axis in response to homeostasis disruption, which may occur due to disturbance of the gut microbiota (Misiak et al., 2020; Turnbull & Rivier, 1995). While current research is limited, these findings suggest an emerging role for the gut microbiota in HPA-axis regulation.

The gut microbiota are also in constant communication with the immune system, which monitors for potential pathogens while retaining a symbiotic relationship with the gut microbiota (Dorrestein, Mazmanian, & Knight, 2014). This is mediated through gut-associated lymphatic tissue (GALT) which contains both innate and adaptive immune cells which maintain homeostasis and immune tolerance to the gut microbiota (Jiao, Wu, Huntington, & Zhang, 2020). Structural studies in the GI tract of GF mice found alterations in GALT formation, suggesting that microbial colonization of the gut is essential for immune function (Jiao et al., 2020). Peyer's patches are secondary lymphoid organs located in the small intestine, containing epithelial M-cells, which monitor for antigens and microbes in the intestinal lumen (Bonnardel et al., 2015), to determine mucosal immune responses to dietary substances to ensure a homeostatic balance of microbiota and prevent bacterial overgrowth (Fung, Olson, & Hsiao, 2017). Epithelial cells and Peyer's patches within the GI tract express the intracellular receptor nucleotide-binding oligomerization domain 2 (NOD2) a widely used biomarker for GI immune functioning (Petnicki-Ocwieja et al., 2009). NOD2 recognizes a component of the bacterial cell wall (Inohara et al., 2003), allowing them to mediate between the gut microbiota and Peyer's patches and guide the development and composition as well as placement of Peyer's patches within the GI tract to maintain immune homeostasis within the gut (Jung, Hugot, & Barreau, 2010).

The gut microbiota and the metabolites it generates are increasingly being recognized for their ability to regulate immune responses (Wu & Wu, 2012). The presence of T helper cells in the GI tract is reduced in GF mice and mice treated with antibiotics, with evidence suggesting that the interaction between the gut microbiota and the innate immune system promotes T helper cells (reviewed in Kamada & Núñez, 2014). Regulatory T (Treg) cells are also involved in managing immune responses and preventing uncontained inflammation (Dorrestein et al., 2014), with SCFAs able to influence the maturation and function of Treg cells in GF mice and prevent colitis in a murine model (Smith et al., 2013). The presence of specialized immune cells in the gut emphasizes the role of the microbiota in regulating gut barrier permeability to maintain an initial line of immune defense from external stimuli.

1.4 Short-chain fatty acids and branched-chain fatty acids

Microbiota composition also influences the production of neurotransmitters and neuroactive metabolites including SCFAs and branched-chain fatty acids (BCFAs). SCFAs and BCFAs are both microbiota-derived metabolites, produced by the fermentation of dietary fiber and amino acids, respectively, and can influence key aspects of gut health such as inflammation and gut barrier permeability (Dalile, Van Oudenhove, Vervliet, & Verbeke, 2019; Gubert, Renoir, & Hannan, 2020; Rios-Covian et al., 2020). Butyrate, acetate and propionate are more widely studied, accounting for approximately 90–95% of SCFAs produced, with other SCFAs and BCFAs making up the remaining 5% (Macfarlane & Macfarlane, 2003).

Histone deacetylase (HDAC) is an enzyme that condenses chromatin structure by removing acetyl groups, resulting in reduced transcriptional availability of genes (Koh, De Vadder, Kovatcheva-Datchary, & Bäckhed, 2016). Butyrate can act as a histone deacetylase (HDAC) inhibitor and block this process (Dalile et al., 2019) suggesting a role of SCFAs in gene transcription. Locally, butyrate is involved in regulating the production of intestinal epithelial cells and regulating the expression of proteins that form tight junctions between these epithelial cells (Peng, Li, Green, Holzman, & Li, 2009; Rakoff-Nahoum, Paglino, Eslami-Varzaneh, Edberg, & Medzhitov, 2004; Wang, Wang, Wang, Wan, & Liu, 2012). Acetate and butyrate increase mucin secretion in rats and butyrate also stimulates mucin secretion in mice by influencing gene expression, helping to protect the epithelial barrier from acidity (Barcelo et al., 2000; Gaudier, Rival, Buisine, Robineau, & Hoebler, 2009). A tight epithelial cell wall and increased mucus production both decrease the permeability of the gut lining, preventing the movement of harmful substances and pathogenic bacteria outside the gut lumen, thus reducing systemic inflammatory responses. Gut microbiota-mediated metabolites contribute to the complex pathways of communication within the microbiota-gut-brain axis which help to maintain homeostasis.

2. Gut microbiota disruption

Gut microbiota are considered stable when they are resistant, resilient or display functional redundancy in response to change (Moya & Ferrer, 2016). If the gut microbiota do not change they are considered resistant, if they can restore the original composition they are considered resilient, and if they can restore original functional capacity they are functionally redundant (Moya & Ferrer, 2016). Microbial stability helps maintain host health and decreases in aging populations (Claesson et al., 2011), highlighting that the gut microbiota are dynamic and their composition can differ over time and between individuals.

Disruption of the gut microbiome occurs when the homeostatic composition of the microbiota becomes imbalanced, resulting in negative effects on overall health (Patterson et al., 2014). A wide range of factors including a poor diet, stress, medication and excessive exercise can all negatively impact the microbiota and lead to this imbalance, harming the metabolic activity of the microbiota and altering their regional distribution along the GI tract (Hawrelak, Hons, & Myers, 2004). Disruption of the gut microbiota has been found alongside changes to the structure and motility of the GI tract, both of which can affect the microbiota's abilities to digest and absorb nutrients from food (Vicentini et al., 2021) which can then alter microbiota-derived by-products and communication to the brain. However, there is an ongoing debate as to the cause or effect relationship between microbial disruption and impairments in GI structure and function (Walker, 2017), suggesting a need to understand the implications of both.

Studies of GI disorders including irritable bowel syndrome (IBS), IBD, obesity and more recently a colon tumor model, have found that disruption of the gut microbiota is a significant aspect of the pathophysiology of these diseases (Greenblum, Turnbaugh, & Borenstein, 2012; Tanoue et al., 2019; Turnbaugh et al., 2006). IBD is characterized by the body attacking cells in the colon and small intestine, resulting in urgent bowel movements, diarrhea, and abdominal pain alongside microbial disruption and inflammation of the GI tract (Assadsangabi & Lobo, 2013). IBS is characterized by bloating, constipation and/or diarrhea as well as abdominal pain, and can be caused by a variety of environmental factors, with clinical and preclinical studies finding microbial disruption (Chassard et al., 2012). Disruption is also seen in obesity studies where the altered abundance of the major gut bacterial phyla, *Firmicutes* and *Bacteroidetes* and the SCFAs they produce are associated with obesity and weight regulation (Chakraborti, 2015; Ley et al., 2005). Studies by Tanoue et al. (2019) also found that the administration of a set of 11 bacterial strains was able to induce intestinal immune cells that inhibit growth in two tumor models. These strains did not leave the gut but instead secreted metabolites that systemically circulate through the body to boost immune cell responses elsewhere, providing evidence of the powerful role microbiota can play in targeting immune responses to fight disease in the body. These studies give insight into the array of impacts that gut microbiota disruption and altered microbiota-derived metabolic function can have on disease pathogenesis.

Disturbances in GI functioning and gut microbiota composition have been found in many brain disorders, from neurodevelopmental disorders such as autism spectrum disorder (ASD) to psychiatric illnesses including schizophrenia (Cryan et al., 2019). Microbial disruption is also found clinically in neurodegenerative diseases such as Parkinson's disease (PD), Alzheimer's disease (AD), amyotrophic lateral sclerosis (ALS) and Huntington's disease (HD) (Rowin, Xia, Jung, & Sun, 2017; Scheperjans et al., 2015; Vogt et al., 2017; Wasser et al., 2020). These findings are also supported by preclinical research, establishing disruption of the gut microbiota in mouse models of PD, AD, ALS and HD (Kong et al., 2018; Lee, Lee, Kim, & Kim, 2019; Perez-Pardo et al., 2018; Wu, Yi, Zhang, Zhou, & Sun, 2015). These neurological disorders consist of neuroinflammation, progressive deterioration of neurons and associated impairments in motor function, and people living with PD, AD, ALS and HD also experience compromised cognition, dementia and behavioral impairments (Spielman, Little, & Klegeris, 2014; Tiryaki & Horak, 2014). Disruption of the gut microbiota can lead to microbiota-mediated inflammation through increased gut barrier permeability and the translocation of microbiota into the systemic circulation; further exacerbating inflammatory responses. This systemic inflammation may also increase BBB permeability, where translocated microbiota can then access the brain (Severance & Yolken, 2019), inducing neuroinflammation, one of the critical characteristics seen in neurodegenerative diseases. Findings of gut microbiota disruption in varied diseases has prompted research into its potential therapeutic role.

2.1 Targeting the gut microbiota

In developing a greater understanding of the complex relationship between host and gut microbiota, in both health and disease, the gut microbiota has emerged as a promising treatment target. This is an area of particular interest for many neurological and neurodegenerative conditions where limited treatment options are available (reviewed in Gubert, Kong, Renoir, & Hannan, 2020). Factors that can lead to the disruption of the gut microbiota, such as diet and exercise, can also be utilized to promote the growth of bacteria associated with stability through lifestyle changes and direct intervention. The Mediterranean diet, high in vegetables, grains and olive oil, was found to alter the gut microbiota in an aging population and increase the abundance of bacterium species positively associated with cognition and decreased those associated with inflammation (Ghosh et al., 2020).

Prebiotics, carbohydrates or dietary fibers which can be used as nutrition by microbiota, also influence the composition of the gut microbiota and can be incorporated through dietary changes or supplementation to specifically modulate the gut microbiota (Valdes, Walter, Segal, & Spector, 2018). Probiotics typically refer to bacterial strains believed to be beneficial to the host and can be specifically introduced into the diet through fermented foods with naturally containing probiotics or supplementation (He & Shi, 2017; Valdes et al., 2018). Prebiotics and probiotics can be used in conjunction to form synbiotics and improve outcomes when manipulating the gut microbiota (Pandey, Naik, & Vakil, 2015). Exercise has also been found to improve gut microbiota outcomes, through the increase of bacterial species associated with the production of butyrate (reviewed in Gubert, Renoir, & Hannan, 2020).

While there is currently limited clinical application of targeting the gut microbiota, antibiotic-resistant *C. difficile* infections are currently treated using fecal microbiota transplant (FMT) from healthy donors. FMT is a technique where the fecal matter from a donor is transferred to a recipient to engraft the donor's gut microbiota and shows promise for its use as a therapeutic tool (Cryan et al., 2019). The microbiota-gut-brain axis is a growing field of research and is being explored as a potential therapeutic target in neurodegenerative diseases such as HD, where we see disrupted microbiota profiles (Wasser et al., 2020). As the gut microbiota is involved in a wide range of host signaling and maintaining homeostasis, its value as a potential modifier of disease outcomes and symptomatic relief holds great potential.

3. Huntington's disease

3.1 HD history and prevalence

Huntington's disease (HD) is a devastating autosomal dominant neurodegenerative disorder that affects both males and females (Bates et al., 2015), resulting in motor, cognitive and psychiatric symptoms as well as disturbances in gut function (Bates et al., 2015; Ramos & Garrett, 2017; Wasser et al., 2020). The first patient record with what we now know as HD dates to 1842, but it was not until a lecture and description of the disease by George Huntington in 1872, that it became known as Huntington's chorea (Bruyn, 1968). In the 1980s, the name was changed to HD, after the extensive non-motor symptoms and link to chromosome 4 were established (reviewed in Roos, 2010), and later in 1993 the 4p16.3 gene for HD was found (The Huntington Disease Research Group, 1993). Prevalence of HD in the Caucasian population is 5 to 10 per 100,000, with a high prevalence in Venezuela, spanning 10 generations, while Japan reports one-tenth of the prevalence of the Caucasian population (Bates, Tabrizi, & Jones, 2014; Wexler, 2004).

3.2 HD etiology

HD is a genetic disorder caused by the expansion of a trinucleotide (CAG) tandem repeat, in the *huntingtin* (*HTT*) gene that encodes for an expanded polyglutamine tract in the huntingtin protein (The Huntington Disease Research Group, 1993). This monogenetic mutation of the *HTT* gene distinguishes HD from many other neurodegenerative disorders which are commonly polygenic and multifactorial (Lipe & Bird, 2009). Non-pathological trinucleotide repeats range from 10 to 35 repeats, mutations ranging between 36 and 39 show incomplete penetrance, while 40 + repeats result in pathological symptoms and the manifestation of HD (Landles & Bates, 2004; Lee et al., 2012). Repeats of 60+ are associated with juvenile-onset HD, which occurs approximately 5% of the time (Nance & Myers, 2001), while most cases are diagnosed later in life with the mean age of symptom onset being between 30 and 50 years (reviewed in Roos, 2010).

The trinucleotide repeat length is not static between generations, and due to genetic anticipation, CAG repeat length tends to increase in subsequent generations, particularly with paternal transmission (Semaka et al., 2013;

Wheeler et al., 2007). The length of the trinucleotide repeat partially determines the variation in onset time but is not the sole predictor, suggesting evidence for the involvement of gene-environment interactions (Langbehn, Brinkman, Falush, Paulsen, & Hayden, 2004). Environmental factors and genetic modifiers are suggested to influence both the disease onset and progression, first established preclinically (Van Dellen, Blakemore, Deacon, York, & Hannan, 2000) and later followed with clinical evidence (Wexler, 2004), leading to differences in presentation despite the disease's monogenic etiology (Bates et al., 2014; Gusella, Macdonald, & Lee, 2014).

3.3 HD pathology

Non-mutated huntingtin protein functions are relatively unknown; however, it does play a role in neurons throughout the brain and is essential for healthy development before birth, with deletion of the *HTT* gene in mouse models resulting in early embryo death (Reiner, Dragatsis, Zeitlin, & Goldowitz, 2003; White et al., 1997). The mutated Htt protein is more widely studied and expressed ubiquitously throughout many cell types in the brain and other areas of the nervous system, and this has also been shown in people with the *HTT* gene mutation (Sassone, Colciago, Cislaghi, Silani, & Ciammola, 2009). Centrally, mutated Htt protein results in a wide range of molecular dysregulation, including brain-derived neurotrophic factor (BDNF) downregulation, white matter atrophy and selective neurodegeneration of striatal medium spiny neurons as well cortical neurons (Ferrer, Goutan, Marín, Rey, & Ribalta, 2000; Halliday et al., 1998; Ross & Tabrizi, 2011; Zuccato et al., 2005).

As neurodegeneration and symptom severity progress, specific regions of the brain are increasingly affected (Bohanna, Georgiou-Karistianis, Hannan, & Egan, 2008; Shaffer et al., 2017). Distinct degeneration can be seen in the striatum where there is an enlargement of the lateral ventricles due to bilateral atrophy (reviewed in Bates et al., 2015). The striatum is predominately composed of GABAergic medium spiny neurons which are characteristically lost due to cytotoxicity in HD (Choi, Choi, & Hong, 2017). Basal ganglia function is also impacted, leading to alterations in motor function and mood (reviewed in Bates et al., 2015). Other areas of the brain are typically impacted by a reduction in cell number and volume loss including the cerebral cortex, thalamus, hypothalamus, substantia nigra, the cerebellum and the brain stem (reviewed in Bates et al., 2015). The exact pathogenic mechanism resulting in neuronal loss is not fully understood, although it is thought that there is a disruption to multiple processes including transcriptional regulation and oxidative stress (reviewed in Bates et al., 2015). HD research has primarily focused on neuropathology and preventing neuronal death, with limited research exploring the peripheral pathology of HD.

Peripherally, the mutated Htt protein results in skeletal muscle atrophy and progressive weight loss (Aziz et al., 2008; Sathasivam, Hobbs, Mangiarini, et al., 1999). Weight loss in people living with HD is frequently reported, increasing with a higher CAG repeat number (Aziz et al., 2008). Beyond this, causation is not well understood with no relationship between weight loss and chorea established in HD or other movement disorders (reviewed in Roos, 2010). It has been suggested that decreased appetite, difficulties chewing and swallowing, as well as GI function and the gut microbiome, may all be involved in HD-induced weight loss (reviewed in Patil et al., 2021).

3.4 HD clinical presentation

HD is typically diagnosed after the onset of distinctive motor symptoms, including involuntary jerking movements known as chorea (Bates et al., 2015; Myrianthopoulos, 1966), changes in gait, speech and difficulties chewing and swallowing (reviewed in Roos, 2010). These motor symptoms can be exhibited anywhere between the ages of 2 and 80 years old, with the mean onset of symptoms occurring between 30 and 50 years of age and a mean symptomatic disease duration of 17–20 years before death, as there is currently no known cure or effective treatment (Gubert, Renoir, & Hannan, 2020; Myers, 2004). Juvenile HD onset also exhibits different motor symptoms compared to adult-onset HD, not showing the typical chorea but rather rigidity and dystonia, with a genome-wide association study (GWAS) identifying genetic modifiers that modulate the onset of clinical symptoms (Lee et al., 2015).

Before the onset of motor deficits, there can be changes to the individual's personality, mood, and cognitive capacity during the prodromal stage of the disease, all of which can significantly interfere with a person's quality of life (Dayalu & Albin, 2015; Tabrizi et al., 2011). Cognitive symptoms include memory deficits, reduced awareness and emotional impairments (Lawrence et al., 1998), with changes in decision making and judgment due to damaged executive functioning involving the prefrontal cortex and striatum circuitry (Murphy et al., 2000). Deficits in memory function present as distinctly different to AD or other forms of dementia, with the underlying pathogenic mechanisms likely to include specific molecular pathways, neuronal synaptopathy and circuitopathy (Hannan, 2018). Psychiatric symptoms include depression, anxiety, irritability, and mood disturbances (Dale & van Duijn, 2015; Lawrence et al., 1998; Pla, Orvoen, Saudou, David, & Humbert, 2014), with 73–98% of people living with HD being affected by at least one psychiatric disorder or symptom (Goh et al., 2018). In both pre-symptomatic and diagnosed people living with HD, many psychiatric symptoms are exacerbated over time (Craufurd, Thompson, & Snowden, 2001; Kirkwood et al., 2002) and a greater prevalence of depression can be seen up to one decade before motor symptoms in pre-symptomatic HD gene carriers (Duff, Paulsen, Beglinger, Langbehn, & Stout, 2007; Epping et al., 2016). Clinical studies also identify mania, anxiety, obsessive-compulsive symptoms and psychosis as other common psychiatric symptoms (Eddy, Parkinson, & Rickards, 2016; Goh et al., 2018), all of which considerably increase disease burden for both the person living with HD and caregivers (Paoli et al., 2017).

HD symptomology is diagnosed, measured, and disease progression is monitored using the Unified Huntington's Disease Rating Scale (UHDRS) (Youssov et al., 2013). Diagnosis can also be based on confirmed family history of HD and subsequent positive genetic testing (McColgan & Tabrizi, 2018). HD is uniformly fatal, resulting in continuous decline of the individual, severely affecting the quality of life, and often resulting in full-time care, with individuals reaching the late stage of the disease 10–25 years after clinical diagnosis (Bates et al., 2015; Edmondson & Goodman, 2017). Interestingly, the onset and severity of HD symptoms and their progressive degeneration vary between people living with HD (Wexler, 2004), reflected in the diverse neuropathology of HD, where abnormalities can appear before the clinical onset of the disease. Genomic background combined with environmental and lifestyle experiences through development and adulthood are also likely to contribute to symptom heterogeneity (reviewed in Gubert, Renoir, & Hannan, 2020) with promising findings in preclinical HD mouse models suggesting that environmental interventions are able to delay disease onset (Van Dellen et al., 2000).

Currently, there are no disease-modifying therapeutic options available for HD, with treatment options limited to a multidisciplinary response to symptoms (Killoran & Biglan, 2014). Pharmaceutical treatment options are available for chorea (with limited efficacy), however, other symptoms experienced including cognitive dysfunction and mood disturbances are treated in a non-HD specific manner (Killoran & Biglan, 2014), or not at all. Intervention strategies such as physical, occupational and speech therapies, as well as dietary consultation, can all assist individuals as the disease progresses. While helpful in their symptomatic relief, these treatment options fail to address the underlying HD pathology and new therapeutic targets are desperately needed to improve the quality of life for people living with HD.

3.5 HD mouse models

HD mouse models are used to enhance our understanding of disease pathology, investigate underlying mechanisms, and identify potential therapeutic targets. They are particularly insightful for gut studies, as they have a controlled environment and diet, both of which can modulate the GI tract and gut microbiota. In 1996, the first transgenic mouse models of HD, the R6/1 and R6/2 lines were engineered to express exon 1 of the human *HTT* gene, with respective transgene expressions of 31% and 75% (Li, Popovic, & Brundin, 2005; Mangiarini et al., 1996). The R6/1 and R6/2 lines are among the most widely studied and used in therapeutic preclinical trials, due to their accurate modeling of adult-onset and juvenile HD, respectively, with replicated neuronal cell dysfunction, formation of mutant huntingtin protein aggregates and regional brain decay (Li et al., 2005). Furthermore, known timepoints for symptom onset and degeneration allows investigation into the onset and modulation of HD pathogenesis.

After the development of the R6/1 and R6/2 mice, numerous other transgenic and knock-in mouse models of HD have been developed, with varying lengths of the CAG repeats, fragment sizes, promotors and background strains (Li et al., 2005). While multiple models can make study comparison challenging, it does allow for rigorous testing across various genetic and symptomatic features, however, no effective therapy in animal models has also shown significant effects in people living with HD (Li et al., 2005).

As HD is incredibly complex, integrating information at molecular, cellular, circuit and systems levels is crucial, so we can establish much needed therapeutic targets to improve the quality of life for people living with HD or even find a cure. HD research is crucial to understanding the mechanisms and pathways leading to selective vulnerability of specific cell populations in the brain and peripheral tissues as well as understanding gene-environment interactions that influence the disorder, all of which may lead to the development of new therapies. Furthermore, within the context of HD, there is a significant knowledge gap within the field of the GI tract and gut microbiome, highlighting a potential new avenue to explore microbiota as targets for novel therapeutic interventions.

4. The gut microbiota in HD

Dysfunction of the GI tract is frequently reported in people living with HD, including symptoms such as constipation and diarrhea, nutrient deficiencies, gastritis, and weight loss (Kobal, Matej, Koželj, & Podnar, 2018; Robbins, Ho, & Barker, 2006; van der Burg et al., 2017). Altered gut-derived metabolites and inflammatory plasma cytokines are found in people living with HD before the clinical onset (Andrich, Wobben, Klotz, Goetze, & Saft, 2009; Lin & Beal, 2006). Despite these presenting factors, knowledge about the peripheral pathology in HD is limited, and it is not known yet whether these GI changes are secondary or causal to other characteristics of HD, such as cognitive deterioration and psychiatric disturbances or loss of mobility (Wasser et al., 2020).

Preclinically, mutant Htt is found in a variety of cell types in the duodenum, rectum and ENS of the stomach in the HdhQ150 knock-in and the R6/2 transgenic mouse models of HD (Sathasivam, Hobbs, Mangiarini, et al., 1999; Van Der Burg et al., 2011). Polyglutamine inclusions, a hallmark neuropathological feature of HD, were also found in both these models throughout the network of nerves between the muscle layers of the GI tract that helps regulate peristalsis (Moffitt, McPhail, Woodman, Hobbs, & Bates, 2009).

As HD is an autosomal dominant disease, the presence of mutant Htt protein in the gut may influence GI tract function and the gut microbiota in early life, which could then modulate disease onset and symptom severity. Preclinical investigation into GI function found impaired gut motility and malabsorption of food in R6/2 mice (Van Der Burg et al., 2011) and the first findings of microbial disruption in HD were published by Kong et al. (2018), who found significant differences in the composition of the gut microbiome between wild-type (WT) and R6/1 HD mice at 12 weeks of age, prior to the onset of motor dysfunction in this model. Continuation of these findings demonstrated that the gut microbiome of the R6/1 model of became unstable during the pre-symptomatic stage and had increased fluctuation in functional capacity when compared to WT littermates (Kong et al., 2020). Microbial disruption was also found in the R6/2 model of HD, alongside increased intestinal permeability, disruption of epithelial cell organization and the reduction of the tight junction protein occludin in the colon mucosa compared to WT controls (Stan et al., 2020).

Wasser et al. (2020) provided the first clinical evidence of gut disruption in people living with HD. The study included 42 HD gene expansion carriers (HDGECs) and 36 healthy controls matched for age and sex. HDGECs consisted of 19 people diagnosed with HD and 23 people in the premanifest stage, classified based on the probability of clinical diagnosis within 5 years, with participant selection also accounting for antibiotics and anti-inflammatory medication. This study reported a significant decrease in alpha diversity and significant differences in beta diversity in the HDGECs compared to healthy controls and did not find any significant differences between HDGECs in the pre-manifest stage and those diagnosed with HD (Wasser et al., 2020). Conversely, Du et al. (2021) found an increase in alpha and beta diversity in 33 people living with HD compared to 33 age and sex matched healthy controls and correlated abundance of bacterium genus with plasma cytokine levels and assessment of clinical HD characteristics. While these two studies are limited by their participation size, they reveal the HD gut microbiome may not have a distinct profile. These studies are the foundation of microbiota-gut-brain axis research in HD and have allowed us to understand the potential role of the GI tract and the gut microbiota in many aspects of HD pathogenesis.

5. Gut microbiota and HD motor symptoms

HD is characterized by severe movement disturbances consisting of bradykinesia, balance and gait disturbances as well as changes in eye movement, speech and hand movements (reviewed in McColgan & Tabrizi, 2018). An association between a lower abundance of *Eubacterium hallii* (*E. hallii*) correlated with more severe motor signs as measured on the UHDRS in symptomatic HDGECs, while in premanifest HDGECs, a significant negative relationship between the abundance of *E. hallii* and estimated proximity to disease onset was observed (Wasser et al., 2020). These findings are consistent with lower levels of *E. hallii* also found in PD (Gerhardt & Mohajeri, 2018), where microbial disruption is thought to promote alpha-synuclein-mediated motor deficits and neuroinflammation (Fitzgerald, Murphy, & Martinson, 2019; Sampson et al., 2016), and we could be seeing a similar microbiota-derived inflammatory response in HD.

In mouse models, HD motor symptoms are typically measured using behavioral tests such as paw clasping, rotarod and Digigait assessments.

Clasping is a phenotypic characteristic of HD transgenic mice, independent of motor coordination and is described as the retraction of paws during tail suspension, where mice are suspended by their tails for a brief period (e.g., 30 s) and observed for the clasping action of the four paws (Gubert et al., 2021). The HD clasping phenotype seen in R6/1 mice reflects striatal dysfunction and is a simple way to observe progressive motor degeneration (Zhang et al., 2014). The rotarod test is widely used to evaluate motor coordination in mouse models, where the mice are placed on an accelerating motorized rotating cylinder and must walk forwards to remain upright without falling off (Deacon, 2013). The Digigait apparatus consists of a transparent testing chamber that sits above a translucent accelerating treadmill belt and a digital camera is mounted underneath the treadmill, recording video footage of the animal's paws for gait analysis (Pallier, Drew, & Morton, 2008).

At 12 weeks of age in the R6/1 HD mouse model, a significant difference in the gut microbiota was seen in comparison to WT mice, which coincided with rotarod deficits, but before the appearance of hind-limb clasping (Kong et al., 2018). At the same time point, metabolomic and metagenomic analysis of the microbiota in R6/1 HD mice showed increased volatility—the degree of compositional change over time (Bastiaanssen et al., 2021)—suggesting possible early effects of HD mutation in the gut before significant motor dysfunction (Kong et al., 2020). Furthermore, significant differences in alpha and beta diversity of R6/1 HD mice coincided with the onset of motor symptoms including performing worse on the rotarod, increased clasping scores and gait deficits (Gubert et al., 2021). These studies demonstrate that gut disruption occurs in HD often before overt motor deficits, suggesting GI factors may be involved in modulating motor symptom presentation and progression.

6. Gut microbiota and HD-induced weight loss

Weight loss is a detrimental complication in HD and a hallmark characteristic of the disorder frequently leading to general weakness and a decline in quality of life for people living with HD (Djoussé et al., 2002). Surprisingly, the development of chorea is unlikely to affect weight loss, with body mass index (BMI) being significantly lower in people living with HD in comparison to control subjects, even at an early stage with a recent diagnosis and minimal involuntary movements (Djoussé et al., 2002). Furthermore, neither disease duration, dystonia or chorea, and its associated disturbances in energy metabolism, were associated with BMI among subjects with HD, suggesting that changes to motor function do not influence weight loss (Djoussé et al., 2002).

HD-induced weight loss is likely a complex interaction between genetic, metabolic and GI factors. Weight loss has been correlated with a higher CAG repeat (Aziz et al., 2008), suggesting the *HTT* gene could have a systemic component associated with a metabolic defect, resulting in a generalized loss of body mass. Weight loss also occurs in people living with HD and R6/2 HD mice despite high dietary intake again suggesting metabolic changes (Liot, Valette, Pépin, Flament, & Brouillet, 2017; Van Der Burg et al., 2011). Cachexia and lack of nutrition due to swallowing difficulties and vomiting develop in people living with HD which also provides challenges in ensuring adequate caloric intake (Trejo et al., 2004), while nutritional deficiencies could also be causing tissue and cellular deficits, leading to weight loss.

Considerable skeletal muscle atrophy, noted in both people living with HD and animal models, occurs alongside progressive weight loss in HD (Sanberg, Fibiger, & Mark, 1981; Sathasivam, Hobbs, Turmaine, et al., 1999). Htt inclusions are found in skeletal muscle cells of R6/2 mice, which could contribute to mitochondrial dysfunction and abnormal energy metabolism resulting in atrophy (Aziz et al., 2008; Sathasivam, Hobbs, Turmaine, et al., 1999). Studies in GF mice, which also show decreased skeletal muscle mass, reduced transcription of skeletal muscle growth genes and increased skeletal muscle atrophy (Lahiri et al., 2019), have found that FMT from specific pathogen-free mice increased skeletal muscle mass (Lahiri et al., 2019). Although yet to be explored in HD, these findings suggest a promising link between gut microbiota and weight loss due to skeletal muscle atrophy.

Microbial dysfunction was also associated with body weight impairment in HDGECs (Wasser et al., 2020), while R6/1 HD mice showed an increase in alpha diversity and lost weight as they aged, despite no differences in food and water intake in comparison to WT (Gubert et al., 2021). R6/1 HD mice also showed impaired weight gain, alongside an increase in *Bacteroidetes* and a proportional decrease in *Firmicutes* (Kong et al., 2018) and this proportional decrease in *Firmicutes* abundance was also seen in male HDGECs (Wasser et al., 2020). In both preclinical and clinical obesity studies, a decrease in *Bacteroidetes* bacteria is associated with weight gain, while a decrease in *Firmicutes* is associated with weight loss (Chakraborti, 2015; Ley et al., 2005), implicating microbiota composition in weight loss. Decreased mucous thickness and villi length also coincided with weight loss in R6/2 HD mice and 12 weeks of age, suggesting structural changes are also affecting nutrient absorption (Van Der Burg et al., 2011). However, malabsorption was inversely associated with body weight after the onset of weight loss, suggesting gut dysfunction may be acceleratory and appear with disease progression and is not a developmental factor (Van Der Burg et al., 2011).

7. Gut microbiota and cognition in HD

Cognitive impairments, culminating in dementia, can have onset in the pre-manifest (pre-motor symptomatic) stages of HD, with cognitive abilities declining over the course of the disease (reviewed in Bates et al., 2015). Investigating cognitive outcomes in HD research can provide insight into potential mechanisms and therapeutic targets (Paulsen, 2011). In their assessment of the relationship between the gut microbiota and HD symptomology, Wasser et al. (2020) found a significant negative relationship between *E. hallii* with disease onset and memory performance, assessed using the Huntington's Disease Cognitive Assessment Battery, in HDGEC subjects (Wasser et al., 2020). This is of particular interest as these bacteria play a role in SCFA production and inflammation, suggesting microbiota-derived inflammation could be contributing to HD pathogenesis (Du et al., 2021; Wasser et al., 2020). Additionally, the abundance of *Lactobacillus* in people living with HD was negatively correlated with Mini-Mental State Examination scores (Du et al., 2021; Wasser et al., 2020), indicating the importance of following up on these intriguing findings linking cognition in HD to the gut microbiota.

Findings in the R6/2 model suggest that impairments to junction proteins result in increased BBB permeability, with a study investigating the BBB in the R6/2 HD model and post-mortem brain tissue of human HD patients observing mutant Htt aggregates throughout neurovascular tissues and changes to the BBB (Di Pardo et al., 2017; Drouin-Ouellet et al., 2015). Preclinically, studies suggest that disturbance to the BBB occurs before overt symptom development, suggesting changes to the BBB may contribute to cognitive decline, similarly to the early compromised BBB seen in AD (van de Haar et al., 2016). The BBB can be modulated by gut microbiota-derived metabolites (Parker, Fonseca, & Carding, 2020). In one study, GF adult mice that received FMT from SPF mice or were administered bacterial strains that produce SCFAs, resulted in positive reinforcement of BBB integrity (Braniste et al., 2014). Further research is needed to investigate the potential effects of the breakdown of the BBB on cognition in HD.

Studies using GF mice have suggested that the gut microbiota plays a role in myelination, and specifically, the myelination abnormalities in HD (Radulescu et al., 2020). Using the bacterial artificial chromosome-mediated transgenic model of HD (BACHD), Radulescu et al. (2020) found that GF BACHD mice had an abnormal myelin thickness that differed from SPF BACHD mice, which possess a gut microbiota. Experimental manipulation of the gut microbiota in BACHD mice was able to modulate specific consequences of HD pathogenesis, highlighting a need for further research into the potential therapeutic outcomes of the gut microbiota in HD.

A study investigating the prebiotic lactoferrin demonstrated an increase in BDNF expression and cognition which may be beneficial in neurodegenerative diseases, including HD, where BDNF expression is impacted (Chen et al., 2015). There is currently limited investigation into how the gut microbiota impacts cognitive impairments in HD, however, research into the effects of the gut microbiota on cognition offers potential therapeutic options to investigate.

8. Gut microbiota and immune function in HD

Studies exploring the role of the immune system in HD have found altered gut-derived metabolites and inflammatory plasma cytokines in people living with HD before disease onset (Andrich et al., 2009; Lin & Beal, 2006). Diseases associated with disturbed gut microbiota often also feature immune dysfunction in their pathology, including IBD, obesity, ASD and neurodegenerative diseases including multiple sclerosis and AD (Bettcher, Tansey, Dorothée, & Heneka, 2021; Dorrestein et al., 2014).

Du et al. (2021) assessed gut microbiota disturbances and host cytokine response in HD by comparing fecal microbiota composition with plasma cytokines and clinical outcomes. Their findings did not conclude any significant differences in the plasma cytokine concentrations between people living with HD and healthy controls, however, they did find correlations between the abundance of bacterial species in the fecal microbiota, cytokine concentrations and HD clinical outcomes. The genus *Intestinimonas* had a higher abundance in people living with HD and was positively correlated with the anti-inflammatory cytokine IL-4 and total functional capacity, a quality of life measure; suggesting this genus may have a positive, anti-inflammatory impact on people living with HD (Du et al., 2021). *Bilophila* was found to have a negative relationship with the pro-inflammatory cytokine IL-6 in this study, conversely, previous murine literature in gut and systemic inflammation has reported a positive correlation between *Bilophila* and pro-inflammatory cytokines including IL-6 (Devkota et al., 2012; Feng et al., 2017). These findings suggest further investigation is needed to understand the role of the gut microbiota in HD immune dysfunction.

Akkermansia is a bacterial genus implicated in anti-inflammatory processes which colonizes the GI tract, using gastric mucin as a nutrient source (Chelakkot et al., 2018). *Akkermansia* has been used to restore intestinal permeability in mice, and a study in elderly Italians aged between 105 and 109 years old found an abundance of *Akkermansia,* associating the genus with maintenance of health and longevity (Biagi et al., 2016; Bodogai et al., 2018; van der Lugt et al., 2018). *Akkermansia muciniphila* was specifically found to influence gut permeability through tight junction proteins (Chelakkot et al., 2018), which is of particular interest in HD as male HDGECs were reported to have a decreased abundance of the *Akkermansiaceae* family, suggesting an microbiota composition may increase epithelial permeability in HD (Wasser et al., 2020). While the results of these studies require further follow-up, they do provide insight into the influence of the gut microbiota on inflammation in HD and potential therapeutic routes.

Alonso, Pisa, and Carrasco (2019) described both fungal and bacterial species in the striatum and frontal cortex of people living with HD, suggesting the accumulation of mutated Htt protein in the brain impairs neural and immune function, creating an environment that encourages microbial colonization of brain tissue. It is important to note that this study should be replicated, with the addition of proper control samples, to avoid potential tissue contamination. Gut microbiota disruption in HD could lead to a breakdown in the homeostasis of the gut which can affect the functioning of the GALT and associated immune functions (Toor et al., 2019), and may allow for bacterial translocation across the intestinal barrier (Jung et al., 2010). It has been proposed that a breakdown in the typical microbiota-gut-brain axis communication pathways, may drive inflammatory pathways in the brain and be present in the clinical presentation of HD (Burokas, Moloney, Dinan, & Cryan, 2015; Wasser et al., 2020).

9. Microbiota-gut-brain axis in HD

While we have previously discussed immune pathways of the microbiota-gut-brain axis in HD, this section will focus on the role of the mucous layer within the GI tract and its communication with the ENS, vagal nerve and HPA axis. Changes in the gut microbiota population in HD can result in the altered production of microbiota-derived metabolites such as neurotransmitters, SCFAs and BCFAs and other neuroactive compounds. Changes to these by-products can in turn affect the bi-directional communication to the brain, influencing the neuroinflammation of HD, and other pathogenic features along the way. We are in the infancy of understanding the connection and communication between the microbiota and the brain along the microbiota-gut-brain axis in HD. The limited research thus far (since the first study by Kong et al., 2018) is predominantly in mouse models and further clinical studies are urgently needed.

9.1 Mucosal function

Disruption of the gut microbiota is thought to affect the integrity of the gut, increase permeability and affect SCFAs, which help regulate both microbes and mucous production at the gut barrier (Maynard & Weinkove, 2018; Venegas et al., 2019). *E. hallii* was found to be reduced in people living with HD (Wasser et al., 2020) which is of particular interest given this bacteria's role in butyrate production and therefore gut permeability and mucosal function, which could lead to increased permeability and inflammation (reviewed in Louis & Flint, 2009). Moreover, the mucosal barrier thickness of the R6/2 HD model was reported to be reduced at 12 weeks of age across the stomach fundus and colon alongside shortened villi length and changes in inhibitory and excitatory neuropeptides in the enteric nervous system (Van Der Burg et al., 2011). These mechanistic changes could affect gut motility and nutrient absorption, with R6/2 HD mice also showing a higher fecal output to food intake ratio (Van Der Burg et al., 2011). Alongside mucosal regulation, butyrate's role as a HDAC inhibitor is pertinent due to transcriptional dysregulation being a pathogenic feature of HD (Sharma & Taliyan, 2015) and has prompted investigations into the use of HDAC inhibitors preclinically and clinically as a therapeutic tool (Sharma & Taliyan, 2015).

9.2 Enteric nervous system and vagal nerve communication

In R6/2 HD mice, loss of ENS neurons was observed alongside an increase in gut transit time and proposed to be associated with small intestine bacterial overgrowth commonly known as SIBO, which can cause inflammation and GI disturbances (Van Der Burg et al., 2011). Mutant Htt protein is also present throughout the ENS of the stomach, duodenum and rectum of the HdhQ150/Q150 and R6/2 HD mice, and could therefore directly interfere with cell function and nutrient absorption (Moffitt et al., 2009).

Furthermore, the timing of changes in these models (R6/2 HD mice mimic juvenile-onset) suggest these changes may occur in the early stages of HD.

9.3 HPA axis

In HD, Htt-immunoreactive inclusions are found in the hypothalamus (Aziz, Swaab, Pijl, & Roos, 2007) and magnetic resonance imaging scans showed significant hypothalamic bilateral atrophy (Kassubek et al., 2004). Furthermore, people living with HD also present with increased corticosteroid levels, suggesting HPA-axis dysregulation (Heuser, Chase, & Mouradian, 1991). Animal models of HD show hypothalamic dysfunction and damage to the hypothalamus, pituitary and adrenal glands (Peterseén et al., 2005; Petersén & Björkqvist, 2006; Sathasivam, Hobbs, Turmaine, et al., 1999), which was correlated with increased thirst in R6/2 mice (Wood et al., 2008). Several regions of the hypothalamus are involved in maintaining energy homeostasis of the body (Aziz et al., 2007) and a dysregulated neuroendocrine system could influence muscle cell metabolism contributing to the weight loss we see in HD. Dysregulation of the HPA axis is thought to contribute to the higher levels of depression and suicidality seen in HD compared to the general population, with both major depressive disorder and HD being associated with HPA axis hyperactivity and disrupted gut microbiota (Cryan et al., 2019; Du & Pang, 2015; Hubers et al., 2015).

10. Gut microbiota and HD sexual dimorphism

Sexual dimorphism occurs when there are distinct features of a disorder between sexes and is a common feature of neurodegenerative diseases such as PD, AD and HD (Farhadi et al., 2017; Jiao et al., 2016; Roos et al., 1991). Clinical studies found that women with HD present more severe symptoms that progress faster (Zielonka, 2018; Zielonka et al., 2013), while in R6/1 mice, females showed exacerbated depressive-like behaviors due to impaired serotonin signaling (Renoir et al., 2011). On the other hand, males presented with impaired spatial memory, weight gain and motor coordination before females (reviewed in Mo, Renoir, & Hannan, 2016). Differences in running wheel activity between males and females have been previously reported in R6/1 mice (Ransome & Hannan, 2013) and the 140 CAG knock-in mouse model of HD (Dorner, Miller, Barton, Brock, & Rebec, 2007) and female HdhQ200/200 mice display earlier deterioration of locomotor and motor coordination

(Cao, Viray, Zweifel, & Stella, 2019). These findings suggest a complex effect of sex on the phenotypic presentation and progression of HD in both mice and humans.

We are also beginning to see sexual dimorphism in the gut microbiota across clinical HD and HD mouse models. In male and female R6/1 mice, microbiota species were similar in abundance, but differed phylogenetically, meaning they were of similar quantity but not closely related, while males also showed an increased microbial disruption in comparison to females (Kong et al., 2018). Clinically, Wasser et al. (2020) also examined the gut microbiome composition between groups and sex at the Phylum and Family level and saw no significant difference between female HDGEC and healthy control groups. In males, there were several differences between HDGECs and healthy controls at a Phylum and family level, with significantly lower abundances of *Firmicutes*, *Lachnospiraceae* and *Akkermansiaceae* in HDGEC males (Wasser et al., 2020). *Akkermansiaceae* plays a role in gut barrier maintenance and inflammation (Chelakkot et al., 2018) and *Lachnospiraceaes* and *Firmicutes* are involved in the production of butyrate (Louis, Young, Holtrop, & Flint, 2010). Across a cohort of R6/1 and WT mice, HD males drank less, weighed less and showed decreased butyrate and valerate concentrations in comparison to females (Gubert et al., 2021). This study also characterized gut microbiome signatures discriminating exercising male and female R6/1 mice, but no sex differences were seen in GI macroscopy and motility (Gubert et al., 2021).

Together, these findings suggest a complex relationship between sex and the gut microbiota, all of which could alter metabolic by-products and subsequent gut-brain axis communication, therefore contributing to the dimorphic presentation of HD. Despite reporting dimorphism in the gut microbiota and the presentation of HD symptomology, these studies are often asynchronous and a clear relationship between the two is yet to be established. Furthermore, understanding how environmental factors can influence the dimorphism of the gut composition and structure could provide insight into the sex differences we see in HD.

11. Environment

The gut lumen and its microbiota are a consistently changing environment, adapting to environmental factors such as diet, medication, stress, and exercise, all of which can result in individual variation in microbial diversity and species richness (Clarke et al., 2014;

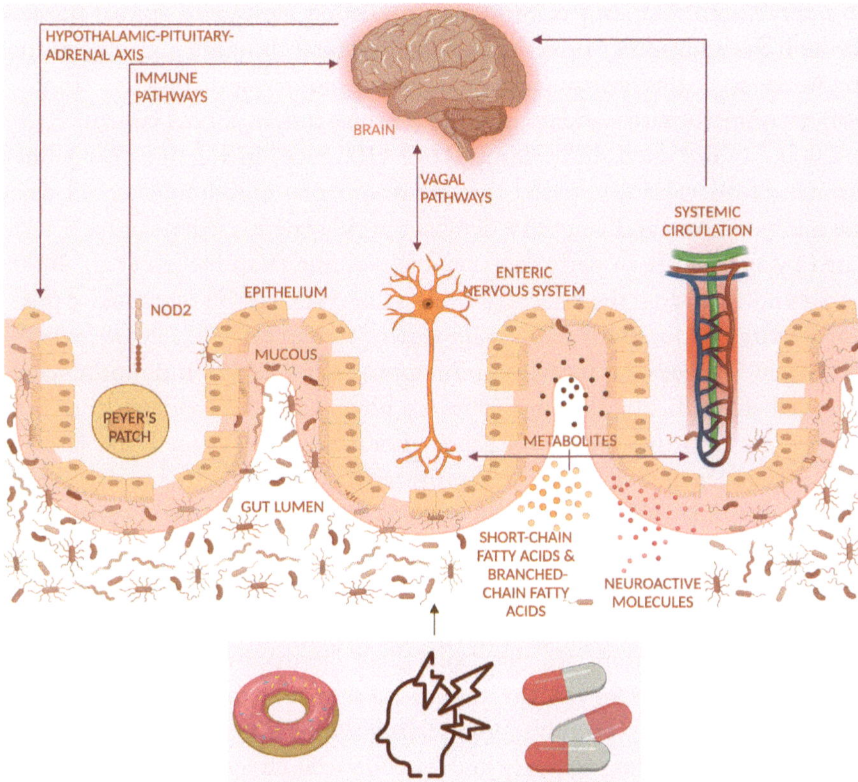

Fig. 2 Environmental influences can lead to altered microbial diversity and increased epithelial permeability, affecting both gastrointestinal function and microbiota-gut-brain axis communication. Created with BioRender.com

Cryan et al., 2019; Hawrelak et al., 2004) (Fig. 2). Despite environmental influences, overall greater microbial diversity has beneficial effects on human health and cognition (Coyte & Schluter, 2015; Le Chatelier et al., 2013).

In 2004, Wexler and colleagues collected 3989 DNA samples from Venezuelan HD relatives and genotyped the samples for HD alleles, finding that variability in the onset of HD consisted of both genetic (40%) and environmental (60%) components, with family relationships and shared environments playing an essential role in HD onset. A pilot study also found that people living with HD who undergo combined speech, physical and occupational therapies over 3 weeks showed cognitive and motor improvements (Zinzi et al., 2007), demonstrating that environmental factors can influence both HD onset and symptoms. Preclinically, environmental enrichment is an intervention that aims to mimic a stimulating lifestyle in animal models, through the addition of novel tunneling, climbing, housing and objects into the home cage, thus enhancing visual motor, sensory and cognitive stimulation in comparison to standard housing (Nithianantharajah & Hannan, 2006). In 2000, the first such study (using R6/1 HD mice) in a genetic model of a human disorder, demonstrated that environmental enrichment could delay the onset of motor and cognitive deficits and delay neurodegeneration as indicated by an increase in peristriatal cerebral volume (Van Dellen et al., 2000).

In understanding the effects of exercise on HD, passivity or lack of physical activity is associated with earlier onset (Trembath et al., 2010), while a structured 12-week exercise program improved cognitive and motor symptoms (Khalil et al., 2013). Furthermore, a physically active video game could moderately improve motor measures after 6 weeks (Kloos, Fritz, Kostyk, Young, & Kegelmeyer, 2013) demonstrating that physical activity can be beneficial and alleviate symptoms in people living with HD. Preclinically, in the R6/1 HD mouse model, voluntary wheel running ameliorated deficits in short-term spatial memory and delayed expression of rear paw clasping (Pang, Stam, Nithianantharajah, Howard, & Hannan, 2006). These findings demonstrate that both these environmental factors can be beneficial in HD, however, the role of exercise and an enrichment equivalent is yet to be explored in people living with HD.

Research exploring exercise and its beneficial effects on the gut microbiota is well established both clinically and preclinically (Gubert & Hannan, 2021; Gubert, Kong, et al., 2020). Male elite rugby players were shown to have higher gut microbiota diversity in comparison to healthy, BMI matched male controls (Clarke et al., 2014) and improved cardiorespiratory fitness correlated with an increase in microbiota diversity (Estaki et al., 2016). A 12-week exercise program also induced changes in the gut microbiota, which correlated with improved glucose homeostasis and insulin sensitivity through the production of SCFAs and BCFAs (Liu et al., 2020). Preclinically, exercise in the form of voluntary wheel running in rats can positively modulate microbiota composition and increase butyrate concentrations (Matsumoto et al., 2008), which can strengthen the gut epithelial barrier (Canani et al., 2011) and voluntary wheel running has also been found to improve microbiota abundance and biodiversity in WT C57BL/6 mice (Choi et al., 2013). It is important to note that intensity and duration of exercise interventions needs to be considered, as strenuous exercise can lead to gut barrier damage, permeability, and inflammation.

The interaction between environmental enrichment and the microbiota is not well understood, with few studies bringing these two ideas together, and predominantly in preclinical research. One study found that environmental enrichment can enhance the lifespan of a colon-tumor mice model by improving receptor signaling involved in tumor wound repair, mucosal barrier functioning and microbiota diversity while decreasing inflammatory responses (Bice et al., 2017). Environmental enrichment also reduced pro-inflammatory cytokines in feces and inflammation-inducing genes in the colon (Singh et al., 2019). Unfortunately, both studies included the addition of running wheels to their EE protocol, making it impossible to separate the effect of enrichment from exercise alone.

In 2021, a link between the beneficial effects of enrichment and exercise in HD mice and the gut microbiome was established, with these housing interventions showing different microbiome compositions to standard-housed conditions (Gubert et al., 2021). Mice were randomly assigned into either standard-housed (SH), environmental enrichment (EE) or exercise (EX) groups according to sex and genotype, due to the coprophagic nature of mice. SH conditions included wood chips and tissues for bedding, EE cages contained tunneling, climbing objects and novel objects, excluding running wheels and EX cages only contained running wheels. EE, EX, and SH conditions differed in beta diversity indicating that the gut microbiota of these housing groups are distinguishable in abundance (Bray-Curtis index) and are phylogenetically different (unweighted Unifrac index). This study identified discriminating bacteria signatures for EE, EX and SH conditions, between male WT and R6/1 HD mice as well as between female WT and R6/1 mice, with the *Bacteroidales*, *Lachnospirales* and *Oscillospirales* being the main discriminatory orders (Gubert et al., 2021). Discriminating bacterial signatures between housing groups in both male and female WT mice demonstrate the importance of studying these housing interventions separately in both sexes and WT mice for translatability to other models (Gubert et al., 2021). Furthermore, these changes in microbiome composition between housing groups, occurred at 12 weeks in the R6/1 model, when microbial disruption was previously established (Kong et al., 2018). This suggests that even a short exposure to these housing interventions can influence the gut microbiota and that the beneficial effects we see from extended EE and EX exposure, could occur via the microbiota-gut-brain axis, but further research is needed to establish underlying pathways and mechanisms.

12. Limitations

As with any animal research, the translatability of these predominantly preclinical findings to humans comes with its challenges. Each mouse model of HD is unique and can therefore only partially mimic the human HD phenotype, with differences in symptoms between transgenic mice and people living with HD to be expected. Furthermore, translatability challenges are also present in assessing GI structure and function, as well as microbiota profiling, metagenomics and metabolomics, due to anatomical differences. To effectively assess GI changes in humans requires complex methods and larger groups over an extended time frame and geographical areas, to account for factors such as BMI, diet, ethnicity, age, exposure to antibiotics, prebiotics and probiotics, as well as exercise and environmental stimulation. Additionally, exploring gut parameters clinically in HD needs to account for the effects of medication, the progressive worsening of symptoms and the impact of the *HTT* gene mutation. Due to these challenges, rodent models (and other valid animal models) are currently an excellent way to explore functional pathways and underlying mechanisms of microbiota-gut-brain axis interactions and for developing novel therapeutic approaches.

13. Conclusion

We are at the beginning of exciting research in the field of HD and the microbiota-gut-brain axis. Understanding the role of the gut microbiota in HD is a crucial step to improving our knowledge of this devastating disease. Collectively, these studies suggest that the microbiota could be involved in HD pathogenesis; however, whether this role is modulatory or comorbid, and whether microbiota dysfunction in the gut could influence brain dysfunction and associated symptoms, is yet to be established. Currently, preclinical models such as the R6 lines provide useful tools to explore this relationship, as they replicate the microbial disruption observed in people living with HD (Kong et al., 2018; Wasser et al., 2020).

HD-induced GI changes emphasize that HD is not only neurodegenerative but also a gut-related disorder and understanding the role of the microbiota and the GI tract in the peripheral pathology of the disease needs to be further explored. Furthermore, understanding the effects of antibiotics, FMT therapies, dietary, prebiotic, and probiotic interventions could help to alleviate GI symptoms in people living with HD. In extending the existing literature on how environmental factors, such as environmental enrichment and exercise, are interacting with the gut microbiota, promising new therapeutic targets may be identified. Novel therapeutics, such as enviromimetics, which mimic the beneficial effects of environmental enrichment (McOmish & Hannan, 2007; Nithianantharajah & Hannan, 2006), or exercise mimetics (Gubert & Hannan, 2021), may hold promise for HD and other disorders. These therapeutic targets could have beneficial effects beyond the scope of HD, extending to other neurodegenerative diseases and psychiatric disorders. HD currently has no disease-modifying treatments and therefore the race is on to find new ways to help HD sufferers and their families.

References

Abautret-Daly, Á., Dempsey, E., Parra-Blanco, A., Medina, C., & Harkin, A. (2018). Gut-brain actions underlying comorbid anxiety and depression associated with inflammatory bowel disease. *Acta Neuropsychiatrica*, *30*(5), 275–296. https://doi.org/10.1017/neu.2017.3.

Agostoni, E., Chinnock, J. E., De Burgh Daly, M., & Murray, J. G. (1957). Functional and histological studies of the vagus nerve and its branches to the heart, lungs and abdominal viscera in the cat. *The Journal of Physiology*, *135*(1), 182–205. https://doi.org/10.1113/jphysiol.1957.sp005703.

Al-Asmakh, M., & Zadjali, F. (2015). Use of germ-free animal models in microbiota-related research. *Journal of Microbiology and Biotechnology*, *25*(10), 1583–1588. https://doi.org/10.4014/jmb.1501.01039.

Alonso, R., Pisa, D., & Carrasco, L. (2019). Brain microbiota in Huntington's disease patients. *Frontiers in Microbiology*, *10*(November), 1–16. https://doi.org/10.3389/fmicb.2019.02622.

Andrich, J. E., Wobben, M., Klotz, P., Goetze, O., & Saft, C. (2009). Upper gastrointestinal findings in Huntington's disease: Patients suffer but do not complain. *Journal of Neural Transmission*, *116*(12), 1607–1611. https://doi.org/10.1007/s00702-009-0310-1.

Assadsangabi, A., & Lobo, A. J. (2013). Diagnosing and managing inflammatory bowel disease. *Practitioner*, *257*(1763), 13–18.

Aziz, N. A., Swaab, D. F., Pijl, H., & Roos, R. A. C. (2007). Hypothalamic dysfunction and neuroendocrine and metabolic alterations in Huntington'S disease: Clinical consequences and therapeutic implications. *Reviews in the Neurosciences*, *18*(3–4), 223–252. https://doi.org/10.1515/REVNEURO.2007.18.3-4.223.

Aziz, N. A., Van Der Burg, J. M. M., Landwehrmeyer, G. B., Brundin, P., Stijnen, T., & Roos, R. A. C. (2008). Weight loss in Huntington disease increases with higher CAG repeat number. *Neurology*, *71*(19), 1506–1513.

Barcelo, A., Claustre, J., Moro, F., Chayvialle, J. A., Cuber, J. C., & Plaisancié, P. (2000). Mucin secretion is modulated by luminal factors in the isolated vascularly perfused rat colon. *Gut*, *46*(2), 218–224. https://doi.org/10.1136/gut.46.2.218.

Bastiaanssen, T. F. S., Gururajan, A., van de Wouw, M., Moloney, G. M., Ritz, N. L., Long-Smith, C. M., et al. (2021). Volatility as a concept to understand the impact of stress on the microbiome. *Psychoneuroendocrinology*, *124*(November 2020), 105047. https://doi.org/10.1016/j.psyneuen.2020.105047.

Bates, G., Tabrizi, S., & Jones, L. (Eds.). (2014). *Huntington's disease (No. 64)* Oxford Monographs on Medical G.
Bates, G. P., Dorsey, R., Gusella, J. F., Hayden, M. R., Kay, C., Leavitt, B. R., et al. (2015). Huntington's disease. *ACTA Paulista de Enfermagem*, *23*(5), 671–676. https://doi.org/10.1590/S0103-21002010000500014.
Bettcher, B. M., Tansey, M. G., Dorothée, G., & Heneka, M. T. (2021). Peripheral and central immune system crosstalk in Alzheimer disease — A research prospectus. *Nature Reviews Neurology*, *17*(11), 689–701. https://doi.org/10.1038/s41582-021-00549-x.
Biagi, E., Franceschi, C., Rampelli, S., Severgnini, M., Ostan, R., Turroni, S., et al. (2016). Gut microbiota and extreme longevity. *Current Biology*, *26*(11), 1480–1485. https://doi.org/10.1016/j.cub.2016.04.016.
Bice, B. D., Stephens, M. R., Georges, S. J., Venancio, A. R., Bermant, P. C., Warncke, A. V., et al. (2017). Environmental enrichment induces pericyte and IgA-dependent wound repair and lifespan extension in a colon tumor model. *Cell Reports*, *19*(4), 760–773. https://doi.org/10.1016/j.celrep.2017.04.006.
Birchenough, G. M. H., Johansson, M. E. V., Gustafsson, J. K., Bergström, J. H., & Hansson, G. C. (2015). New developments in goblet cell mucus secretion and function. *Mucosal Immunology*, *8*(4), 712–719. https://doi.org/10.1038/mi.2015.32.
Bodogai, M., O'Connell, J., Kim, K., Kim, Y., Moritoh, K., Chen, C., et al. (2018). Commensal bacteria contribute to insulin resistance in aging by activating innate B1a cells. *Science Translational Medicine*, *10*(468). https://doi.org/10.1126/scitranslmed.aat4271.
Bohanna, I., Georgiou-Karistianis, N., Hannan, A. J., & Egan, G. F. (2008). Magnetic resonance imaging as an approach towards identifying neuropathological biomarkers for Huntington's disease. *Brain Research Reviews*, *58*(1), 209–225. https://doi.org/10.1016/j.brainresrev.2008.04.001.
Bonaz, B., Bazin, T., & Pellissier, S. (2018). The Vagus nerve at the Interface of the microbiota-gut-brain Axis. *Frontiers in Neuroscience*, *12*, 49. https://doi.org/10.3389/fnins.2018.00049.
Bonnardel, J., Da Silva, C., Masse, M., Montañana-Sanchis, F., Gorvel, J. P., & Lelouard, H. (2015). Gene expression profiling of the Peyer's patch mononuclear phagocyte system. *Genomics Data*, *5*, 21–24. https://doi.org/10.1016/j.gdata.2015.05.002.
Braniste, V., Al-Asmakh, M., Kowal, C., Anuar, F., & Abbaspour, A. (2014). The gut microbiota influences blood-brain barrier permeability in mice. *Science Translational Medicine*, *17*(1), 53–55. https://doi.org/10.1126/scitranslmed.3009759.The.
Brown, E. G., Tanner, C. M., & Goldman, S. M. (2018). The microbiome in neurodegenerative disease. *Current Geriatrics Reports*, 7(2), 81–91. https://doi.org/10.1007/s13670-018-0240-6.
Brüssow, H. (2020). Problems with the concept of gut microbiota dysbiosis. *Microbial Biotechnology*. https://doi.org/10.1111/1751-7915.13479.
Bruyn, G. W. (1968). Huntington's chorea: Historical, clinical and laboratory synopsis. *6. Handbook of Clinical Neurology* (pp. 298–378). Amsterdam: Elsevier.
Burokas, A., Moloney, R. D., Dinan, T. G., & Cryan, J. F. (2015). Microbiota regulation of the mammalian gut–brain axis. *Advances in Applied Microbiology*, *91*, 1–62.
Canani, R. B., Di Costanzo, M., Leone, L., Pedata, M., Meli, R., & Calignano, A. (2011). Potential beneficial effects of butyrate in intestinal and extraintestinal diseases. *World Journal of Gastroenterology*, *17*(12), 1519–1528. https://doi.org/10.3748/wjg.v17.i12.1519.
Cao, J. K., Viray, K., Zweifel, L., & Stella, N. (2019). Sex-dependent impaired locomotion and motor coordination in the HdhQ200/200 mouse model of Huntington's disease. *Neurobiology of Disease*, *132*(May), 104607. https://doi.org/10.1016/j.nbd.2019.104607.
Chakraborti, C. K. (2015). New-found link between microbiota and obesity. *World Journal of Gastrointestinal Pathophysiology*, *6*(4), 110. https://doi.org/10.4291/wjgp.v6.i4.110.

Chassard, C., Dapoigny, M., Scott, K. P., Crouzet, L., Del'Homme, C., Marquet, P., et al. (2012). Functional dysbiosis within the gut microbiota of patients with constipated-irritable bowel syndrome. *Alimentary Pharmacology and Therapeutics*, *35*(7), 828–838. https://doi.org/10.1111/j.1365-2036.2012.05007.x.

Chelakkot, C., Choi, Y., Kim, D. K., Park, H. T., Ghim, J., Kwon, Y., et al. (2018). Akkermansia muciniphila-derived extracellular vesicles influence gut permeability through the regulation of tight junctions. *Experimental and Molecular Medicine*, *50*(2). e450-11 https://doi.org/10.1038/emm.2017.282.

Chen, Y., Zheng, Z., Zhu, X., Shi, Y., Tian, D., Zhao, F., et al. (2015). Lactoferrin promotes early neurodevelopment and cognition in postnatal piglets by upregulating the BDNF signaling pathway and polysialylation. *Molecular Neurobiology*, *52*(1), 256–269. https://doi.org/10.1007/s12035-014-8856-9.

Choi, K., Choi, Y., & Hong, S. (2017). Stem cell transplantation for Huntington's diseases. *Methods*, *133*(2018), 104–112. https://doi.org/10.1016/j.ymeth.2017.08.017.

Choi, J. J., Eum, S. Y., Rampersaud, E., Daunert, S., Abreu, M. T., & Toborek, M. (2013). Exercise attenuates PCB-induced changes in the mouse gut microbiome. *Environmental Health Perspectives*, *121*(6), 725–730. https://doi.org/10.1289/ehp.1306534.

Claesson, M. J., Cusack, S., O'Sullivan, O., Greene-Diniz, R., De Weerd, H., Flannery, E., et al. (2011). Composition, variability, and temporal stability of the intestinal microbiota of the elderly. *Proceedings of the National Academy of Sciences of the United States of America*, *108*(SUPPL. 1), 4586–4591. https://doi.org/10.1073/pnas.1000097107.

Clarke, S. F., Murphy, E. F., O'Sullivan, O., Lucey, A. J., Humphreys, M., Hogan, A., et al. (2014). Exercise and associated dietary extremes impact on gut microbial diversity. *Gut*, *63*(12), 1913–1920. https://doi.org/10.1136/gutjnl-2013-306541.

Coyte, K. Z., & Schluter, J. (2015). The ecology of the microbiome: Networks, competition and stability. *Science*, *350*(6261), 663–666.

Craufurd, D., Thompson, J. C., & Snowden, J. S. (2001). Behavioral changes in Huntington disease. *Cognitive and Behavioral Neurology*, *14*(4), 219–226.

Cresci, G. A. M., & Izzo, K. (2019). Gut Microbiome. *Adult Short Bowel Syndrome*, 45–54. https://doi.org/10.1016/b978-0-12-814330-8.00004-4.

Cryan, J. F., O'Riordan, K. J., Cowan, C. S. M., Sandhu, K. V., Bastiaanssen, T. F. S., Boehme, M., et al. (2019). The microbiota-gut-brain Axis. *Physiological Reviews*, *99*(4), 1877–2013. https://doi.org/10.1152/physrev.00018.2018.

Dale, M., & van Duijn, E. (2015). Anxiety in Huntington's disease. *Journal of Neuropsychiatry and Clinical Neurosciences*, *27*(4), 262–271. https://doi.org/10.1176/appi.neuropsych.14100265.

Dalile, B., Van Oudenhove, L., Vervliet, B., & Verbeke, K. (2019). The role of short-chain fatty acids in microbiota–gut–brain communication. *Nature Reviews. Gastroenterology & Hepatology*, *16*(8), 461–478. https://doi.org/10.1038/s41575-019-0157-3.

Dayalu, P., & Albin, R. L. (2015). Huntington disease: Pathogenesis and treatment. *Neurologic Clinics*, *33*(1), 101–114.

Deacon, R. M. J. (2013). Measuring motor coordination in mice. *Journal of Visualized Experiments*, *May*, 1–8. https://doi.org/10.3791/2609.

Devkota, S., Wang, Y., Musch, M. W., Leone, V., Fehlner-Peach, H., Nadimpalli, A., et al. (2012). Dietary-fat-induced taurocholic acid promotes pathobiont expansion and colitis in Il10 −/− mice. *Nature*, *487*(7405), 104–108. https://doi.org/10.1038/nature11225.

Di Pardo, A., Amico, E., Scalabrì, F., Pepe, G., Castaldo, S., Elifani, F., et al. (2017). Impairment of blood-brain barrier is an early event in R6/2 mouse model of Huntington disease. *Scientific Reports*, 7(January), 1–8. https://doi.org/10.1038/srep41316.

Djoussé, L., Knowlton, B., Cupples, L. A., Marder, K., Shoulson, I., & Myers, R. H. (2002). Weight loss in early stage of Huntington's disease. *Neurology*, *59*(9), 1325–1330. https://doi.org/10.1212/01.WNL.0000031791.10922.CF.

Dominguez-Bello, M. G., Godoy-Vitorino, F., Knight, R., & Blaser, M. J. (2018). Role of the microbiome in human development. *Gut*. https://doi.org/https://doi.org/10.1136/gutjnl-2018-317503.

Donaldson, G. P., Lee, S. M., & Mazmanian, S. K. (2015). Gut biogeography of the bacterial microbiota. *Nature Reviews Microbiology*, *14*(1), 20–32. https://doi.org/10.1038/nrmicro3552.

Dorner, J. L., Miller, B. R., Barton, S. J., Brock, T. J., & Rebec, G. V. (2007). Sex differences in behavior and striatal ascorbate release in the 140 CAG knock-in mouse model of Huntington's disease. *Behavioural Brain Research*, *178*(1), 90–97. https://doi.org/10.1016/j.bbr.2006.12.004.

Dorrestein, P. C., Mazmanian, S. K., & Knight, R. (2014). Finding the missing links among metabolites, microbes, and the host. *Immunity*, *40*(6), 824–832. https://doi.org/10.1016/j.immuni.2014.05.015.

Drouin-Ouellet, J., Sawiak, S. J., Cisbani, G., Lagacé, M., Kuan, W. L., Saint-Pierre, M., et al. (2015). Cerebrovascular and blood-brain barrier impairments in Huntington's disease: Potential implications for its pathophysiology. *Annals of Neurology*, *78*(2), 160–177. https://doi.org/10.1002/ana.24406.

Du, G., Dong, W., Yang, Q., Yu, X., Ma, J., Gu, W., et al. (2021). Altered gut microbiota related to inflammatory responses in patients with Huntington's disease. *Frontiers in Immunology*, *11*(February), 1–12. https://doi.org/10.3389/fimmu.2020.603594.

Du, X., & Pang, T. Y. (2015). Is dysregulation of the HPA-axis a core pathophysiology mediating co-morbid depression in neurodegenerative diseases? *Frontiers in Psychiatry*, *6*(MAR), 1–33. https://doi.org/10.3389/fpsyt.2015.00032.

Duff, K., Paulsen, J. S., Beglinger, L. J., Langbehn, D. R., & Stout, J. C. (2007). Psychiatric symptoms in Huntington's disease before diagnosis: The predict-HD study. *Biological Psychiatry*, *62*(12), 1341–1346. https://doi.org/10.1016/j.biopsych.2006.11.034.

Eddy, C. M., Parkinson, E. G., & Rickards, H. E. (2016). Changes in mental state and behaviour in Huntington's disease. *The Lancet Psychiatry*, *3*(11), 1079–1086. https://doi.org/10.1016/S2215-0366(16)30144-4.

Edmondson, M. C., & Goodman, L. (2017). *Contemporary health care for Huntington disease. Huntington disease. Vol. 144* (1st ed.). Elsevier B.V. https://doi.org/10.1016/B978-0-12-801893-4.00014-6.

Epping, E. A., Ph, D., Kim, J., Ph, D., Craufurd, D., Brashers-krug, T. M., et al. (2016). *Longitudinal psychiatric symptoms in prodromal Huntington's disease: A decade of data.* https://doi.org/https://doi.org/10.1176/appi.ajp.2015.14121551.

Estaki, M., Pither, J., Baumeister, P., Little, J. P., Gill, S. K., Ghosh, S., et al. (2016). Cardiorespiratory fitness as a predictor of intestinal microbial diversity and distinct metagenomic functions. *Microbiome*, *4*, 1–13. https://doi.org/10.1186/s40168-016-0189-7.

Farhadi, F., Vosoughi, K., Shahidi, G. A., Delbari, A., Lökk, J., & Fereshtehnejad, S. M. (2017). Sexual dimorphism in Parkinson's disease: Differences in clinical manifestations, quality of life and psychosocial functioning between males and females. *Neuropsychiatric Disease and Treatment*, *13*, 329–338. https://doi.org/10.2147/NDT.S124984.

Feng, Z., Long, W., Hao, B., Ding, D., Ma, X., Zhao, L., et al. (2017). A human stool-derived Bilophila wadsworthia strain caused systemic inflammation in specific-pathogen-free mice. *Gut Pathogens*, *9*(1), 1–10. https://doi.org/10.1186/s13099-017-0208-7.

Ferrer, I., Goutan, E., Marín, C., Rey, M. J., & Ribalta, T. (2000). Brain-derived neurotrophic factor in Huntington disease. *Brain Research*, *866*(1–2), 257–261. https://doi.org/10.1016/S0006-8993(00)02237-X.

Fitzgerald, E., Murphy, S., & Martinson, H. A. (2019). Alpha-synuclein pathology and the role of the microbiota in Parkinson's disease. *Frontiers in Neuroscience*, *13*(APR), 1–13. https://doi.org/10.3389/fnins.2019.00369.

Frankiensztajn, L. M., Elliott, E., & Koren, O. (2020). The microbiota and the hypothalamus-pituitary-adrenocortical (HPA) axis, implications for anxiety and stress disorders. *Current Opinion in Neurobiology, 62*, 76–82. https://doi.org/10.1016/j.conb.2019.12.003.

Fung, T. C., Olson, C. A., & Hsiao, E. Y. (2017). Interactions between the microbiota, immune and nervous systems in health and disease. *Nature Neuroscience, 20*(2), 145–155. https://doi.org/10.1038/nn.4476.

Furness, J. B. (2006). *The enteric nervous system*. Wiley.

Gaudier, E., Rival, M., Buisine, M. P., Robineau, I., & Hoebler, C. (2009). Butyrate enemas upregulate Muc genes expression but decrease adherent mucus thickness in mice colon. *Physiological Research, 58*(1), 111–119.

Gerhardt, S., & Mohajeri, M. H. (2018). Changes of colonic bacterial composition in Parkinson's disease and other Neurodegenarative diseases. *Nutrients, 10*(6), 708. https://doi.org/10.3390/nu10060708.

Gershon, M. D. (2004). Review article: Serotonin receptors and transporters — Roles in normal and abnormal gastrointestinal motility. *Alimentary Pharmacology & Therapeutics, 20*, 3–14.

Ghosh, T. S., Rampelli, S., Jeffery, I. B., Santoro, A., Neto, M., Capri, M., et al. (2020). Mediterranean diet intervention alters the gut microbiome in older people reducing frailty and improving health status: The NU-AGE 1-year dietary intervention across five European countries. *Gut, 69*(7), 1218–1228. https://doi.org/10.1136/gutjnl-2019-319654.

Goh, A. M., Wibawa, P., Loi, S. M., Walterfang, M., Velakoulis, D., & Looi, J. C. (2018). Psychiatry Australasian Huntington's disease: Neuropsychiatric manifestations of Huntington 's disease. *Australasian Psychiatry, 26*(4), 366–375. https://doi.org/10.1177/1039856218791036.

Greenblum, S., Turnbaugh, P. J., & Borenstein, E. (2012). Metagenomic systems biology of the human gut microbiome reveals topological shifts associated with obesity and inflammatory bowel disease. *Proceedings of the National Academy of Sciences of the United States of America, 109*(2), 594–599. https://doi.org/10.1073/pnas.1116053109.

Gubert, C., & Hannan, A. J. (2021). Exercise mimetics: Harnessing the therapeutic effects of physical activity. *Nature Reviews Drug Discovery, 20*(11), 862–879. https://doi.org/10.1038/s41573-021-00217-1.

Gubert, C., Kong, G., Renoir, T., & Hannan, A. J. (2020). Exercise, diet and stress as modulators of gut microbiota: Implications for neurodegenerative diseases. *Neurobiology of Disease, 134*(April 2019), 104621. https://doi.org/10.1016/j.nbd.2019.104621.

Gubert, C., Love, C. J., Kodikara, S., Mei Liew, J. J., Renoir, T., Lê Cao, K.-A., et al. (2021). Gene-environment-gut interactions in Huntington's disease mice are associated with environmental modulation of the gut microbiome. *IScience, 25*(1), 103687. https://doi.org/10.1016/j.isci.2021.103687.

Gubert, C., Renoir, T., & Hannan, A. J. (2020). Why Woody got the blues: The neurobiology of depression in Huntington's disease. *Neurobiology of Disease*. https://doi.org/10.1016/j.nbd.2020.104958.

Gusella, J. F., Macdonald, M. E., & Lee, J. M. (2014). Genetic modifiers of Huntington's disease. *Movement Disorders, 29*(11), 1359–1365. https://doi.org/10.1002/mds.26001.

Halliday, G. M., McRitchie, D. A., Macdonald, V., Double, K. L., Trent, R. J., & McCusker, E. (1998). Regional specificity of brain atrophy in Huntington's disease. *Experimental Neurology, 154*(2), 663–672. https://doi.org/10.1006/exnr.1998.6919.

Hannan, A. J. (2018). Synaptopathy, circuitopathy and the computational biology of Huntington 's disease. *BMC Biology*.

Hawrelak, J. A., Hons, B., & Myers, S. P. (2004). The causes of intestinal dysbiosis - a review. *Alternative Medicine Review, 9*(2), 180–197.

He, M., & Shi, B. (2017). Gut microbiota as a potential target of metabolic syndrome: The role of probiotics and prebiotics. *Cell & Bioscience*, 7(1), 1–14. https://doi.org/10.1186/s13578-017-0183-1.

Herath, M., Hosie, S., Bornstein, J. C., & Franks, A. E. (2020). The role of the gastrointestinal mucus system in intestinal homeostasis: Implications for neurological disorders. *Frontiers in Cellular and Infection Microbiology*, *10*, 248. https://doi.org/10.3389/fcimb.2020.00248.

Heuser, I. J. E., Chase, T. N., & Mouradian, M. M. (1991). The limbic-hypothalamic-pituitary-adrenal axis in Huntington 's disease. *Biological Psychiatry*, *30*, 943–952.

Hosoi, T., Okuma, Y., & Nomura, Y. (2000). Electrical stimulation of afferent vagus nerve induces IL-1β expression in the brain and activates HPA axis. *American Journal of Physiology. Regulatory, Integrative and Comparative Physiology*, *279*(1), 141–147. https://doi.org/10.1152/ajpregu.2000.279.1.r141.

Hubers, A. A. M., van der Mast, R. C., Pereira, A. M., Roos, R. A. C., Veen, L. J., Cobbaert, C. M., et al. (2015). Hypothalamic-pituitary-adrenal axis functioning in Huntington's disease and its association with depressive symptoms and suicidality. *Journal of Neuroendocrinology*, *27*(3), 234–244. https://doi.org/10.1111/jne.12255.

Inohara, N., Ogura, Y., Fontalba, A., Gutierrez, O., Pons, F., Crespo, J., et al. (2003). Host recognition of bacterial muramyl dipeptide mediated through NOD2: Implications for Crohn's disease. *Journal of Biological Chemistry*, *278*(8), 5509–5512. https://doi.org/10.1074/jbc.C200673200.

Jandhyala, S. M., Talukdar, R., Subramanyam, C., Vuyyuru, H., Sasikala, M., & Reddy, D. N. (2015). Role of the normal gut microbiota. *World Journal of Gastroenterology*, *21*(29), 8836–8847. https://doi.org/10.3748/wjg.v21.i29.8787.

Jiao, S. S., Le Bu, X., Liu, Y. H., Zhu, C., Wang, Q. H., Shen, L. L., et al. (2016). Sex dimorphism profile of Alzheimer's disease-type pathologies in an APP/PS1 mouse model. *Neurotoxicity Research*, *29*(2), 256–266. https://doi.org/10.1007/s12640-015-9589-x.

Jiao, Y., Wu, L., Huntington, N. D., & Zhang, X. (2020). Crosstalk between gut microbiota and innate immunity and its implication in autoimmune diseases. *Frontiers in Immunology*, *11*(February), 1–15. https://doi.org/10.3389/fimmu.2020.00282.

Johansson, M. E. V., Phillipson, M., Petersson, J., Velcich, A., Holm, L., & Hansson, G. C. (2008). The inner of the two Muc2 mucin-dependent mucus layers in colon is devoid of bacteria. *Proceedings of the National Academy of Sciences of the United States of America*, *105*(39), 15064–15069. https://doi.org/10.1073/pnas.0803124105.

Jung, C., Hugot, J.-P., & Barreau, F. (2010). Peyer's patches: The immune sensors of the intestine. *International Journal of Inflammation*, *2010*, 1–12. https://doi.org/10.4061/2010/823710.

Kamada, N., & Núñez, G. (2014). Regulation of the immune system by the resident intestinal bacteria. *Gastroenterology*, *146*(6), 1477–1488. https://doi.org/10.1053/j.gastro.2014.01.060.

Kassubek, J., Juengling, F. D., Kioschies, T., Henkel, K., Karitzky, J., Kramer, B., et al. (2004). Topography of cerebral atrophy in early Huntington's disease: A voxel based morphometric MRI study. *Journal of Neurology, Neurosurgery and Psychiatry*, *75*(2), 213–220.

Khalil, H., Quinn, L., Van Deursen, R., Dawes, H., Playle, R., Rosser, A., et al. (2013). What effect does a structured home-based exercise programme have on people with Huntington's disease? A randomized, controlled pilot study. *Clinical Rehabilitation*, *27*(7), 646–658. https://doi.org/10.1177/0269215512473762.

Killoran, A., & Biglan, K. M. (2014). Current therapeutic options for Huntington's disease: Good clinical practice versus evidence-based approaches? *Movement Disorders*, *29*(11), 1404–1413. https://doi.org/10.1002/mds.26014.

Kirkwood, S. C., Siemers, E., Viken, R., Hodes, M. E., Conneally, P. M., Christian, J. C., & Foroud, T. (2002). Longitudinal personality changes among presymptomatic Huntington disease gene carriers. *Cognitive and Behavioral Neurology*, *15*(3), 192–197.

Kloos, A. D., Fritz, N. E., Kostyk, S. K., Young, G. S., & Kegelmeyer, D. A. (2013). Video game play (dance dance revolution) as a potential exercise therapy in Huntington's disease: A controlled clinical trial. *Clinical Rehabilitation*, *27*(11), 972–982. https://doi.org/10.1177/0269215513487235.

Kobal, J., Matej, K., Koželj, M., & Podnar, S. (2018). Anorectal dysfunction in Presymptomatic mutation carriers and patients with Huntington's disease. *Journal of Huntington's Disease*, 7(3), 259–267. https://doi.org/10.3233/JHD-170280.

Koenig, J. E., Spor, A., Scalfone, N., Fricker, A. D., Stombaugh, J., Knight, R., et al. (2011). Succession of microbial consortia in the developing infant gut microbiome. *Proceedings of the National Academy of Sciences of the United States of America*, *108*(SUPPL. 1), 4578–4585. https://doi.org/10.1073/pnas.1000081107.

Koh, A., De Vadder, F., Kovatcheva-Datchary, P., & Bäckhed, F. (2016). From dietary fiber to host physiology: Short-chain fatty acids as key bacterial metabolites. *Cell*, *165*(6), 1332–1345. https://doi.org/10.1016/j.cell.2016.05.041.

Kong, G., Cao, K. A. L., Judd, L. M., Li, S. S., Renoir, T., & Hannan, A. J. (2018). Microbiome profiling reveals gut dysbiosis in a transgenic mouse model of Huntington's disease. *Neurobiology of Disease*. https://doi.org/10.1016/j.nbd.2018.09.001.

Kong, G., Ellul, S., Narayana, V., Kanojia, K., Ha, H. T. T., Li, S., et al. (2020). *An integrated metagenomics and metabolomics approach implicates the microbiome-gut-brain-axis in the pathogenesis of Huntington's disease transgenic mice*. https://doi.org/10.21203/rs.2.23316/v1.

Lahiri, S., Kim, H., Garcia-Perez, I., Reza, M. M., Martin, K. A., Kundu, P., et al. (2019). The gut microbiota influences skeletal muscle mass and function in mice. *Science Translational Medicine*, *11*(502). eaan5662 https://doi.org/10.1126/scitranslmed.aan5662.

Landles, C., & Bates, G. P. (2004). Huntingtin and the molecular pathogenesis of Huntington's disease. Fourth in molecular medicine review series. *EMBO Reports*, *5*(10), 958–963. https://doi.org/10.1038/sj.embor.7400250.

Langbehn, D. R., Brinkman, R. R., Falush, D., Paulsen, J. S., & Hayden, M. R. (2004). A new model for prediction of the age of onset and penetrance for Huntington's disease based on CAG length. *Clinical Genetics*, *65*(4), 267–277. https://doi.org/10.1111/j.1399-0004.2004.00241.x.

Lawrence, A. D., Hodges, J. R., Rosser, A. E., Kershaw, A., Ffrench-Constant, C., Rubinsztein, D. C., et al. (1998). Evidence for specific cognitive deficits in preclinical Huntington's disease. *Brain*, *121*(7), 1329–1341. https://doi.org/10.1093/brain/121.7.1329.

Le Chatelier, E., Nielsen, T., Qin, J., Prifti, E., Hildebrand, F., Falony, G., et al. (2013). Richness of human gut microbiome correlates with metabolic markers. *Nature*, *500*(7464), 541–546. https://doi.org/10.1038/nature12506.

Lee, H. J., Lee, K. E., Kim, J. K., & Kim, D. H. (2019). Suppression of gut dysbiosis by Bifidobacterium longum alleviates cognitive decline in 5XFAD transgenic and aged mice. *Scientific Reports*, *9*(1), 1–12. https://doi.org/10.1038/s41598-019-48342-7.

Lee, J. M., Ramos, E. M., Lee, J. H., Gillis, T., Mysore, J. S., Hayden, M. R., et al. (2012). CAG repeat expansion in Huntington disease determines age at onset in a fully dominant fashion. *Neurology*, *78*(10), 690–695. https://doi.org/10.1212/WNL.0b013e318249f683.

Lee, J. M., Wheeler, V. C., Chao, M. J., Vonsattel, J. P. G., Pinto, R. M., Lucente, D., et al. (2015). Identification of genetic factors that modify clinical onset of Huntington's disease. *Cell*, *162*(3), 516–526. https://doi.org/10.1016/j.cell.2015.07.003.

Ley, R. E., Bäckhed, F., Turnbaugh, P., Lozupone, C. A., Knight, R. D., & Gordon, J. I. (2005). Obesity alters gut microbial ecology. *Proceedings of the National Academy of Sciences of the United States of America, 102*(31), 11070–11075. https://doi.org/10.1073/pnas.0504978102.

Li, J. Y., Popovic, N., & Brundin, P. (2005). The use of the R6 transgenic mouse models of Huntington's disease in attempts to develop novel therapeutic strategies. *NeuroRx, 2*(3), 447–464. https://doi.org/10.1602/neurorx.2.3.447.

Lin, M. T., & Beal, F. M. (2006). Mitochondrial dysfunction and oxidative stress in neurodegenerative disease. *Nature Review, 443.*

Liot, G., Valette, J., Pépin, J., Flament, J., & Brouillet, E. (2017). Energy defects in Huntington's disease: Why "in vivo" evidence matters. *Biochemical and Biophysical Research Communications, 483*(4), 1084–1095. https://doi.org/10.1016/j.bbrc.2016.09.065.

Lipe, H., & Bird, T. (2009). Journal of the neurological sciences late onset Huntington disease: Clinical and genetic characteristics of 34 cases. *Journal of the Neurological Sciences, 276*(1–2), 159–162. https://doi.org/10.1016/j.jns.2008.09.029.

Liu, Y. X., Qin, Y., Chen, T., Lu, M., Qian, X., Guo, X., et al. (2021). A practical guide to amplicon and metagenomic analysis of microbiome data. *Protein & Cell, 12*(5), 315–330. https://doi.org/10.1007/s13238-020-00724-8.

Liu, Y., Wang, Y., Ni, Y., Tse, M. A., Panagiotou, G., Xu, A., et al. (2020). Gut microbiome fermentation determines the efficacy of exercise for diabetes prevention. *Cell Metabolism, 1–15.* https://doi.org/10.1016/j.cmet.2019.11.001.

Louis, P., & Flint, H. J. (2009). Diversity, metabolism and microbial ecology of butyrate-producing bacteria from the human large intestine. *FEMS Microbiology Letters, 294*(1), 1–8. https://doi.org/10.1111/j.1574-6968.2009.01514.x.

Louis, P., Young, P., Holtrop, G., & Flint, H. J. (2010). Diversity of human colonic butyrate-producing bacteria revealed by analysis of the butyryl-CoA:Acetate CoA-transferase gene. *Environmental Microbiology, 12*(2), 304–314. https://doi.org/10.1111/j.1462-2920.2009.02066.x.

Luczynski, P., Neufeld, K. A. M. V., Oriach, C. S., Clarke, G., Dinan, T. G., & Cryan, J. F. (2016). Growing up in a bubble: Using germ-free animals to assess the influence of the gut microbiota on brain and behavior. *International Journal of Neuropsychopharmacology, 19*(8), 1–17. https://doi.org/10.1093/ijnp/pyw020.

Ma, Z., & Li, W. (2019). How and why men and women differ in their microbiomes: Medical ecology and network analyses of the Microgenderome. *Advanced. Science, 6*(23). https://doi.org/10.1002/advs.201902054.

Macfarlane, S., & Macfarlane, G. T. (2003). Regulation of short-chain fatty acid production. *Proceedings of the Nutrition Society, 62*(1), 67–72. https://doi.org/10.1079/pns2002207.

Mangiarini, L., Sathasivam, K., Seller, M., Cozens, B., Harper, A., Hetherington, C., et al. (1996). Exon 1 of the HD gene with an expanded CAG repeat is sufficient to cause a progressive neurological phenotype in transgenic mice. *Cell, 87*, 493–506. Retrieved from http://ac.els-cdn.com/S0092867400813690/1-s2.0-S0092867400813690-main.pdf?_tid=7db7695e-7afe-11e7-94f8-00000aacb35d&acdnat=1502062090_1e45eb383db00141a08e5daa67072892.

Matsumoto, M., Inoue, R., Tsukahara, T., Ushida, K., Chiji, H., Matsubara, N., et al. (2008). Voluntary running exercise alters microbiota composition and increases n-butyrate concentration in the rat cecum. *Bioscience, Biotechnology, and Biochemistry, 72*(2), 572–576. https://doi.org/10.1271/bbb.70474.

Mawdsley, J. E., & Rampton, D. S. (2005). Psychological stress in IBD: New insights into pathogenic and therapeutic implications. *Gut, 54*(10), 1481–1491. https://doi.org/10.1136/gut.2005.064261.

Maynard, C., & Weinkove, D. (2018). The gut microbiota and ageing. *Biochemistry and Cell Biology of Ageing: Part I Biomedical Science*, 351–371.

McColgan, P., & Tabrizi, S. J. (2018). Huntington's disease: A clinical review. *European Journal of Neurology, 25*(1), 24–34. https://doi.org/10.1111/ene.13413.

McOmish, C. E., & Hannan, A. J. (2007). Enviromimetics: Exploring gene-environment interactions to identify therapeutic targets for brain disorders. *Expert Opinion on Therapeutic Targets*, *11*(7), 899–913. https://doi.org/10.1517/14728222.11.7.899.

Misiak, B., Łoniewski, I., Marlicz, W., Frydecka, D., Szulc, A., Rudzki, L., et al. (2020). The HPA axis dysregulation in severe mental illness: Can we shift the blame to gut microbiota? *Progress in Neuro-Psychopharmacology and Biological Psychiatry*, *102*(February). https://doi.org/10.1016/j.pnpbp.2020.109951.

Mo, C., Renoir, T., & Hannan, A. J. (2016). What's wrong with my mouse cage? Methodological considerations for modeling lifestyle factors and gene-environment interactions in mice. *Journal of Neuroscience Methods*, *265*, 99–108. https://doi.org/10.1016/j.jneumeth.2015.08.008.

Moffitt, H., McPhail, G. D., Woodman, B., Hobbs, C., & Bates, G. P. (2009). Formation of polyglutamine inclusions in a wide range of non-CNS tissues in the HdhQ150 Knock-in mouse model of Huntington's disease. *PLoS One*, *4*(11). https://doi.org/10.1371/journal.pone.0008025.

Moya, A., & Ferrer, M. (2016). Functional redundancy-induced stability of gut microbiota subjected to disturbance. *Trends in Microbiology*, *24*(5), 402–413. https://doi.org/10.1016/j.tim.2016.02.002.

Murphy, K. P. S. J., Carter, R. J., Lione, L. A., Mangiarini, L., Mahal, A., Bates, G. P., et al. (2000). Abnormal synaptic plasticity and impaired spatial cognition in mice transgenic for exon 1 of the human Huntington's disease mutation. *The Journal of Neuroscience*, *20*(13), 5115–5123.

Myers, R. H. (2004). Huntington's disease genetics. *NeuroRx*, *1*, 255–262.

Myrianthopoulos, N. C. (1966). Huntington's chorea. *Journal of Medical Genetics*. https://doi.org/10.1136/jmg.3.4.298.

Nance, M. A., & Myers, R. H. (2001). Juvenile onset Huntington's disease - clinical and research perspectives. *Mental Retardation and Developmental Disabilities Research Reviews*. https://doi.org/10.1002/mrdd.1022.

Nithianantharajah, J., & Hannan, A. J. (2006). Enriched environments, experience-dependent plasticity and disorders of the nervous system. *Nature Reviews Neuroscience*, *7*(9), 697–709. https://doi.org/10.1038/nrn1970.

Pallier, P. N., Drew, C. J. G., & Morton, A. J. (2008). The detection and measurement of locomotor deficits in a transgenic mouse model of Huntington's disease are task- and protocol-dependent: Influence of non-motor factors on locomotor function. *Brain Research Bulletin*, *78*(6), 347–355. https://doi.org/10.1016/j.brainresbull.2008.10.007.

Pandey, K. R., Naik, S. R., & Vakil, B. V. (2015). Probiotics, prebiotics and synbiotics- a review. *Journal of Food Science and Technology*, *52*(12), 7577–7587. https://doi.org/10.1007/s13197-015-1921-1.

Pang, T. Y. C., Stam, N. C., Nithianantharajah, J., Howard, M. L., & Hannan, A. J. (2006). Differential effects of voluntary physical exercise on behavioral and brain-derived neurotrophic factor expression deficits in huntington's disease transgenic mice. *Neuroscience*, *141*(2), 569–584. https://doi.org/10.1016/j.neuroscience.2006.04.013.

Pannaraj, P. S., Li, F., Cerini, C., Bender, J. M., Yang, S., Rollie, A., et al. (2017). Association between breast milk bacterial communities and establishment and development of the infant gut microbiome. *JAMA Pediatrics*, *171*(7), 647–654. https://doi.org/10.1001/jamapediatrics.2017.0378.

Paoli, R. A., Botturi, A., Ciammola, A., Silani, V., Prunas, C., Lucchiari, C., et al. (2017). Neuropsychiatric burden in Huntington's disease. *Brain Sciences*. https://doi.org/10.3390/brainsci7060067.

Parker, A., Fonseca, S., & Carding, S. R. (2020). Gut microbes and metabolites as modulators of blood-brain barrier integrity and brain health. *Gut Microbes*, *11*(2), 135–157. https://doi.org/10.1080/19490976.2019.1638722.

Patil, R. S., Vyas, S. G., Quazi, W. T., Tembhurnikar, H. J., Milmile, P. S., & Umekar, M. J. (2021). The gut microbiome in Huntington disease: A review. *GSC Biological and Pharmaceutical Sciences*, *15*(3), 317–326. https://doi.org/10.30574/gscbps.2021.15.3.0180.

Patterson, E., Cryan, J. F., Fitzgerald, G. F., Ross, R. P., Dinan, T. G., & Stanton, C. (2014). Gut microbiota, the pharmabiotics they produce and host health. *Proceedings of the Nutrition Society*, *760*(September 2014), 477–489. https://doi.org/10.1017/S0029665114001426.

Paulsen, J. S. (2011). Cognitive impairment in Huntington disease: Diagnosis and treatment. *Current Neurology and Neuroscience Reports*, *11*(5), 474–483. https://doi.org/10.1007/s11910-011-0215-x.

Peng, L., Li, Z.-R., Green, R. S., Holzman, I. R., & Li, J. (2009). Butyrate enhances the intestinal barrier by facilitating tight junction assembly via activation of AMP-activated protein kinase in Caco-2 cell monolayers. *The Journal of Nutrition*. https://doi.org/10.3945/jn.109.104638.1619.

Perez-Pardo, P., Dodiya, H. B., Engen, P. A., Forsyth, C. B., Huschens, A. M., Shaikh, M., et al. (2018). Role of TLR4 in the gut-brain axis in Parkinson's disease: A translational study from men to mice. *Gut*, 829–843. https://doi.org/10.1136/gutjnl-2018-316844.

Peterseén, Å., Gil, J., Maat-Schieman, M. L. C., Björkqvist, M., Tanila, H., Araújo, I. M., et al. (2005). Orexin loss in Huntington's disease. *Human Molecular Genetics*, *14*(1), 39–47. https://doi.org/10.1093/hmg/ddi004.

Petersén, Å., & Björkqvist, M. (2006). Hypothalamic-endocrine aspects in Huntington's disease. *European Journal of Neuroscience*, *24*(4), 961–967. https://doi.org/10.1111/j.1460-9568.2006.04985.x.

Petnicki-Ocwieja, T., Hrncir, T., Liu, Y. J., Biswas, A., Hudcovic, T., Tlaskalova-Hogenova, H., et al. (2009). Nod2 is required for the regulation of commensal microbiota in the intestine. *Proceedings of the National Academy of Sciences of the United States of America*, *106*(37), 15813–15818. https://doi.org/10.1073/pnas.0907722106.

Pla, P., Orvoen, S., Saudou, F., David, D. J., & Humbert, S. (2014). Mood disorders in Huntington's disease: From behavior to cellular and molecular mechanisms. *Frontiers in Behavioral Neuroscience*, *8*(APR), 1–15. https://doi.org/10.3389/fnbeh.2014.00135.

Radulescu, C. I., Garcia-Miralles, M., Sidik, H., Bardile, C. F., Yusof, N. A. B. M., Lee, H. U., et al. (2020). Reprint of: Manipulation of microbiota reveals altered callosal myelination and white matter plasticity in a model of Huntington disease. *Neurobiology of Disease*, *135*(September 2018), 104744. https://doi.org/10.1016/j.nbd.2020.104744.

Rakoff-Nahoum, S., Paglino, J., Eslami-Varzaneh, F., Edberg, S., & Medzhitov, R. (2004). Recognition of commensal microflora by toll-like receptors is required for intestinal homeostasis Seth. *Journal of the Torrey Botanical Society*, *130*(3), 147. https://doi.org/10.2307/3557550.

Ramos, A. R. S., & Garrett, C. (2017). Huntington's disease: Premotor phase. *Neurodegenerative Diseases*, *17*(6), 313–322.

Ranjan, R., Rani, A., Metwally, A., McGee, H. S., & Perkins, D. L. (2016). Analysis of the microbiome: Advantages of whole genome shotgun versus 16S amplicon sequencing. *Biochemical and Biophysical Research Communications*, *469*(4), 967–977. https://doi.org/10.1016/j.bbrc.2015.12.083.

Ransome, M. I., & Hannan, A. J. (2013). Impaired basal and running-induced hippocampal neurogenesis coincides with reduced Akt signaling in adult R6/1 HD mice. *Molecular and Cellular Neurosciences*, *54*, 93–107. https://doi.org/10.1016/j.mcn.2013.01.005.

Reiner, A., Dragatsis, I., Zeitlin, S., & Goldowitz, D. (2003). Wild-type huntingtin plays a role in brain development and neuronal survival. *Molecular Neurobiology*, *28*(3), 259–275. https://doi.org/10.1385/MN:28:3:259.

Renoir, T., Pang, T. Y. C., Zajac, M. S., Chan, G., Du, X., Leang, L., et al. (2011). Treatment of depressive-like behaviour in Huntington's disease mice by chronic

sertraline and exercise. *British Journal of Pharmacology*, *165*(5), 1375–1389. https://doi.org/10.1111/j.1476-5381.2011.01567.x.
Rios-Covian, D., González, S., Nogacka, A. M., Arboleya, S., Salazar, N., Gueimonde, M., et al. (2020). An overview on fecal branched short-chain fatty acids along human life and as related with body mass index: Associated dietary and anthropometric factors. *Frontiers in Microbiology*, *11*(May), 1–9. https://doi.org/10.3389/fmicb.2020.00973.
Robbins, A. O., Ho, A. K., & Barker, R. A. (2006). Weight changes in Huntington's disease [6]. *European Journal of Neurology*, *13*(8), 2006. https://doi.org/10.1111/j.1468-1331.2006.01319.x.
Roos, R. A. (2010). Huntington's disease: A clinical review. *Orphanet Journal of Rare Diseases*, *5*. https://doi.org/10.1186/1750-1172-5-40.
Roos, R. A. C., Vegter-van der Vlis, M., Hermans, J., Elshove, H. M., Moll, A. C., van de Kamp, J. J. P., et al. (1991). Age at onset in Huntington's disease: Effect of line of inheritance and patient's sex. *Journal of Medical Genetics*, *28*(8), 515–519. https://doi.org/10.1136/jmg.28.8.515.
Ross, C. A., & Tabrizi, S. J. (2011). Huntington's disease: From molecular pathogenesis to clinical treatment. *The Lancet Neurology*, *10*(1), 83–98. https://doi.org/10.1016/S1474-4422(10)70245-3.
Rowin, J., Xia, Y., Jung, B., & Sun, J. (2017). Gut inflammation and dysbiosis in human motor neuron disease. *Physiological Reports*, *5*(18), 1–6. https://doi.org/10.14814/phy2.13443.
Sampson, T. R., Debelius, J. W., Thron, T., Janssen, S., Shastri, G. G., Ilhan, Z. E., et al. (2016). Gut microbiota regulate motor deficits and neuroinflammation in a model of Parkinson's disease. *Cell*, *167*(6), 1469–1480. https://doi.org/10.1016/j.cell.2016.11.018.
Sanberg, P. R., Fibiger, H. C., & Mark, R. F. (1981). Body weight and dietary factors in Huntington's disease patients compared with matched controls. *Medical Journal of Australia*, *1*(8), 407–409.
Sassone, J., Colciago, C., Cislaghi, G., Silani, V., & Ciammola, A. (2009). Huntington's disease: The current state of research with peripheral tissues. *Experimental Neurology*, *219*(2), 385–397. https://doi.org/10.1016/j.expneurol.2009.05.012.
Sathasivam, K., Hobbs, C., Mangiarini, L., Mahal, A., Turmaine, M., Doherty, P., et al. (1999). Transgenic models of Huntington's disease. *Philosophical Transactions of the Royal Society of London. Series B, Biological Sciences*, *354*, 963–969.
Sathasivam, K., Hobbs, C., Turmaine, M., Mangiarini, L., Mahal, A., Bertaux, F., et al. (1999). Formation of polyglutamine inclusions in non-CNS tissue. *Human Molecular Genetics*, *8*(5), 813–822.
Scheperjans, F., Aho, V., Pereira, P. A. B., Koskinen, K., Paulin, L., Pekkonen, E., et al. (2015). Gut microbiota are related to Parkinson's disease and clinical phenotype. *Movement Disorders*, *30*(3), 350–358. https://doi.org/10.1002/mds.26069.
Schwiertz, A., Gruhl, B., Löbnitz, M., Michel, P., Radke, M., & Blaut, M. (2003). Development of the intestinal bacterial composition in hospitalized preterm infants in comparison with breast-fed, full-term infants. *Pediatric Research*, *54*(3), 393–399. https://doi.org/10.1203/01.PDR.0000078274.74607.7A.
Semaka, A., Kay, C., Doty, C., Collins, J. A., Bijlsma, E. K., Richards, F., et al. (2013). CAG size-specific risk estimates for intermediate allele repeat instability in Huntington disease. *Journal of Medical Genetics*, *50*(10), 696–703. https://doi.org/10.1136/jmedgenet-2013-101796.
Severance, E. G., & Yolken, R. H. (2019). Tracking a dysregulated gut-brain axis with biomarkers of the microbiome. *Biomarkers in Neuropsychiatry*. https://doi.org/10.1016/j.bionps.2019.100009.
Shaffer, J. J., Ghayoor, A., Long, J. D., Kim, R. E. Y., Lourens, S., O'Donnell, L. J., et al. (2017). Longitudinal diffusion changes in prodromal and early HD: Evidence of

white-matter tract deterioration. *Human Brain Mapping*, *38*(3), 1460–1477. https://doi.org/10.1002/hbm.23465.

Shanahan, F. (2002). The host-microbe interface within the gut. *Baillière's Best Practice & Research. Clinical Gastroenterology*, *16*(6), 915–931. https://doi.org/10.1053/bega.2002.0342.

Sharma, S., & Taliyan, R. (2015). Transcriptional dysregulation in Huntington's disease: The role of histone deacetylases. *Pharmacological Research*, *100*, 157–169. https://doi.org/10.1016/j.phrs.2015.08.002.

Singh, Y., El-Hadidi, M., Admard, J., Wassouf, Z., Schulze-Hentrich, J. M., Kohlhofer, U., et al. (2019). Enriched environmental conditions modify the gut microbiome composition and fecal markers of inflammation in parkinson's disease. *Frontiers in Neuroscience*, *13*(OCT), 1–18. https://doi.org/10.3389/fnins.2019.01032.

Smith, P. M., Howitt, M. R., Panikov, N., Michaud, M., Gallini, C. A., Bohlooly-Y, M., et al. (2013). The microbial metabolites, short-chain fatty acids, regulate colonic T reg cell homeostasis. *Science*, *341*(6145), 569–573. https://doi.org/10.1126/science.1241165.

Spielman, L. J., Gibson, D. L., & Klegeris, A. (2018). Unhealthy gut, unhealthy brain: The role of the intestinal microbiota in neurodegenerative diseases. *Neurochemistry International*, *120*(May), 149–163. https://doi.org/10.1016/j.neuint.2018.08.005.

Spielman, L. J., Little, J. P., & Klegeris, A. (2014). Inflammation and insulin/IGF-1 resistance as the possible link between obesity and neurodegeneration. *Journal of Neuroimmunology*, *273*(1–2), 8–21. https://doi.org/10.1016/j.jneuroim.2014.06.004.

Stan, T. L., Soylu-Kucharz, R., Burleigh, S., Prykhodko, O., Cao, L., Franke, N., et al. (2020). Increased intestinal permeability and gut dysbiosis in the R6/2 mouse model of Huntington's disease. *Scientific Reports*, *10*(1), 1–9. https://doi.org/10.1038/s41598-020-75229-9.

Sudo, N., Chida, Y., Aiba, Y., Sonoda, J., Oyama, N., Yu, X. N., et al. (2004). Postnatal microbial colonization programs the hypothalamic-pituitary-adrenal system for stress response in mice. *Journal of Physiology*, *558*(1), 263–275. https://doi.org/10.1113/jphysiol.2004.063388.

Tabrizi, S. J., Scahill, R. I., Durr, A., Roos, R. A. C., Leavitt, B. R., Jones, R., et al. (2011). Biological and clinical changes in premanifest and early stage Huntington's disease in the TRACK-HD study: The 12-month longitudinal analysis. *The Lancet Neurology*, *10*(1), 31–42. https://doi.org/10.1016/S1474-4422(10)70276-3.

Tanoue, T., Morita, S., Plichta, D. R., Skelly, A. N., Suda, W., Sugiura, Y., et al. (2019). A defined commensal consortium elicits CD8 T cells and anti-cancer immunity. *Nature*, *565*(7741), 600–605. https://doi.org/10.1038/s41586-019-0878-z.

The Huntington Disease Research Group. (1993). A novel gene containing a trinucleotide repeat that is expanded and unstable on Huntington's disease chromosomes. *Cell*, *72*(6), 971–983. https://doi.org/10.1016/0092-8674(93)90585-E.

Tiryaki, E., & Horak, H. A. (2014). ALS and other motor neuron diseases. *CONTINUUM Lifelong Learning in Neurology*, *20*(5), 1185–1207. https://doi.org/10.1212/01.CON.0000455886.14298.a4.

Toor, D., Wasson, M. K., Kumar, P., Karthikeyan, G., Kaushik, N. K., Goel, C., et al. (2019). Dysbiosis disrupts gut immune homeostasis and promotes gastric diseases. *International Journal of Molecular Sciences*, *20*(10), 1–14. https://doi.org/10.3390/ijms20102432.

Trejo, A., Tarrats, R. M., Alonso, M. E., Boll, M. C., Ochoa, A., & Velásquez, L. (2004). Assessment of the nutrition status of patients with Huntington's disease. *Nutrition*, *20*(2), 192–196. https://doi.org/10.1016/j.nut.2003.10.007.

Trembath, M. K., Horton, Z. A., Tippett, L., Hogg, V., Collins, V. R., Churchyard, A., et al. (2010). A retrospective study of the impact of lifestyle on age at onset of Huntington disease. *Movement Disorders*, *25*(10), 1444–1450. https://doi.org/10.1002/mds.23108.

Turnbaugh, P. J., Ley, R. E., Mahowald, M. A., Magrini, V., Mardis, E. R., & Gordon, J. I. (2006). An obesity-associated gut microbiome with increased capacity for energy harvest. *Nature, 444*(7122), 1027–1031. https://doi.org/10.1038/nature05414.

Turnbull, A. V., & Rivier, C. (1995). Regulation of the hpa axis by cytokines. *Brain, Behavior, and Immunity*. https://doi.org/10.1006/brbi.1995.1026.

Valdes, A. M., Walter, J., Segal, E., & Spector, T. D. (2018). Role of the gut microbiota in nutrition and health. *BMJ (Online), 361*, 36–44. https://doi.org/10.1136/bmj.k2179.

van de Haar, H. J., Jansen, J. F. A., van Osch, M. J. P., van Buchem, M. A., Muller, M., Wong, S. M., et al. (2016). Neurovascular unit impairment in early Alzheimer's disease measured with magnetic resonance imaging. *Neurobiology of Aging, 45*, 190–196.

van de Wouw, M., Boehme, M., Lyte, J. M., Wiley, N., Strain, C., O'Sullivan, O., et al. (2018). Short-chain fatty acids: Microbial metabolites that alleviate stress-induced brain–gut axis alterations. *Journal of Physiology, 596*(20), 4923–4944. https://doi.org/10.1113/JP276431.

Van Dellen, A., Blakemore, C., Deacon, R., York, D., & Hannan, A. J. (2000). Delaying the onset of Huntington's in mice. *Nature, 404*(6779), 721–722. https://doi.org/10.1038/35008142.

van der Burg, J. M. M., Gardiner, S. L., Ludolph, A. C., Landwehrmeyer, G. B., Roos, R. A. C., & Aziz, N. A. (2017). Body weight is a robust predictor of clinical progression in Huntington disease. *Annals of Neurology, 82*(3), 479–483. https://doi.org/10.1002/ana.25007.

Van Der Burg, J. M. M., Winqvist, A., Aziz, N. A., Maat-Schieman, M. L. C., Roos, R. A. C., Bates, G. P., et al. (2011). Gastrointestinal dysfunction contributes to weight loss in Huntington's disease mice. *Neurobiology of Disease, 44*(1), 1–8. https://doi.org/10.1016/j.nbd.2011.05.006.

van der Lugt, B., Rusli, F., Lute, C., Lamprakis, A., Salazar, E., Boekschoten, M. V., et al. (2018). Integrative analysis of gut microbiota composition, host colonic gene expression and intraluminal metabolites in aging C57BL/6J mice. *Aging, 10*(5), 930–950. https://doi.org/10.18632/aging.101439.

Venegas, D. P., De La Fuente, M. K., Landskron, G., González, M. J., Quera, R., Dijkstra, G., et al. (2019). Short chain fatty acids (SCFAs)mediated gut epithelial and immune regulation and its relevance for inflammatory bowel diseases. *Frontiers in Immunology, 10*(MAR). https://doi.org/10.3389/fimmu.2019.00277.

Vicentini, F. A., Keenan, C. M., Wallace, L. E., Woods, C., Cavin, J. B., Flockton, A. R., et al. (2021). Intestinal microbiota shapes gut physiology and regulates enteric neurons and glia. *Microbiome, 9*(1), 1–24. https://doi.org/10.1186/s40168-021-01165-z.

Vogt, N. M., Kerby, R. L., Dill-McFarland, K. A., Harding, S. J., Merluzzi, A. P., Johnson, S. C., et al. (2017). Gut microbiome alterations in Alzheimer's disease. *Scientific Reports*, 7(1), 1–11. https://doi.org/10.1038/s41598-017-13601-y.

Walker, W. A. (2017). *Dysbiosis. The microbiota in gastrointestinal pathophysiology*. Elsevier Inc. https://doi.org/10.1016/B978-0-12-804024-9/00025-2.

Wang, H. B., Wang, P. Y., Wang, X., Wan, Y. L., & Liu, Y. C. (2012). Butyrate enhances intestinal epithelial barrier function via up-regulation of tight junction protein claudin-1 transcription. *Digestive Diseases and Sciences, 57*(12), 3126–3135. https://doi.org/10.1007/s10620-012-2259-4.

Wasser, C. I., Mercieca, E.-C., Kong, G., Hannan, A. J., McKeown, S. J., Glikmann-Johnston, Y., et al. (2020). Gut dysbiosis in Huntington's disease: Associations among gut microbiota, cognitive performance and clinical outcomes. *Brain Communications, 2*(2), 1–13. https://doi.org/10.1093/braincomms/fcaa110.

Wexler, N. S. (2004). Venezuelan kindreds reveal that genetic and environmental factors modulate Huntington's disease age od onset. *Fetal Diagnosis and Therapy, 19*(3), 251–260. https://doi.org/10.1159/000076707.

Wheeler, V. C., Persichetti, F., McNeil, S. M., Mysore, J. S., Mysore, S. S., MacDonald, M. E., et al. (2007). Factors associated with HD CAG repeat instability in Huntington disease. *Journal of Medical Genetics*, *44*(11), 695–701. https://doi.org/10.1136/jmg.2007.050930.

White, J. K., Auerbach, W., Duyao, M. P., Vonsattel, J.-P., Gusella, J. F., Joyner, A. L., et al. (1997). Huntingtin is required for neurogenesis and is not impaired by the Huntington's disease CAG expansion. *Nature Genetics*, *17*(4), 404–410.

Wood, N. I., Goodman, A. O. G., van der Burg, J. M. M., Gazeau, V., Brundin, P., Björkqvist, M., et al. (2008). Increased thirst and drinking in Huntington's disease and the R6/2 mouse. *Brain Research Bulletin*, *76*(1–2), 70–79. https://doi.org/10.1016/j.brainresbull.2007.12.007.

Wu, H. J., & Wu, E. (2012). The role of gut microbiota in immune homeostasis and autoimmunity. *Gut Microbes*, *3*(1). https://doi.org/10.4161/gmic.19320.

Wu, S., Yi, J., Zhang, Y. G., Zhou, J., & Sun, J. (2015). Leaky intestine and impaired microbiome in an amyotrophic lateral sclerosis mouse model. *Physiological Reports*, *3*(4), 1–10. https://doi.org/10.14814/phy2.12356.

Youssov, K., Dolbeau, G., Maison, P., Boissé, M. F., De Langavant, L. C., Roos, R. A. C., et al. (2013). Unified Huntington's disease rating scale for advanced patients: Validation and follow-up study. *Movement Disorders*, *28*(12), 1717–1723. https://doi.org/10.1002/mds.25654.

Zhang, J., Saur, T., Duke, A. N., Grant, S. G. N., Platt, D. M., Rowlett, J. K., et al. (2014). Motor impairments, striatal degeneration, and altered dopamine-glutamate interplay in mice lacking PSD-95. *Journal of Neurogenetics*, *28*(1–2), 98–111. https://doi.org/10.3109/01677063.2014.892486.

Zielonka, D. (2018). Factors contributing to clinical picture and progression of Huntington's disease. *Neural Regeneration Research*, *13*(8), 1364–1365. https://doi.org/10.4103/1673-5374.235241.

Zielonka, D., Marinus, J., Roos, R. A. C., De Michele, G., Di Donato, S., Putter, H., et al. (2013). The influence of gender on phenotype and disease progression in patients with Huntington's disease. *Parkinsonism & Related Disorders*, *19*(2), 192–197. https://doi.org/10.1016/j.parkreldis.2012.09.012.

Zinzi, P., Salmaso, D., De Grandis, R., Graziani, G., Maceroni, S., Bentivoglio, A., et al. (2007). Effects of an intensive rehabilitation programme on patients with Huntington's disease: A pilot study. *Clinical Rehabilitation*, *21*(7), 603–613. https://doi.org/10.1177/0269215507075495.

Zuccato, C., Liber, D., Ramos, C., Tarditi, A., Rigamonti, D., Tartari, M., et al. (2005). Progressive loss of BDNF in a mouse model of Huntington's disease and rescue by BDNF delivery. *Pharmacological Research*, *52*(2), 133–139. https://doi.org/10.1016/j.phrs.2005.01.001.

CHAPTER SEVEN

Role of the gut microbiome in multiple sclerosis: From etiology to therapeutics

Shailesh K. Shahi[a,b], Meeta Yadav[a,b], Sudeep Ghimire[a,b], and Ashutosh K. Mangalam[a,b,*]
[a]Department of Pathology, University of Iowa, Iowa City, IA, United States
[b]Iowa City VA Health System, Iowa City, IA, United States
*Corresponding author: e-mail address: ashutosh-mangalam@uiowa.edu

Contents

Abstract

Multiple sclerosis (MS) is a chronic inflammatory disease of the CNS that affects around one million people in the United States. Predisposition or protection from this disease is linked with both genetic and environmental factors. In recent years, gut microbiome has emerged as an important environmental factor in the pathobiology of MS. The gut microbiome supports various physiologic functions, including the development and maintenance of the host immune system, the perturbation of which is known as dysbiosis and has been linked with multiple diseases including MS. We and others have shown that people with MS (PwMS) have gut dysbiosis that is characterized by specific gut bacteria being enriched or depleted. Consequently, there is an emphasis on determining the mechanism(s) through which gut bacteria and/or their metabolites alter the course of MS through their ability to provide protection, predispose individuals, or promote disease progression. Improving our understanding of these mechanisms

International Review of Neurobiology, Volume 167
ISSN 0074-7742
https://doi.org/10.1016/bs.irn.2022.06.001

will allow us to harness the enormous potential of the gut microbiome as a diagnostic and/or therapeutic agent. In this chapter, we will discuss current advances in microbiome research in the context of MS, including a review of specific bacteria that are currently linked with this disease, potential mechanisms of disease pathogenesis, and the utility of microbiome-based therapy for PwMS.

Multiple sclerosis (MS) is a chronic autoimmune inflammatory disease of the central nervous system (CNS) that affects around one million people in the United States (US) (Briggs & Hill, 2019; Culpepper, Marrie et al., 2019). The economic impact of MS on the US economy amounts to billions of dollars spent per year on medical care and indirect expenses, which includes lost wages and productivity (Campbell et al., 2014; Jones, Pike, Marshall, & Ye, 2016). Similar to other autoimmune diseases, both genetic and environmental factors are linked with affording resistance to, and protection from, the disease. Among genetic factors, the HLA region on chromosome six is strongly linked with susceptibility and resistance to MS (Alvarado-de la Barrera et al., 2000; Barcellos et al., 2003; DeLuca et al., 2007; Dyment et al., 2005; Oksenberg et al., 2004; Olerup & Hillert, 1991; Ramagopalan et al., 2007, 2008). However, a concordance rate for MS of only 30% among monozygotic twins suggests environmental factor(s) are also crucial for predisposing individuals to this disease (Kuusisto et al., 2008). Currently, there is no consensus on environmental contributors, as multiple factors such as infectious agents (e.g., Epstein-Barr Virus), bacterial infection, lack of vitamin D, and geographic location (higher incidence in area closer to polar region) have all been linked with susceptibility and resistance to MS.

One of the proposed hypotheses focused on the effect of the environment on disease incidence is the hygiene hypothesis, which posits a link between the increased incidence of inflammatory diseases in industrialized/developed nations. According to this hypothesis, the strong emphasis on cleanliness and hygiene aimed at reducing the incidence of infectious diseases in developed countries might contribute to an increased predisposition to inflammatory diseases. In recent years, the hygiene hypothesis has been revised to suggest that industrialization has also resulted in the loss or alteration of trillions of microbes living in and on our body (i.e., the microbiome), with the maximum density of these microbes observed in the gut. The microbiome plays a crucial role in maintaining a healthy state by contributing to a number of physiological processes such as food digestion, nutrient absorption, thermogenesis, regulation of the immune system, generation of neurotransmitters, among others. Factors such as acidification of water, urban living, loss of green space, increased use of chemicals in our daily life, including in foods, and heavy use of antibiotics all disrupt homeostasis of the gut microbiota, thus inducing a state of dysbiosis. Consequently, this can result in the predisposition and propagation of inflammatory diseases. Indeed, a number of studies, including the one from our group, have shown that people with MS (PwMS) exhibit an altered gut microbiota compared to age- and sex-matched healthy controls (HC). Similar microbial dysbiosis has been reported for other inflammatory and neurological diseases such as inflammatory bowel disease (IBD) (Shaw et al., 2016; Tamboli, Neut, Desreumaux, & Colombel, 2004; Willing et al., 2010), rheumatoid arthritis (RA) (Chen et al., 2016; Scher & Abramson, 2013; Zhang et al., 2015), Ankylosing Spondylitis (AS) (Wen et al., 2017; Zhou et al., 2020), Lupus (He, Shao, Li, Xie, & Wen, 2016; Hevia et al., 2014; Luo et al., 2018), ASD (Finegold et al., 2002; Ha et al., 2021; Kang et al., 2017; Levi Mortera et al., 2022), Alzheimer's disease (AD) (Liu et al., 2019; Vogt et al., 2017; Wu et al., 2021; Zhuang et al., 2018) and Parkinson disease (PD) (Qian et al., 2018; Vascellari et al., 2020; Zhou et al., 2020). In this chapter, we will review the microbiome association studies that have been conducted in PwMS and the animal model of this disease, experimental autoimmune encephalomyelitis (EAE). This will be followed by a discussion of the potential mechanisms by which the gut microbiome can influence disease outcomes in inflammatory disorders. Finally, we will explore the use of bacteria as a potential treatment option for MS.

1. Microbiome associations studies in MS

The role of the microbiome in MS was first shown by Miyake et. al., who compared fecal microbiota from patients with relapsing remitting MS (RRMS) (n = 20) to that of healthy controls (HC, n = 40) in Japan. The study reported a dysbiosis in RRMS patients compared to HC, with depletion of *Clostridium* as well as *Prevotella, Fecalibacterium, Sutterella* and *Bacteroides* genera (Miyake et al., 2015). RRMS patients also showed an enrichment of *Bifidobacterium* and *Streptococcus*. Additionally, a collaborative study by Dr. Weiner's group at Harvard and our group (Dr. Mangalam, Mayo Clinic Rochester and University of Iowa) showed that PwMS had gut dysbiosis with enrichment and depletion of certain gut bacteria. Similar to the findings reported previously in PwMS from Japan, we also observed a decreased abundance of *Prevotella* and increased abundance of *Blautia*. The abundance of other bacteria such as *Parabacteroides, Adlercreutzia, Collinsella, and Lactobacillus* was also increased. Similar additional studies also showed gut dysbiosis in PwMS, the findings of which are summarized in Table 1. Despite variability among these studies, which were conducted across different geographical regions (North America, Europe, and Asia), there were some commonalities. For example, several studies reported that bacteria such as *Prevotella, Parabacteroides, Adlercreutzia, Collinsella, Fecalibacterium, Lactobacillus, Blautia,* and *Akkermansia* were all increased in abundance among PwMS. However, these studies used fecal samples as a surrogate for measuring bacteria levels in the gut. Thus, the findings in a study from Italy are important to consider as the gut microbiota was profiled from mucosal biopsies obtained from PwMS (Cosorich et al., 2017). This study showed an inverse correlation between the relative abundance of *Prevotella* and disease severity. Additionally, this study also showed an inverse correlation between the abundance of *Prevotella* and IL-17 expressing CD4+ T helper cells (Th17 cells). This was the first study to directly correlate disease severity with the frequency of Th17 cells or the abundance of a specific gut bacteria (*Prevotella* spp). Four additional studies also similarly showed a reduced abundance of *Prevotella* genera in PwMS (Table 1).

In addition of microbiome profiling, germ-free (GF) mice have been used as an important tool for studying the human microbiome. In addition to profiling the microbiome of PwMS, two studies published in the same issue of PNAS reported possible mechanisms through which gut bacteria might contribute to pathobiology of MS. It is important to note that these studies used germ-free (GF) mice, which lack any bacteria and are thus an important tool for studying the human microbiome in isolation. One study from the Branzini laboratory at the University of California at San Francisco, PwMS had an increased abundance of *Akkermansia muciniphila* and *Acinetobacter calcoaceticus* in addition to a reduced abundance of *Parabacteroides distasonis* (Cekanaviciute et al., 2017). An accompanying study from the Wekerle laboratory in Germany analyzed the fecal microbiome of 34 monozygotic twin pairs discordant for MS and showed that those with the disease had enrichment of *Akkermansia*. When the fecal microbiota was transferred from healthy twins into those with MS, *Adlercreutzia* and *Sutterella* both became enriched (Berer et al., 2017). Most importantly, both studies showed that the fecal microbiota from PwMS but not HC could induce severe disease in GF mice. These were the first studies to show an expansion of distinct bacteria in GF mice

Table 1 Comparison of MS microbiome studies.

# Samples (Country) Reference	Type of sample analysis method	Lower abundance in MS patients	Increased abundance in MS patients
RRMS (n=31) HC (n=36) (USA) (Chen et al., 2016)	Fecal Bacterial 16S rRNA V3–V5	*Prevotella* *Parabacteroides* *Adlercreutzia* *Collinsella* *Lactobacillus*	*Flavobacterium* *Pedobacter* *Pseudomonas* *Mycoplasma* *Haemophilus* *Blautia* *Dorea*
RRMS (n=60) HC (n=43) (USA) (Jangi et al., 2016)	Fecal Bacterial 16S rRNA V3–5 and V4	*Butyricimona* *Prevotella* *Parabacteroides*	*Methanobrevibacter* *Akkermansia*
RRMS (n=20) HC (n=40) (Japan) (Miyake et al., 2015)	Fecal Bacterial 16S rRNA V1-V2	*Bacteroides* (*B. stercoris*, *B. coprocola*, *B. coprophilus*) *Fecalibacterium* *Prevotella* (*P. copri*) *Anaerostipes* *Clostridium* *Sutterella* (*S. wadsworthensis*)	*Bifidobacterium* *Streptococcus* *Streptococcus thermophilus* *Eggerthella lenta*
RRMS (n=30) HC (n=14) (UK) (Yu, Peng, Zhang, Yin, & Zhou, 2022)	Fecal Bacterial 16S rRNA V4	*Eubacterium eligens* *Prevotella copri* uncultured *Bacteroides* sp. uncultured *alpha Proteobacterium* uncultured *Pseudomonas* sp	*Faecalibacterium* sp. *Ruminococcus* sp. uncultured *Oscillospiraceae bacterium* uncultured *Blautia* sp. uncultured *Sinorhizobium* sp *Anaerostipes* sp. *Clostridium bolteae* uncultured *Dialister* sp. *Alistipes onderdonkii* *Bifidobacterium longum* *Coriobacterium* sp

Continued

Table 1 Comparison of MS microbiome studies.—cont'd

# Samples (Country) Reference	Type of sample analysis method	Lower abundance in MS patients	Increased abundance in MS patients
RRMS (n=71) HC (n=71) (USA) (, Cekanaviciute et al., 2017)	Fecal Bacterial 16S rRNA V4	*Parabacteroides distasonis*	*Akkermansia muciniphila* *Acinetobacter calcoaceticus*
RRMS (n=24) HC (n=25) (USA) (Cantoni et al., 2022)	Fecal Bacterial 16S rRNA V1–V3	*Faecalibacterium* *Prevotella* unclassified *Lachnospiraceae* *Anaerostipes*	
RRMS (n=24) HC (n=25) (USA) (Cantoni et al., 2022)	Fecal Metagenomic	*Bifidobacterium longum* *Clostridium leptum* *Faecalibacterium prausnitzii* Bacteroides thetaiotaomicron unclassified *Parabacteroides* unclassified *Escherichia*	
RRMS (n=19) HC (n=17) (Italy) (Cosorich et al., 2017)	Intestinal tissue 16S rRNA V3-V5	*Prevotella*	*Streptococcus*
Monozygotic twin pairs: MS (n=34) Unaffected twin (n=34) (Germany) (Berer et al., 2017)	Fecal Bacterial 16S rRNA V3-V5		*Akkermansia*

POMS (n = 20) HC (n = 20) (Canada) (Mirza et al., 2022)	Fecal Metagenomic	*Pseudomonas Corrgata* *Haemophilus Influenzae* *Kochuria* *Spingopyxis sp.QXT-31*	*Methanobrevibacter smithi* *Moribacter* *Mycoplasma bovoculi* *Diaphorobacter polyhydrobutyrativorans* *Methanobrevibacter millerae* *Acanobacterium sp 2701*
pWMS (n = 25) HC (n = 22) (USA) (Shah et al., 2021)	Fecal Fungal 18S rRNA ITS 1		*Saccharomyces* *Aspergillus*
RRMS (n = 20) HC (n = 33) (USA) (Yadav et al., 2021)	Fecal Fungal 18S rRNA ITS 2	*Ascomycota* *Saccharomyces*	*Basidiomycota* *Candida* *Epicoccum*

RRMS, relapsing-remitting multiple sclerosis; HC, healthy controls; n, number; PwMS, people with MS; POMS, pediatric onset MS; rRNA, ribosomal RNA; V, variable region of 16S rRNA; ITS, internal transcribed spacer.

transplanted with fecal microbiota from PwMS vs. HCs and a causal role of the gut microbiota from patients with this disease.

Although there were some common bacteria among different studies across geographical regions, as described above, this was accompanied by a high degree of variability. This variability could be a true reflection of the differences in the gut microbiota that arise in different locations and patient populations. However, differences in the abundance of specific gut bacteria could also be due to the use of 16S primers that are specific to variable regions of rRNA. There are 9 variable regions in $\sim$1500 nucleotide-long 16S rRNA and current high-throughput sequencing technology can only sequence 500–600 bases without significant error. Therefore, primers specific to different V-regions have been used in some microbiome profiling studies and could be responsible for some of the variability in the bacteria taxa reported among published studies. It is worth noting that as marker-based profiling, such as 16S rRNA metagenomic sequencing, lacks information on functional/metabolic capabilities of the microbiome, more recent studies (including two summarized in Table 1) have focused instead on using shotgun metagenomic sequencing to identify bacterial strains and pathways that are enriched or depleted in PwMS. In this regard, we recently analyzed the fecal microbiome from PwMS in Iowa and observed an increase in *Eggerthella* spp. in addition to *Akkermansia* and *Dorea* (unpublished observation). Thus, we anticipate more studies will use shotgun metagenomic sequencing in next few years to improve our understanding of the bacterial metabolic pathways that are linked with predisposition to, or protection from, MS.

Although the gut microbiome data of fungi and viruses are not as advanced as that for bacteria, recent studies, including from our group, profiled the fungal microbiome (mycobiome) in PwMS and showed fungi including *Saccharomyces* and *Aspergillus* were enriched, whereas Ascomycota phyla and *Saccharomyces* genera were reduced (Shah et al., 2021; Yadav et al., 2021). We also reported an increased abundance of *Candida* and *Epicoccum*. Of note, studies of other inflammatory diseases such as IBD and RA also showed changes in the mycobiome of these patients, characterized by an enrichment of *Candida* and a reduction in *Saccharomyces* genera. However, mycobiome profiling is still in its infancy and additional studies are required to further understand its role in MS pathogenesis.

Collectively, data analyzing the microbiome from PwMS suggests that this disease is associated with gut dysbiosis, characterized by a loss of *Prevotella, Parabacteroides, Adlercreutzia and Lactobacillus* and/or enrichment of *Akkermansia, Dorea, Eggerthella* and *Archaea-Methanobrevibacter* (summarized in Table 1). Additionally, PwMS exhibit fungal dysbiosis, and, an altered abundance of *Candida* and *Saccharomyces*. However, further studies are needed to identify and characterize the role of fungi in this patient population. These changes in the abundance of gut bacteria in PwMS could predispose and/or exacerbate disease by inducing and/or propagating a pro-inflammatory response in the intestine, which can compromise gut barrier integrity and induce a systemic inflammatory state that ultimately drives CNS autoimmunity. In the subsequent sections, we will discuss the potential mechanisms through which the gut microbiota influences the pathogenesis of MS.

2. Mechanisms of gut dysbiosis

The human gut is colonized by a large number of microorganisms ($\sim 10^{13}$ bacteria) that support various physiologic functions. Although there is an appreciation for the link between the gut microbiome changes and MS, the mechanisms that drive this association are still unclear. In a healthy individual, there is a diverse community of bacteria that helps maintain gut homeostasis, either by interacting with intestinal epithelial cells (IECs) or by generating beneficial metabolites that keep the gut barrier intact and induce immunoregulatory cells. It is hypothesized that perturbations in a healthy gut microbiome lead to the development of inflammatory diseases, including MS (Berer et al., 2017; Cekanaviciute et al., 2017; Chung & Kasper, 2010; Colpitts & Kasper, 2017; Jangi et al., 2016; Rothhammer & Quintana, 2016; Wu & Wu, 2012). Indeed, a number of factors including genetics, diet, modern lifestyles, and excessive hygiene can induce gut dysbiosis and consequently shift the balance between beneficial bacteria (i.e., symbionts) and bacteria that disrupt normal homeostasis (i.e., pathobionts), thus promoting an inflammatory phenotype (Abdelhamid & Luo, 2018; Almeida, Silvernale, Kuo, & Staller, 2019; Dibley, Coggrave, McClurg, Woodward, & Norton, 2017). Specifically with regard to the pathogenesis of MS, gut dysbiosis can cause several changes that predispose individuals to this disease, which include: (i) disrupting tight-junctions between IECs, which increases gut permeability, translocation of bacterial products into systemic circulation, and induction of systemic inflammation; (ii) depleting bacteria that are responsible for digesting dietary compounds into beneficial metabolites, such as short chain fatty acids (SCFA) and equol; (iii) suppressing pathways that induce immunoregulatory cells such as CD4

+FoxP3+ regulatory T cells (Tregs); and (iv) inducing pro-inflammatory Th1/Th17 cells either directly or indirectly through molecular mimicry and bystander activation of autoreactive CD4+ T cells.

2.1 Leaky gut and systemic inflammation

The intestinal lining is made up of millions of epithelial cells that form tight junctions, which allows communication between gut bacteria and the host as well as the passage of selective metabolites. Emerging evidence suggests that gut microbiota composition affects the gut permeability and plays an important role in the pathogenesis of diseases such as MS. Not surprisingly, diet has the most significant influence on shaping the gut microbial community. (Singh et al., 2017) Under homeostatic conditions, symbionts (beneficial bacteria) maintain a balance between the pro- and anti-inflammatory immune responses through conversion of ingested food (fiber, poly unsaturated fatty acid-containing food, tryptophan and phytoestrogens) into metabolites such as SCFA, specialized proresolving mediators (SPM) precursors (e.g. ω-3 fatty acids), indoles and equol, which promote an anti-inflammatory state. Specifically, gut bacteria such as *Roseburia* can induce production of SCFA and SPMs whereas bacteria such as *Prevotella, Parabacteroides, Adlercreutzia, Lactobacillus* produce equol by metabolizing phytoestrogen from dietary compounds. Moreover, these beneficial metabolites help maintain an anti-inflammatory environment at mucosal surfaces by inducing immunoregulatory cells (Freedman, Shahi, & Mangalam, 2018). Thus, symbionts within the gut microbiota play an important role in regulating the immune response, in part by producing beneficial metabolites to maintain an intact gut barrier and their depletion in PwMS provide support for the contributions of gut dysbiosis in MS (De Weirdt et al., 2017; Devillard, McIntosh, Duncan, & Wallace, 2007; Hiippala et al., 2018).

In contrast, altering the composition of the gut microbiota in a manner that enriches pathobionts and depletes beneficial symbionts is linked with gut barrier dysfunction. Increased intestinal permeability (i.e., a leaky gut) enables bacterial contents and toxins to pass from the gut lumen into the systemic circulation, which, in turn, may induce the production of pro-inflammatory mediators such as IL-1β, IL-6, IL-8, and TNF-α, and thus initiate an inflammatory cascade. Notably, numerous studies show that IL-1β, IL-6, IL-8, and TNF-α are modulated in the sera of PwMS (Dujmovic et al., 2009; Goyal et al., 2019; Ireland, Monson, & Davis, 2015). Additionally, gut dysbiosis allows the translocation of bacterial toxins such as lipopolysaccharide (LPS) into systemic circulation, which is another potent inducer of pro-inflammatory cytokines (Raetz & Whitfield, 2002). The development of a leaky gut and the associated systemic inflammation that arises due to an increase in inflammatory mediators is termed *leaky gut syndrome*, which is linked with multiple inflammatory diseases, including MS (Abdelhamid & Luo, 2018; Poisson et al., 2015; Simeonova, Ivanovska, Murdjeva, Carvalho, & Maes, 2018). Indeed, multiple studies have reported that PwMS have increased gut permeability (Buscarinu et al., 2018; Pellizoni et al., 2021; Sjostrom et al., 2021). Enhanced immune activation induced by a leaky gut can propagate a myelin-specific auto-reactive immune response in genetically-susceptible individuals. In addition, increased systemic inflammation can also promote trafficking of activated inflammatory cells to the CNS causing inflammation, demyelination and neuroaxonal damage. Additionally, in the animal model of MS, experimental autoimmune encephalomyelitis (EAE), the development of disease is associated with increased intestinal permeability whereas animals that were treated with specific gut bacteria to suppress disease displayed reduced gut permeability (Mangalam et al., 2017; Nouri, Bredberg, Westrom, & Lavasani, 2014). Thus, this evidence provides further support that increased intestinal permeability due to gut dysbiosis is a mechanism by which pathobionts influence the pathobiology of MS.

2.2 Modulation of the immune response by the gut microbiota

The initial process of autoimmunity is the peripheral activation of autoreactive CD4+ T cells via the presentation of autoantigens by antigen presenting cells (APCs) that express disease-susceptible MHC class-II molecule(s) (called Human leukocyte antigen (HLA) molecule(s) in humans). Collective evidence suggests that the onset of MS is the result of an aberrant T cell mediated immune response to a number of myelin antigens. In addition, the ability of APCs to produce pro-inflammatory cytokines such as IL-1β, IL-6, IL-10, IL-12, IL-23, and TNF-α plays a crucial role in activating encephalitogenic CD4+ T cell responses and in determining disease susceptibility/severity. Once activated, myelin antigen-specific CD4+ T cells produce pro-inflammatory cytokines such as IL-1β, IL-6, IL-17A, IFNγ and GM-CSF. The encephalitogenic capacity of autoreactive T cells to cause disease is dependent on the ability of the cell to produce some or all of these cytokines. At the same time, tolerance to self-antigen is maintained by regulatory cells such as Treg cells. During homeostasis there is a balance between Treg and Th17 cells. Thus, any factors (cell/molecules) with ability to activate APCs or autoreactive CD4 + T cells or suppress the Treg population can break tolerance to self-antigen and activate myelin-specific autoreactive T and B cell responses, thus precipitating disease. Interestingly, seminal studies performed in GF mice lacking any microbes have established the importance of the microbiota in developing a functional immune system, including in the induction of Tregs, Th17 and mucosal IgA response (Round & Mazmanian, 2009).

2.2.1 Suppression of the immunoregulatory response

The gut microbiota has emerged as an important player in the induction of immunoregulatory responses at mucosal surfaces. As such, there is a large quantity of bacteria in the gut lumen that increases in number from the stomach to the colon, with the colon having the maximum density of microbes. Generally, the host immune system mounts an immune response when it encounters microbes such as bacteria. However, despite the large number of bacteria in the gut, the host uses multiple mechanisms to prevent immune cell activation upon interaction with these organisms and maintains a symbiotic relationship with the gut microbiota.

The interaction among gut microbiota, intestinal epithelial, and immune cells plays an important role in host immunity. Physical separation due to the tight junctions of IECs keeps bacteria from entering into circulation, and failure to maintain these junctions results in a leaky gut, as discussed above. Besides this, there are other mechanisms in place that maintain a tolerogenic response at mucosal surfaces, which is driven by the interaction between IECs, the gut microbiota, and their metabolites. Discussion of these mechanisms are beyond the scope of this chapter and readers are encouraged to read the reviews published elsewhere (Allaire et al., 2018; Brown, Kenny, & Xavier, 2019; Kayama, Okumura, & Takeda, 2020). Instead, here we focus on the ability of the gut microbiota to modulate Tregs, which primarily occurs in the colon. It is important to highlight that there are at least two types of Treg cell populations: one that is naturally produced in the thymus and is defined by its expression of FoxP3, Helios, and Neuropilin-1, and a population that is induced in the periphery or intestines (referred to as iTregs). Most bacteria-induced iTregs arise in the large intestine. They can be induced by a number of different bacteria, such as spore forming *Clostridium* spp., *Bacteroide* spp., *Prevotella* spp., *Lactobacillus* spp., and *Parabacteroides* spp. GF mice that lack any bacteria are characterized by an underdeveloped immune system and significantly reduced Treg population. However, colonization of GF mice with a single bacterium,

Bacteroides fragilis, or a mixture of chloroform-resistant *Clostridium* bacteria induces Tregs in GF mice (Atarashi et al., 2011, 2013; Round & Mazmanian, 2010; Telesford et al., 2015). The capability of *Clostridium* species to induce Tregs seems to be dependent on the production of SCFAs (Smith et al., 2013). Notably, bacteria that are absent in PwMS can be administered to mice with EAE (the animal model of MS), which induces Tregs and suppresses disease (discussed further in the section Bacteria as Drugs). Thus, the absence or reduction of bacteria that induce Tregs is another mechanism by which the gut microbiota influences MS disease.

2.2.2 Induction of a pro-inflammatory Th1/Th17 response by gut bacteria

The gut microbiota plays a critical role in developing and regulating innate and adaptive immunity of the host, including Treg and Th17 responses. Segmentous filamentous bacterium (SFB) as well as other bacteria that harbor the ability to adhere to intestinal epithelium cells in the ileum can induce Th17 cells in GF mice (Ivanov II et al., 2009, 2008; Ivanov & Littman, 2010). Thus, gut bacteria that are enriched in PwMS can influence the pathobiology of MS by inducing Th1/Th17 cells or suppressing the Treg population.

Multiple studies using EAE have validated the importance of the gut microbiota in the pathogenesis of this disease. Utilizing broad spectrum antibiotics to deplete the gut flora, Ochoa-Repáraz et al showed the microbiota is required for EAE to develop, as mice treated with antibiotics exhibited impaired disease development. These findings were validated by Lee et al who utilized germ-free mice to show they develop an attenuated form of EAE (Lee, Menezes, Umesaki, & Mazmanian, 2011). However, colonizing GF mice with SFB, which have capacity to induce IL-17, resulted in severe EAE, suggesting that Th17 cells that are induced by gut bacteria play an important role in CNS autoimmunity (Lee et al., 2011). Besides directly activating Th17 cells, SFB also activated pro-inflammatory dendritic cells with better antigen presentation capacity, which in turn can further enhance antigen presentation and activation of autoreactive T cells. The importance of the gut microbiota for the development of EAE was further shown by Berer et al., who used GF T-cell receptor (TCR) transgenic SJL mice expressing myelin oligodendrocyte glycoprotein (MOG)92-106 peptide specific CD4. All CD4 T cells in these TCR transgenic mice are specific for $MOG_{92\text{-}104}$ peptide and develop spontaneous relapsing-remitting

EAE in conventional mouse colony (Berer et al., 2011). However, when these TCR transgenic mice are raised in a germ-free environment lacking any gut bacteria, they are protected from EAE. Colonization with gut microbiota from conventional mice restore disease susceptibility in GF TCR transgenic mice which suggest a critical role of gut microbiota in the development of EAE. The reduced level of disease in these GF mice was due to impaired T and B cell activation, especially production of IL-17 and MOG-specific antibodies, respectively. Interestingly, in this study SFB alone failed to activate Th17 cells to initiate full blown disease. Although, both Lee et al. (2011) and Berer et al. (2011) provide evidence that the gut microbiota is critical in the pathobiology of EAE, the discrepancies between the two studies suggest it plays a highly complex and complicated role. Alternatively, it is possible that a specific gut bacterium such as SFB or combination of gut bacteria with ability to stimulate Th17 cells can trigger autoimmune disease.

Gut bacteria can also influence inflammation through molecular mimicry due to the structural similarity between bacterial antigens and encephalitogenic autoantigens. Bacterial antigens can trigger bystander activation of autoreactive T and/or B cells that are directed against CNS antigens. Once activated these encephalitogen T cells can traffic to the CNS and induce an inflammatory response, which results in the destruction of myelin and release of autoantigens. These autoantigens then can activate additional autoreactive T cells and propagate the inflammation and demyelination in CNS. A number of microbial pathogens are proposed to have either antigen similarity to antigens derived from myelin proteins or an ability to cross-react with myelin antigen. For example, Epstein-Barr virus antigens are cross-reactive with myelin-specific CD4 T, CD8 T and B cells (Gabibov et al., 2011; Lang et al., 2002; Lindsey, 2017; Morecraft, Blair, & Maynard, 1988; Tengvall et al., 2019). However, disease-initiating antigens might not be myelin specific, as shown by a recent study from Roland Martin's laboratory (Planas et al., 2018). In this study, they used brain-infiltrating CD4 T cell clones from PwMS and identified guanosine diphosphate (GDP)–L-fucose synthase as a potential autoantigen, which was cross-reactive with the myelin antigen. Most importantly, these autoreactive CD4 T cells can also be activated by homologous peptides derived from the gut microbiota, especially from *Akkermansia*. This study demonstrates that protein derived from the gut microbiota can directly stimulate autoreactive T cell repertoires.

3. Influence of diet on the pathobiology of MS

The gut of newly born babies is mostly sterile; in the first few days after birth, the majority of bacteria that are present belong to the Proteobacteria phyla; subsequently their gut is colonized by milk-digesting bacteria, such as *Lactobacillus* and *Bifidobacterium* species (Derrien, Alvarez, & de Vos, 2019). Thus, early in life the composition of the microbiome is simple and dominated by only a few bacterial species. However, the shift from a milk-based diet to solid food leads to a large expansion of the gut microbiota, which is dominated with bacteria from Firmicutes and Bacteroidetes phyla and ultimately results in the establishment of a complex microbiome. As changes in the microbiome occur based on the types of food consumed by the host, it is not surprising that an individual's diet is one of the strongest factors influencing the composition of gut microbiome.

At the same time, there are diets which can predispose and/or worsen the disease. The best example of this is a western diet that is rich in fat and protein, and which has been linked with an increased incidence of obesity in developed countries. Correspondingly, obesity significantly increases the risk and severity of MS and obese PwMS show a reduced responsiveness to current MS therapies and receive a worse prognosis than non-obese patients. In mice, a high-fat diet (HFD) induces gut dysbiosis, and this is characterized by an enrichment in pathobionts, in particular bacteria that are linked with induction of pro-inflammatory Th17 response. Additionally, gut bacteria that are altered by a HFD can increase disease severity by allowing the expansion of methane and H_2S-producing bacteria, which induce constipation and reduce gut motility. Similarly, a diet that is high in fat and sugar/fructose can also influence MS; this type of diet induces inflammation in other models of autoimmune disease, such as inflammatory bowel disease. However, data on the precise mechanism of diet-induced inflammation and disease exacerbation is lacking in MS.

4. Targeting gut bacteria to treat multiple sclerosis

Autoimmune diseases are only observed in a very small fraction of any given population. Taking into consideration the existing data on microbiome signatures, it is fair to postulate that microbiota predominantly have a protective role against autoimmunity. Thus, understanding how this protection is afforded will enable us to harness the power of the microbiome and utilize the beneficial components as therapeutic agents to treat various autoimmune diseases. In this regard, a number of groups, including ours, are working on bacteriotherapy studies (summarized in Table 2). The gut commensal-based therapies can be divided into two groups: those that use probiotics and those that use bacteria as drugs (BRUGS).

4.1 Probiotic-based therapy to treat autoimmune diseases

According to National Center for Complimentary and Integrative health (NCCIH) in the US, probiotics are defined as live organisms that can provide health benefits when consumed or applied to the body. A number of different probiotic combinations are currently being tried as therapies for various autoimmune diseases; however at present there is no consensus on the efficacy of probiotics in MS. Although the role of the gut microbiota in the pathogenesis of MS is increasingly appreciated, the few studies that have tested the therapeutic effect of probiotics in PwMS have been inclusive. In a randomized double-blind placebo-controlled clinical trial of Iranian PwMS, *Kouchaki et al* showed that taking a probiotic mixture containing *Lactobacillus acidophilus, Lactobacillus casei,* and *Bifidobacterium bifidum* improved scores on the expanded disability status scale (EDSS) and decreased levels of inflammatory markers (Kouchaki et al., 2017). A recent study from the Weiner group showed that treatment for eight weeks with VSL3, a probiotics containing four strains of *Lactobacillus*, three strains of *Bifidobacterium* and one strain of *Streptococcus* modulated the gut microbiota (Tankou, Regev, Healy, Cox, et al., 2018; Tankou, Regev, Healy, Tjon, et al., 2018) and induced an anti-inflammatory immune response. Thus, although the effect of probiotic therapy for autoimmune diseases is still inconclusive, future studies with different probiotic combinations and/or a longer course of treatment might provide an answer on the effectiveness of this approach.

4.2 Bacteria as drugs (BRUGS)

As probiotic-based therapies have been inconclusive in their effectiveness in treating autoimmune diseases, a number of groups, including ours, have focused on colonizing patients with a specific single bacterium that is otherwise absent in these individuals. The majority of probiotics used in clinical trials contain a mixture of *Bifidobacterium*, *Lactobacillus,* and *Streptococcus;* however microbiome studies in PwMS have failed to report that these

Table 2 Specific bacteriotherapy and MS.

Study design	Intervention used	Therapy effect on host	Therapy effect on gut microbiota
Human fecal study MS = 9, Control = 13 (Tankou, Regev, Healy, Cox, et al., 2018; Tankou, Regev, Healy, Tjon, et al., 2018)	LBS/VSL3/Vivomixx (3600 billion CFU/day for 2 months) containing *Lactobacillus acidophilus (DSM 24735), L. plantarum* (DSM 24730), *L. paracasei* (DSM 24733), *L. delbrueckii subsp. bulgaricus* (DSM 24734), *Bifidobacterium longum* (DSM 24736), *B. breve* (DSM 24732), *B. infantis* (DSM 24737), and *Streptococcus thermophilus* (DSM 24731)	Induced and anti-inflammatory peripheral innate immune response characterized by decreased frequency of intermediate monocytes ($CD14^{high}CD16^{low}$) in PwMS as well as decreased MFI of costimulatory marker CD80 on classical monocytes in controls In controls, trend toward decreased frequency of inflammatory monocytes following VSL3 supplementation. Administration of VSL3 was also associated with decreased MFI of HLA-DR on myeloid-derived dendritic cells (CD45 + LIN − CD11c +) in PwMS	Increased relative abundance of *Lactobacillus* and *Bifidobacterium* species in PwMS Increased relative of *Veillonellaceae* and genus *Collinsella* in healthy controls Decrease abundance of *Akkermansia*, *Blautia* and *Dorea* in PwMS
Human fecal study MS (on glatiramer acetate) = 7, Untreated = 2 HC = 13 (Tankou, Regev, Healy, Cox, et al., 2018; Tankou, Regev, Healy, Tjon, et al., 2018)	VSL3/Vivomixx (3600 billion CFU/day for 2 months) containing *Lactobacillus acidophilus (DSM 24735), L. plantarum* (DSM 24730), *L. paracasei* (DSM 24733), *L. delbrueckii subsp. bulgaricus* (DSM 24734), *Bifidobacterium longum* (DSM 24736), *B. breve* (DSM 24732), *B. infantis* (DSM 24737), and *Streptococcus thermophilus* (DSM 24731)	Discontinuation of VSL3 induced a pro-inflammatory immune response characterized by increased frequency of inflammatory monocytes ($CD14^{low}CD16^{high}$) in controls. Decreased frequency of IL-10 + Tregs in controls as well as a trend toward decreased frequency of LAP + Tregs following discontinuation of VSL3	VSL3 administration increased relative abundance of several species including *Lactobacillus*, *Streptococcus* and *Bifidobactetrium* in both PwMS and control patients but the relative abundance of these bacteria returned to normal after discontinuing the therapy
Human clinical trial MS = 30 HC = 30 (Kouchaki et al., 2017)	"Probiotic capsule" containing *Lactobacillus acidophilus*, *Lactobacillus casei*, *Bifidobacterium bifidum* and *Lactobacillus fermentum*	Improved expanded disability status scale (EDSS), beck depression inventory, general health questionnaire, and depression anxiety and stress scale Decreased serum insulin, homeostasis model of assessment-estimated insulin resistance, beta cell function and total-/HDL-cholesterol, and significantly increased quantitative insulin sensitivity check index and HDL-cholesterol levels compared with the placebo	

Continued

Table 2 Specific bacteriotherapy and MS.—cont'd

Study design	Intervention used	Therapy effect on host	Therapy effect on gut microbiota
Human Clinical trial (diet study) (Saresella et al., 2017)	High-vegetable/low-protein diet (HV/LP diet group; $N = 10$) or a "Western Diet" (WD group; $N = 10$) given to PwMS	In HV/LP group, IL-17-producing CD4+ T lymphocytes and PD-1-expressing CD4+ T lymphocytes were significantly decreased PD-L1-expressing monocytes were significantly increased In the HV/LP diet group, positive correlations between *Lachnospiraceae* and both CD14+/IL-10+ and CD14+/TGFβ+monocytes, as well as between *Lachnospiraceae* and CD4+/CD25+/FoxP3+ T lymphocytes were observed	Increased *Lachnospiraceae* in HV/LP compared to WD
C57BL/6JOlaHsd mice (Calvo-Barreiro et al., 2020)	Lactibiane ike (Pileje, Paris, France) containing *Bifidobacterium lactis* LA 304, L. *acidophilus* LA 201, *L. salivarius* LA 302 Vivomixx (Frifols, Barcelona, Spain) containing *L. s acidophilus* (DSM 24735), *L. plantarum* (DSM 24730), *L. paracasei* (DSM 24733), *L. delbrueckii subsp. bulgaricus* (DSM 24734), *Bifidobacterium longum* (DSM 24736), *B. breve* (DSM 24732), *B. infantis* (DSM 24737), and *Streptococcus thermophilus* (DSM 24731)	Lactibiane ike improved clinical outcome while Vivomixx did not Both treatments improved motor coordination skills Lower demyelination and T cell inflammatory infiltrate density in white matter of spinal cord in both Lactibiane treatment decreased RORγt expression by 4-fold, increased Tregs, and reduced peripheral plasma cells Both increased percentage of myeloid DCs in periphery compared to control	Lactibiane caused an increase in relative abundance of *Lachnoclostridium* and *Bifidobacterium* Vivomixx caused the increase in Streptococcus
C57BL/6 mice (Mestre et al., 2020)	Vivomixx containing *L. paracasei* DSM 24734, *L. plantarum* DSM 24730, *L. acidophilus* DSM 24735, *L. delbrueckii* ssp. *bulgaricus* DSM 24734, *B. longum* DSM 24736, *B. infantis* DSM 24737, *B. breve* DSM 24732, *S. thermophilus* DSM 24731	Improved motor disability. Reduced microgliosis, astrogliosis and leukocyte infiltration in the brain. Expression of IL-1β and IL-6 was diminished and promoted IL-10 gene expression Increased plasma levels of butyrate and acetate levels in mice that received Vivomixx Restricted IL-17 production by Th17-polarized $CD4^+$ T cells purified from the mesenteric lymph nodes	Abundance of *Anaerostipe, Dorea, Oscillospira, Enterobacteraceae* or *Ruminococcus* were decreased whereas *Bacteroides, Odoribacter, Lactobacillus, or Sutterella* were increased

C57BL/6 mice (Salehipour et al., 2017)	*Lactobacillus plantarum* A7 or *Bifidobacterium animalis* PTCC 1631 or mixture of both	*L. plantarum* A7 and *B. animalis* ameliorated EAE condition by favoring Th2 and Treg differentiation via upregulation of Foxp3 and GATA3 in the brain and spleen as well as inhibited the differentiation of Th1 and Th17 cells.	
C57BL/6 mice (Secher et al., 2017)	*Escherichia coli* strain Nissle 1917 (ECN, 10^8 cfu/animal)	ECN treatment reduced disease severity ECN protection is because of dampened inflammatory cytokine production and reduced migration of MOG-specific T cells from periphery to CNS ECN treatment repaired intestinal permeability dysfunction	
HLA-DR3.DQ8 mice (Mangalam et al., 2017)	*Prevotella histicola*	*P. histicola* challenge led to a suppressed EAE along with a decrease in proinflammatory Th1 and Th17 cells, and an increase in the frequencies of CD4 + FoxP3 + regulatory T cells, tolerogenic dendritic cells, and suppressive macrophages	Increased abundance of *Prevotella, Lactobacillus*, and *Sutterella* in *P. histicola*-treated group
HLA-DR3.DQ8 mice (Shahi et al., 2019)	*Prevotella histicola* or Copaxone	*P. histicola* suppressed EAE as effectively as Copaxone treatment *P. histicola*-treated mice had higher frequency and number of FoxP3 + T cells in periphery as well as in the gut along with decreased frequency of pro-inflammatory IFN-γ and IL17-producing CD4 T cells in the CNS	*P. histicola*-treated group also showed higher relative abundance of *Lactobacillus*, which was lost in EAE control group

Continued

Table 2 Specific bacteriotherapy and MS.—cont'd

Study design	Intervention used	Therapy effect on host	Therapy effect on gut microbiota
C57BL/6 mice (He et al., 2019)	*Lactobacillus reuteri* DSM 17938	Ameliorated EAE. *L. reuteri* treatment reduced Th1/Th17 cells and their associated cytokines IFN-γ/IL-17 in EAE mice	Relative abundances of *Proteobacteria* and *Deferribacteres* were significantly increased in the colonic contents of EAE mice, while the relative abundance of *Bacteroidetes* was reduced. *L. reuteri* treatment reversed the effects of EAE on the relative abundance of these phyla in EAE mice At the genus level, *Anaeroplasma* and *Rikenellaceae* were increased while *Prevotella* was decreased in EAE mice. After *L. reuteri* treatment to EAE mice, the percentage of these genera was improved
C57BL/6 mice (Chen et al., 2019)	1. *Clostridium butyricum* (5×10^6 CFU/mL) 2. Norfloxacin (5 mg/kg)	Both treatments significantly ameliorated severity of EAE *C. butyricum* reduced levels of Th17 and Th1 in CNS, inguinal lymph nodes, spleen small intestine and colon MAPK signaling was suppressed by using *C. butyricum* and Norfloxacin	Both significantly increased *Bacteroidetes* and decreased *Firmicutes* Norfloxacin increased *Veruucomicrobia* Both treatments reduced abundance of *Desulfovibroneceae* and *Ruminococcus*

PwMS, people with MS; EAE, experimental autoimmune encephalomyelitis; EDSS, expanded disability status scale; MFI, mean fluorescence intensity; HC, healthy control.

bacteria are lacking in patients compared to HCs (with the exception of *Lactobacillus* being absent in one study). Thus, a better approach is to identify the bacteria that are lacking in specific autoimmune diseases and use these as a therapy. Our argument for this approach is that providing a specific bacterium that is lacking in a patient will be better at treating the disease than providing a non-specific commensal bacterium. Thus, we coined the term BRUGS (bacteria as drugs) to differentiate this treatment from that of probiotics, as the latter are harmless bacteria which may or may not be reduced during gut dysbiosis. This approach would involve the use of a specific BRUGS, designed from analyzing microbial signatures, as a therapy for a particular autoimmune disorder. This aligns with the therapeutic philosophy that is progressing towards precision and personalized medicine as experimental biologists and clinicians are increasingly appreciative of the fact that a one-size-fits-all therapeutic approach is not always effective in treating disease.

Multiple studies in MS have validated the importance of the *Prevotella* genus in disease as this bacterium displays reduced abundance in PwMS and is increased after treatment with disease modifying therapies (DMTs) (Table 1). Based on these findings, we successfully isolated a strain of *Prevotella, P histicola,* from human subjects and administered this as a therapy in an animal model of MS. Notably, this suppressed disease and was as efficient as currently used first line MS therapies, including Copaxone and Betaseron (Chen et al., 2016; Mangalam et al., 2017; Shahi et al., 2019). Interestingly, *P histicola* was also able to suppress disease in the collagen-induced arthritis model in HLA-DR4 transgenic mice (Marietta et al., 2016) and type-1-diabetes mouse model (Marietta et al., 2022).

There are other microbiota candidates such as *Bacteroides fragilis (B. fragilis),* which also suppress disease in other animal models, including those for MS (Ochoa-Reparaz et al., 2010), colitis (Mazmanian, Round, & Kasper, 2008) and autism spectrum disorder (Hsiao et al., 2013). *B. fragilis* was the first gut commensal that has been proposed as a potential therapeutic by the Kasper group. Dennis Kasper, Sarkis Mazmanian, and Llyod Kasper as well as others have published extensively on the therapeutic ability and potential mechanism of *B. fragilis* in modulating immune responses (Mazmanian, Liu, Tzianabos, & Kasper, 2005), preventing colonization of pathobionts, and suppressing inflammatory and neurological diseases, including MS, IBD and autism spectrum disorder. A study by Mazamanian's group showed that *B. fragilis* helps to promote maturation of the host immune system. Moreover, they showed that the immunomodulatory ability of

B. fragilis is dependent on bacterial polysaccharide (PSA) (Mazmanian et al., 2005), as a mutant strain of *B. fragilis* lacking PSA failed to direct maturation of the host immune response. This group next showed that *B. fragilis* expressing PSA, but not the mutant lacking PSA, can prevent development of colitis (Mazmanian et al., 2008) and EAE (Ochoa-Reparaz et al., 2010). The ability of PSA to suppress disease was due to its ability to induce Tregs (Telesford et al., 2015) as well as IL-10-producing B and T cells (Mazmanian et al., 2008; Ramakrishna et al., 2019). Collectively, these studies highlight the ability of *B. fragilis* to perform immunomodulatory functions and therefore its potential as a therapeutic agent to treat inflammatory autoimmune diseases.

4.3 Prebiotics/diet-based therapy

Prebiotics are foods that are degraded into metabolites by gut bacteria and are beneficial to human health (Davani-Davari et al., 2019). Given that diet has the strongest impact on the gut microbiome, even more than genetic effects (Ridaura et al., 2013), it is not surprising that the scientific community is slowly becoming more receptive to its use as potential therapeutic agent. Indeed, the importance of the diet in human disease is highlighted by multiple ongoing clinical trials based on modulating the diet, which include those that mimic fasting, the Mediterranean diet, the keto diet, the paleo diet, and the Wahls diet, among others (Chehade et al., 2019; Claflin, van der Mei, & Taylor, 2018; Dahan, Segal, & Shoenfeld, 2017; Esposito, Bonavita, Sparaco, Gallo, & Tedeschi, 2018; Forsyth et al., 2018; Vieira, Pagovich, & Kriegel, 2014; Winkvist, Barebring, Gjertsson, Ellegard, & Lindqvist, 2018; Yadav et al., 2016).

We have recently shown the importance of a phytoestrogen diet in modulating disease in an animal model of MS. Under anaerobic conditions in the gut, intestinal bacteria convert phytoestrogen daidzein (an isoflavone) to equol and this metabolite is suggested to possess immunomodulatory activity due to its structural similarity with endogenous estrogen. Further, our group showed that PwMS lack phytoestrogen-metabolizing bacteria (i.e., *Parabacteroides* (Chen et al., 2016; Cekanaviciute et al., 2017) and *Adlercreutzia* (Berer et al., 2017; Chen et al., 2016), and we have also recently shown that dietary phytoestrogen (isoflavone) can protect from disease in a chronic-EAE model as well as in a relapsing remitting-EAE (RR-EAE) model (Jensen et al., 2021). Most importantly, depletion of gut microbiota abolished the disease-protective effect, whereas colonization of mice with

Parabacteroides and Adlercreutzia restored protection from disease. This indicates that the protection from disease that is afforded by phytoestrogen requires the presence of gut bacteria. In addition to a phytoestrogen diet, Choi et al. showed that a diet that mimics fasting can suppress EAE by inducing the gut microbiota, Tregs, and corticosterone, which is shown to be immune-suppressive. Additionally, a number of diets such as a diet rich in SCFA (propionic acid), resveratrol, or those that impose calorie restrictions suppress disease in animal models of MS. Thus, these studies highlight the importance of modulating the diet and gut bacteria as potential treatment options for MS. As this is an emerging field and multiple clinical trials are underway, future discussions will include assessment of the effectiveness of these and other diet-based therapies. In the meantime, those who are interested in further information on this topic are encouraged to read the excellent reviews mentioned above.

5. Conclusions

As our understanding of microbial variation and corresponding genetic parameters continues to grow, remodeling of the human gut microbiota and their associated functions has reached an intriguing milestone in the development of therapeutic options for MS. Data from prospective, randomized, multi-centric, longitudinal studies is on the horizon and will enable us to determine the efficacy of microbiota-targeted approaches to treat various autoimmune disorders. In an ideal setting, a personalized strategy could be designed for each patient, which will enable the care team consisting of clinicians, basic scientists, and nutritionists to develop dietary interventions that promote increases in specific microbial populations that could influence the outcome of autoimmune disorders. These interventions could be in the form of custom diets and/or probiotics or targeted antibiotics. The use of such treatments, in individual or combined regimens, may manipulate the gut microbiota in a way that will prove to fundamentally change our approach to modern medicine. Although this sounds simple, there is still a need to carefully investigate the complex interactions occurring in the context of host genetic backgrounds in order to identify optimal treatment strategies for MS. Although early studies show promise, the complexity of interactions between the host and the gut microbiota still warrants the question of whether therapeutic interventions will be successful for a majority of the individuals suffering from this complex disease of immune dysregulation.

Acknowledgments

The authors acknowledge funding from the National Multiple Sclerosis Society (RG 5138A1/1T), National Institute of Health (NIH)/NIAID (1R01AI137075-01), VA Merit Award (I01CX002212-01), a Carver Trust Medical Research Initiative Grant, and the University of Iowa Environmental Health Sciences Research Center, NIEHS/NIH (P30 ES005605). MY was supported on an institutional training grant National Research Service Award (NRSA T90) in Oral Health Research from the National Institute of Dental & Craniofacial Research (T90-DE 23520-7 to Dr. Scott Levy).

Conflict of interest statement

AM is inventor of a technology claiming the use of *Prevotella histicola* for the treatment of autoimmune diseases. The patent for the technology is owned by Mayo Clinic, who has given exclusive license to Evelo Biosciences. AM received royalties from Mayo Clinic (paid by Evelo Biosciences).

References

Abdelhamid, L., & Luo, X. M. (2018). Retinoic acid, leaky gut, and autoimmune diseases. *Nutrients*, *10*(8), 1016.

Allaire, J. M., Crowley, S. M., Law, H. T., Chang, S. Y., Ko, H. J., & Vallance, B. A. (2018). The intestinal epithelium: Central coordinator of mucosal immunity. *Trends in Immunology*, *39*(9), 677–696.

Almeida, M. N., Silvernale, C., Kuo, B., & Staller, K. (2019). Bowel symptoms predate the diagnosis among many patients with multiple sclerosis: A 14-year cohort study. *Neurogastroenterology and Motility*, *31*(6), e13592.

Alvarado-de la Barrera, C., Zuniga-Ramos, J., Ruiz-Morales, J. A., Estanol, B., Granados, J., & Llorente, L. (2000). HLA class II genotypes in Mexican Mestizos with familial and nonfamilial multiple sclerosis. *Neurology*, *55*(12), 1897–1900.

Atarashi, K., Tanoue, T., Oshima, K., Suda, W., Nagano, Y., Nishikawa, H., et al. (2013). Treg induction by a rationally selected mixture of Clostridia strains from the human microbiota. *Nature*, *500*(7461), 232–236.

Atarashi, K., Tanoue, T., Shima, T., Imaoka, A., Kuwahara, T., Momose, Y., et al. (2011). Induction of colonic regulatory T cells by indigenous Clostridium species. *Science*, *331*(6015), 337–341.

Barcellos, L. F., Oksenberg, J. R., Begovich, A. B., Martin, E. R., Schmidt, S., Vittinghoff, E., et al. (2003). HLA-DR2 dose effect on susceptibility to multiple sclerosis and influence on disease course. *American Journal of Human Genetics*, *72*(3), 710–716.

Berer, K., Gerdes, L. A., Cekanaviciute, E., Jia, X., Xiao, L., Xia, Z., et al. (2017). Gut microbiota from multiple sclerosis patients enables spontaneous autoimmune encephalomyelitis in mice. *Proceedings of the National Academy of Sciences of the United States of America*, *114*(40), 10719–10724.

Berer, K., Mues, M., Koutrolos, M., Rasbi, Z. A., Boziki, M., Johner, C., et al. (2011). Commensal microbiota and myelin autoantigen cooperate to trigger autoimmune demyelination. *Nature*, *479*(7374), 538–541.

Briggs, F. B., & Hill, E. (2019). Estimating the prevalence of multiple sclerosis using 56.6 million electronic health records from the United States. *Multiple Sclerosis*, *26*, 1948–1952, 1352458519864681.

Brown, E. M., Kenny, D. J., & Xavier, R. J. (2019). Gut microbiota regulation of T cells during inflammation and autoimmunity. *Annual Review of Immunology*, *37*, 599–624.

Buscarinu, M. C., Romano, S., Mechelli, R., Pizzolato Umeton, R., Ferraldeschi, M., Fornasiero, A., et al. (2018). Intestinal permeability in relapsing-remitting multiple sclerosis. *Neurotherapeutics*, *15*(1), 68–74.

Calvo-Barreiro, L., Eixarch, H., Ponce-Alonso, M., Castillo, M., Lebron-Galan, R., Mestre, L., et al. (2020). A commercial probiotic induces tolerogenic and reduces pathogenic responses in experimental autoimmune encephalomyelitis. *Cell*, *9*(4).

Campbell, J. D., Ghushchyan, V., Brett McQueen, R., Cahoon-Metzger, S., Livingston, T., Vollmer, T., et al. (2014). Burden of multiple sclerosis on direct, indirect costs and quality of life: National US estimates. *Multiple Sclerosis and Related Disorders*, *3*(2), 227–236.

Cantoni, C., Lin, Q., Dorsett, Y., Ghezzi, L., Liu, Z., Pan, Y., et al. (2022). Alterations of host-gut microbiome interactions in multiple sclerosis. *eBioMedicine*, *76*, 103798.

Cekanaviciute, E., Yoo, B. B., Runia, T. F., Debelius, J. W., Singh, S., Nelson, C. A., et al. (2017). Gut bacteria from multiple sclerosis patients modulate human T cells and exacerbate symptoms in mouse models. *Proceedings of the National Academy of Sciences of the United States of America*, *114*, 10713–10718.

Chehade, L., Jaafar, Z. A., El Masri, D., Zmerly, H., Kreidieh, D., Tannir, H., et al. (2019). Lifestyle modification in rheumatoid arthritis: Dietary and physical activity recommendations based on evidence. *Current Rheumatology Reviews*, *15*(3), 209–214.

Chen, J., Chia, N., Kalari, K. R., Yao, J. Z., Novotna, M., Paz Soldan, M. M., et al. (2016). Multiple sclerosis patients have a distinct gut microbiota compared to healthy controls. *Scientific Reports*, *6*, 28484.

Chen, H., Ma, X., Liu, Y., Ma, L., Chen, Z., Lin, X., et al. (2019). Gut microbiota interventions with Clostridium butyricum and norfloxacin modulate immune response in experimental autoimmune encephalomyelitis mice. *Frontiers in Immunology*, *10*, 1662.

Chen, J., Wright, K., Davis, J. M., Jeraldo, P., Marietta, E. V., Murray, J., et al. (2016). An expansion of rare lineage intestinal microbes characterizes rheumatoid arthritis. *Genome Medicine*, *8*(1), 43.

Chung, H., & Kasper, D. L. (2010). Microbiota-stimulated immune mechanisms to maintain gut homeostasis. *Current Opinion in Immunology*, *22*(4), 455–460.

Claflin, S. B., van der Mei, I. A. F., & Taylor, B. V. (2018). Complementary and alternative treatments of multiple sclerosis: a review of the evidence from 2001 to 2016. *Journal of Neurology, Neurosurgery, and Psychiatry*, *89*(1), 34–41.

Colpitts, S. L., & Kasper, L. H. (2017). Influence of the gut microbiome on autoimmunity in the central nervous system. *Journal of Immunology*, *198*(2), 596–604.

Cosorich, I., Dalla-Costa, G., Sorini, C., Ferrarese, R., Messina, M. J., Dolpady, J., et al. (2017). High frequency of intestinal TH17 cells correlates with microbiota alterations and disease activity in multiple sclerosis. *Science Advances*, *3*(7), e1700492.

Culpepper, W. J., Marrie, R. A., Langer-Gould, A., Wallin, M. T., Campbell, J. D., Nelson, L. M., et al. (2019). Validation of an algorithm for identifying MS cases in administrative health claims datasets. *Neurology*, *92*(10), e1016–e1028.

Dahan, S., Segal, Y., & Shoenfeld, Y. (2017). Dietary factors in rheumatic autoimmune diseases: A recipe for therapy? *Nature Reviews Rheumatology*, *13*(6), 348–358.

Davani-Davari, D., Negahdaripour, M., Karimzadeh, I., Seifan, M., Mohkam, M., Masoumi, S. J., et al. (2019). Prebiotics: Definition, types, sources, mechanisms, and clinical applications. *Food*, *8*(3).

De Weirdt, R., Hernandez-Sanabria, E., Fievez, V., Mees, E., Geirnaert, A., Van Herreweghen, F., et al. (2017). Mucosa-associated biohydrogenating microbes protect the simulated colon microbiome from stress associated with high concentrations of poly-unsaturated fat. *Environmental Microbiology*, *19*(2), 722–739.

DeLuca, G. C., Ramagopalan, S. V., Herrera, B. M., Dyment, D. A., Lincoln, M. R., Montpetit, A., et al. (2007). An extremes of outcome strategy provides evidence that multiple sclerosis severity is determined by alleles at the HLA-DRB1 locus. *Proceedings of the National Academy of Sciences of the United States of America*, *104*(52), 20896–20901.
Derrien, M., Alvarez, A. S., & de Vos, W. M. (2019). The gut microbiota in the first decade of life. *Trends in Microbiology*, *27*(12), 997–1010.
Devillard, E., McIntosh, F. M., Duncan, S. H., & Wallace, R. J. (2007). Metabolism of linoleic acid by human gut bacteria: Different routes for biosynthesis of conjugated linoleic acid. *Journal of Bacteriology*, *189*(6), 2566–2570.
Dibley, L., Coggrave, M., McClurg, D., Woodward, S., & Norton, C. (2017). It's just horrible: A qualitative study of patients' and carers' experiences of bowel dysfunction in multiple sclerosis. *Journal of Neurology*, *264*(7), 1354–1361.
Dujmovic, I., Mangano, K., Pekmezovic, T., Quattrocchi, C., Mesaros, S., Stojsavljevic, N., et al. (2009). The analysis of IL-1 beta and its naturally occurring inhibitors in multiple sclerosis: The elevation of IL-1 receptor antagonist and IL-1 receptor type II after steroid therapy. *Journal of Neuroimmunology*, *207*(1-2), 101–106.
Dyment, D. A., Herrera, B. M., Cader, M. Z., Willer, C. J., Lincoln, M. R., Sadovnick, A. D., et al. (2005). Complex interactions among MHC haplotypes in multiple sclerosis: Susceptibility and resistance. *Human Molecular Genetics*, *14*(14), 2019–2026.
Esposito, S., Bonavita, S., Sparaco, M., Gallo, A., & Tedeschi, G. (2018). The role of diet in multiple sclerosis: A review. *Nutritional Neuroscience*, *21*(6), 377–390.
Finegold, S. M., Molitoris, D., Song, Y., Liu, C., Vaisanen, M. L., Bolte, E., et al. (2002). Gastrointestinal microflora studies in late-onset autism. *Clinical Infectious Diseases*, *35*(Suppl. 1), S6–S16.
Forsyth, C., Kouvari, M., D'Cunha, N. M., Georgousopoulou, E. N., Panagiotakos, D. B., Mellor, D. D., et al. (2018). The effects of the Mediterranean diet on rheumatoid arthritis prevention and treatment: a systematic review of human prospective studies. *Rheumatology International*, *38*(5), 737–747.
Freedman, S. N., Shahi, S. K., & Mangalam, A. K. (2018). The "gut feeling": Breaking down the role of gut microbiome in multiple sclerosis. *Neurotherapeutics*, *15*(1), 109–125.
Gabibov, A. G., Belogurov, A. A., Jr., Lomakin, Y. A., Zakharova, M. Y., Avakyan, M. E., Dubrovskaya, V. V., et al. (2011). Combinatorial antibody library from multiple sclerosis patients reveals antibodies that cross-react with myelin basic protein and EBV antigen. *The FASEB Journal*, *25*(12), 4211–4221.
Goyal, M., Khanna, D., Rana, P. S., Khaibullin, T., Martynova, E., Rizvanov, A. A., et al. (2019). Computational intelligence technique for prediction of multiple sclerosis based on serum cytokines. *Frontiers in Neurology*, *10*, 781.
Ha, S., Oh, D., Lee, S., Park, J., Ahn, J., Choi, S., et al. (2021). Altered gut microbiota in Korean children with autism spectrum disorders. *Nutrients*, *13*(10).
He, B., Hoang, T. K., Tian, X., Taylor, C. M., Blanchard, E., Luo, M., et al. (2019). Lactobacillus reuteri reduces the severity of experimental autoimmune encephalomyelitis in mice by modulating gut microbiota. *Frontiers in Immunology*, *10*, 385.
He, Z., Shao, T., Li, H., Xie, Z., & Wen, C. (2016). Alterations of the gut microbiome in Chinese patients with systemic lupus erythematosus. *Gut Pathogens*, *8*, 64.
Hevia, A., Milani, C., Lopez, P., Cuervo, A., Arboleya, S., Duranti, S., et al. (2014). Intestinal dysbiosis associated with systemic lupus erythematosus. *MBio*, *5*(5), e01548–01514.
Hiippala, K., Jouhten, H., Ronkainen, A., Hartikainen, A., Kainulainen, V., Jalanka, J., et al. (2018). The potential of gut commensals in reinforcing intestinal barrier function and alleviating inflammation. *Nutrients*, *10*(8).

Hsiao, E. Y., McBride, S. W., Hsien, S., Sharon, G., Hyde, E. R., McCue, T., et al. (2013). Microbiota modulate behavioral and physiological abnormalities associated with neurodevelopmental disorders. *Cell*, *155*(7), 1451–1463.

Ireland, S. J., Monson, N. L., & Davis, L. S. (2015). Seeking balance: Potentiation and inhibition of multiple sclerosis autoimmune responses by IL-6 and IL-10. *Cytokine*, *73*(2), 236–244.

Ivanov, K., II, Atarashi, N., Manel, E. L., Brodie, T., Shima, U., Karaoz, D., et al. (2009). Induction of intestinal Th17 cells by segmented filamentous bacteria. *Cell*, *139*(3), 485–498.

Ivanov, L., II, Frutos Rde, N., Manel, K., Yoshinaga, D. B., Rifkin, R. B., Sartor, B. B., et al. (2008). Specific microbiota direct the differentiation of IL-17-producing T-helper cells in the mucosa of the small intestine. *Cell Host & Microbe*, *4*(4), 337–349.

Ivanov, I. I., & Littman, D. R. (2010). Segmented filamentous bacteria take the stage. *Mucosal Immunology*, *3*(3), 209–212.

Jangi, S., Gandhi, R., Cox, L. M., Li, N., von Glehn, F., Yan, R., et al. (2016). Alterations of the human gut microbiome in multiple sclerosis. *Nature Communications*, *7*, 12015.

Jensen, S. N., Cady, N. M., Shahi, S. K., Peterson, S. R., Gupta, A., Gibson-Corley, K. N., et al. (2021). Isoflavone diet ameliorates experimental autoimmune encephalomyelitis through modulation of gut bacteria depleted in patients with multiple sclerosis. *Science Advance*, 7(28), eabd4595.

Jones, E., Pike, J., Marshall, T., & Ye, X. (2016). Quantifying the relationship between increased disability and health care resource utilization, quality of life, work productivity, health care costs in patients with multiple sclerosis in the US. *BMC Health Services Research*, *16*, 294.

Kang, D. W., Adams, J. B., Gregory, A. C., Borody, T., Chittick, L., Fasano, A., et al. (2017). Microbiota transfer therapy alters gut ecosystem and improves gastrointestinal and autism symptoms: An open-label study. *Microbiome*, *5*(1), 10.

Kayama, H., Okumura, R., & Takeda, K. (2020). Interaction between the microbiota, epithelia, and immune cells in the intestine. *Annual Review of Immunology*, *38*, 23–48.

Kouchaki, E., Tamtaji, O. R., Salami, M., Bahmani, F., Daneshvar Kakhaki, R., Akbari, E., et al. (2017). Clinical and metabolic response to probiotic supplementation in patients with multiple sclerosis: A randomized, double-blind, placebo-controlled trial. *Clinical Nutrition*, *36*(5), 1245–1249.

Kuusisto, H., Kaprio, J., Kinnunen, E., Luukkaala, T., Koskenvuo, M., & Elovaara, I. (2008). Concordance and heritability of multiple sclerosis in Finland: Study on a nationwide series of twins. *European Journal of Neurology*, *15*(10), 1106–1110.

Lang, H. L., Jacobsen, H., Ikemizu, S., Andersson, C., Harlos, K., Madsen, L., et al. (2002). A functional and structural basis for TCR cross-reactivity in multiple sclerosis. *Nature Immunology*, *3*(10), 940–943.

Lee, Y. K., Menezes, J. S., Umesaki, Y., & Mazmanian, S. K. (2011). Proinflammatory T-cell responses to gut microbiota promote experimental autoimmune encephalomyelitis. *Proceedings of the National Academy of Sciences of the United States of America*, *108*(Suppl. 1), 4615–4622.

Levi Mortera, S., Vernocchi, P., Basadonne, I., Zandona, A., Chierici, M., Durighello, M., et al. (2022). A metaproteomic-based gut microbiota profiling in children affected by autism spectrum disorders. *Journal of Proteomics*, *251*, 104407.

Lindsey, J. W. (2017). Antibodies to the Epstein-Barr virus proteins BFRF3 and BRRF2 cross-react with human proteins. *Journal of Neuroimmunology*, *310*, 131–134.

Liu, P., Wu, L., Peng, G., Han, Y., Tang, R., Ge, J., et al. (2019). Altered microbiomes distinguish Alzheimer's disease from amnestic mild cognitive impairment and health in a Chinese cohort. *Brain, Behavior, and Immunity*, *80*, 633–643.

Luo, X. M., Edwards, M. R., Mu, Q., Yu, Y., Vieson, M. D., Reilly, C. M., et al. (2018). Gut microbiota in human systemic lupus erythematosus and a mouse model of lupus. *Applied and Environmental Microbiology*, *84*(4).

Mangalam, A., Shahi, S. K., Luckey, D., Karau, M., Marietta, E., Luo, N., et al. (2017). Human gut-derived commensal bacteria suppress CNS inflammatory and demyelinating disease. *Cell Reports*, *20*(6), 1269–1277.

Marietta, E., Horwath, I., Meyer, S., Khaleghi-Rostamkolaei, S., Norman, E., Luckey, D., et al. (2022). Administration of human derived upper gut commensal Prevotella histicola delays the onset of type 1 diabetes in NOD mice. *BMC Microbiology*, *22*(1), 8.

Marietta, E. V., Murray, J. A., Luckey, D. H., Jeraldo, P. R., Lamba, A., Patel, R., et al. (2016). Suppression of inflammatory arthritis by human gut-derived Prevotella histicola in humanized mice. *Arthritis & Rhematology*, *68*(12), 2878–2888.

Mazmanian, S. K., Liu, C. H., Tzianabos, A. O., & Kasper, D. L. (2005). An immunomodulatory molecule of symbiotic bacteria directs maturation of the host immune system. *Cell*, *122*(1), 107–118.

Mazmanian, S. K., Round, J. L., & Kasper, D. L. (2008). A microbial symbiosis factor prevents intestinal inflammatory disease. *Nature*, *453*(7195), 620–625.

Mestre, L., Carrillo-Salinas, F. J., Feliu, A., Mecha, M., Alonso, G., Espejo, C., et al. (2020). How oral probiotics affect the severity of an experimental model of progressive multiple sclerosis? Bringing commensal bacteria into the neurodegenerative process. *Gut Microbes*, *12*(1), 1813532.

Mirza, A. I., Zhu, F., Knox, N., Forbes, J. D., Van Domselaar, G., Bernstein, C. N., et al. (2022). Metagenomic analysis of the pediatric-onset multiple sclerosis gut microbiome. *Neurology*, *98*(10), e1050–e1063.

Miyake, S., Kim, S., Suda, W., Oshima, K., Nakamura, M., Matsuoka, T., et al. (2015). Dysbiosis in the gut microbiota of patients with multiple sclerosis, with a striking depletion of species belonging to Clostridia XIVa and IV clusters. *PLoS One*, *10*(9), e0137429.

Morecraft, R. J., Blair, W. F., & Maynard, J. A. (1988). Suture extrusion in small arteries and veins. *Scandinavian Journal of Plastic and Reconstructive Surgery and Hand Surgery*, *22*(2), 121–126.

Nouri, M., Bredberg, A., Westrom, B., & Lavasani, S. (2014). Intestinal barrier dysfunction develops at the onset of experimental autoimmune encephalomyelitis, and can be induced by adoptive transfer of auto-reactive T cells. *PLoS One*, *9*(9), e106335.

Ochoa-Reparaz, J., Mielcarz, D. W., Ditrio, L. E., Burroughs, A. R., Begum-Haque, S., Dasgupta, S., et al. (2010). Central nervous system demyelinating disease protection by the human commensal Bacteroides fragilis depends on polysaccharide A expression. *Journal of Immunology*, *185*(7), 4101–4108.

Oksenberg, J. R., Barcellos, L. F., Cree, B. A., Baranzini, S. E., Bugawan, T. L., Khan, O., et al. (2004). Mapping multiple sclerosis susceptibility to the HLA-DR locus in African Americans. *American Journal of Human Genetics*, *74*(1), 160–167.

Olerup, O., & Hillert, J. (1991). HLA class II-associated genetic susceptibility in multiple sclerosis: A critical evaluation. *Tissue Antigens*, *38*(1), 1–15.

Pellizoni, F. P., Leite, A. Z., Rodrigues, N. C., Ubaiz, M. J., Gonzaga, M. I., Takaoka, N. N. C., et al. (2021). Detection of dysbiosis and increased intestinal permeability in Brazilian patients with relapsing-remitting multiple sclerosis. *International Journal of Environmental Research and Public Health*, *18*(9).

Planas, R., Santos, R., Tomas-Ojer, P., Cruciani, C., Lutterotti, A., Faigle, W., et al. (2018). GDP-l-fucose synthase is a CD4(+) T cell-specific autoantigen in DRB3*02:02 patients with multiple sclerosis. *Science Translational Medicine*, *10*(462).

Poisson, L. M., Suhail, H., Singh, J., Datta, I., Denic, A., Labuzek, K., et al. (2015). Untargeted plasma metabolomics identifies endogenous metabolite with drug-like properties in chronic animal model of multiple sclerosis. *The Journal of Biological Chemistry*, *290*(52), 30697–30712.

Qian, Y., Yang, X., Xu, S., Wu, C., Song, Y., Qin, N., et al. (2018). Alteration of the fecal microbiota in Chinese patients with Parkinson's disease. *Brain, Behavior, and Immunity, 70*, 194–202.

Raetz, C. R., & Whitfield, C. (2002). Lipopolysaccharide endotoxins. *Annual Review of Biochemistry, 71*, 635–700.

Ramagopalan, S. V., Deluca, G. C., Morrison, K. M., Herrera, B. M., Dyment, D. A., Lincoln, M. R., et al. (2008). Analysis of 45 candidate genes for disease modifying activity in multiple sclerosis. *Journal of Neurology, 255*(8), 1215–1219.

Ramagopalan, S. V., Morris, A. P., Dyment, D. A., Herrera, B. M., DeLuca, G. C., Lincoln, M. R., et al. (2007). The inheritance of resistance alleles in multiple sclerosis. *PLoS Genetics, 3*(9), 1607–1613.

Ramakrishna, C., Kujawski, M., Chu, H., Li, L., Mazmanian, S. K., & Cantin, E. M. (2019). Bacteroides fragilis polysaccharide A induces IL-10 secreting B and T cells that prevent viral encephalitis. *Nature Communications, 10*(1), 2153.

Ridaura, V. K., Faith, J. J., Rey, F. E., Cheng, J., Duncan, A. E., Kau, A. L., et al. (2013). Gut microbiota from twins discordant for obesity modulate metabolism in mice. *Science, 341*(6150), 1241214.

Rothhammer, V., & Quintana, F. J. (2016). Environmental control of autoimmune inflammation in the central nervous system. *Current Opinion in Immunology, 43*, 46–53.

Round, J. L., & Mazmanian, S. K. (2009). The gut microbiota shapes intestinal immune responses during health and disease. *Nature Reviews. Immunology, 9*(5), 313–323.

Round, J. L., & Mazmanian, S. K. (2010). Inducible Foxp3 + regulatory T-cell development by a commensal bacterium of the intestinal microbiota. *Proceedings of the National Academy of Sciences of the United States of America, 107*(27), 12204–12209.

Salehipour, Z., Haghmorad, D., Sankian, M., Rastin, M., Nosratabadi, R., Soltan Dallal, M. M., et al. (2017). Bifidobacterium animalis in combination with human origin of Lactobacillus plantarum ameliorate neuroinflammation in experimental model of multiple sclerosis by altering CD4+ T cell subset balance. *Biomedicine & Pharmacotherapy, 95*, 1535–1548.

Saresella, M., Mendozzi, L., Rossi, V., Mazzali, F., Piancone, F., LaRosa, F., et al. (2017). Immunological and clinical effect of diet modulation of the gut microbiome in multiple sclerosis patients: A pilot study. *Frontiers in Immunology, 8*, 1391.

Scher, J. U., & Abramson, S. B. (2013). Periodontal disease, Porphyromonas gingivalis, and rheumatoid arthritis: what triggers autoimmunity and clinical disease? *Arthritis Research & Therapy, 15*(5), 122.

Secher, T., Kassem, S., Benamar, M., Bernard, I., Boury, M., Barreau, F., et al. (2017). Oral administration of the probiotic strain Escherichia coli Nissle 1917 reduces susceptibility to neuroinflammation and repairs experimental autoimmune encephalomyelitis-induced intestinal barrier dysfunction. *Frontiers in Immunology, 8*, 1096.

Shah, S., Locca, A., Dorsett, Y., Cantoni, C., Ghezzi, L., Lin, Q., et al. (2021). Alterations of the gut mycobiome in patients with MS. *eBioMedicine, 71*, 103557.

Shahi, S. K., Freedman, S. N., Murra, A. C., Zarei, K., Sompallae, R., Gibson-Corley, K. N., et al. (2019). Prevotella histicola, a human gut commensal, is as potent as COPAXONE(R) in an animal model of multiple sclerosis. *Frontiers in Immunology, 10*, 462.

Shaw, K. A., Bertha, M., Hofmekler, T., Chopra, P., Vatanen, T., Srivatsa, A., et al. (2016). Dysbiosis, inflammation, and response to treatment: A longitudinal study of pediatric subjects with newly diagnosed inflammatory bowel disease. *Genome Medicine, 8*(1), 75.

Simeonova, D., Ivanovska, M., Murdjeva, M., Carvalho, A. F., & Maes, M. (2018). Recognizing the leaky gut as a trans-diagnostic target for neuroimmune disorders using clinical chemistry and molecular immunology assays. *Current Topics in Medicinal Chemistry, 18*(19), 1641–1655.

Singh, R. K., Chang, H. W., Yan, D., Lee, K. M., Ucmak, D., Wong, K., et al. (2017). Influence of diet on the gut microbiome and implications for human health. *Journal of Translational Medicine, 15*(1), 73.
Sjostrom, B., Bredberg, A., Mandl, T., Alonso-Magdalena, L., Ohlsson, B., Lavasani, S., et al. (2021). Increased intestinal permeability in primary Sjogren's syndrome and multiple sclerosis. *Journal of Translational Autoimmunity, 4*, 100082.
Smith, P. M., Howitt, M. R., Panikov, N., Michaud, M., Gallini, C. A., Bohlooly, Y. M., et al. (2013). The microbial metabolites, short-chain fatty acids, regulate colonic Treg cell homeostasis. *Science, 341*(6145), 569–573.
Tamboli, C. P., Neut, C., Desreumaux, P., & Colombel, J. F. (2004). Dysbiosis in inflammatory bowel disease. *Gut, 53*(1), 1–4.
Tankou, S. K., Regev, K., Healy, B. C., Cox, L. M., Tjon, E., Kivisakk, P., et al. (2018). Investigation of probiotics in multiple sclerosis. *Multiple Sclerosis, 24*(1), 58–63.
Tankou, S. K., Regev, K., Healy, B. C., Tjon, E., Laghi, L., Cox, L. M., et al. (2018). A probiotic modulates the microbiome and immunity in multiple sclerosis. *Annals of Neurology, 83*(6), 1147–1161.
Telesford, K. M., Yan, W., Ochoa-Reparaz, J., Pant, A., Kircher, C., Christy, M. A., et al. (2015). A commensal symbiotic factor derived from Bacteroides fragilis promotes human CD39(+)Foxp3(+) T cells and Treg function. *Gut Microbes, 6*(4), 234–242.
Tengvall, K., Huang, J., Hellstrom, C., Kammer, P., Bistrom, M., Ayoglu, B., et al. (2019). Molecular mimicry between Anoctamin 2 and Epstein-Barr virus nuclear antigen 1 associates with multiple sclerosis risk. *Proceedings of the National Academy of Sciences of the United States of America, 116*(34), 16955–16960.
Vascellari, S., Palmas, V., Melis, M., Pisanu, S., Cusano, R., Uva, P., et al. (2020). Gut microbiota and metabolome alterations associated with Parkinson's disease. *mSystems, 5*(5).
Vieira, S. M., Pagovich, O. E., & Kriegel, M. A. (2014). Diet, microbiota and autoimmune diseases. *Lupus, 23*(6), 518–526.
Vogt, N. M., Kerby, R. L., Dill-McFarland, K. A., Harding, S. J., Merluzzi, A. P., Johnson, S. C., et al. (2017). Gut microbiome alterations in Alzheimer's disease. *Scientific Reports*, 7(1), 13537.
Wen, C., Zheng, Z., Shao, T., Liu, L., Xie, Z., Le Chatelier, E., et al. (2017). Quantitative metagenomics reveals unique gut microbiome biomarkers in ankylosing spondylitis. *Genome Biology, 18*(1), 142.
Willing, B. P., Dicksved, J., Halfvarson, J., Andersson, A. F., Lucio, M., Zheng, Z., et al. (2010). A pyrosequencing study in twins shows that gastrointestinal microbial profiles vary with inflammatory bowel disease phenotypes. *Gastroenterology, 139*(6), 1844–1854 e1841.
Winkvist, A., Barebring, L., Gjertsson, I., Ellegard, L., & Lindqvist, H. M. (2018). A randomized controlled cross-over trial investigating the effect of anti-inflammatory diet on disease activity and quality of life in rheumatoid arthritis: The anti-inflammatory diet in rheumatoid arthritis (ADIRA) study protocol. *Nutrition Journal, 17*(1), 44.
Wu, L., Han, Y., Zheng, Z., Peng, G., Liu, P., Yue, S., et al. (2021). Altered gut microbial metabolites in amnestic mild cognitive impairment and Alzheimer's disease: signals in host-microbe interplay. *Nutrients, 13*(1).
Wu, H. J., & Wu, E. (2012). The role of gut microbiota in immune homeostasis and autoimmunity. *Gut Microbes, 3*(1), 4–14.
Yadav, M., Ali, S., Shrode, R. L., Shahi, S. K., Jensen, S. N., Hoang, J., et al. (2021). Multiple sclerosis patients have an altered gut mycobiome and increased fungal to bacterial richness. *bioRxiv*. preprint.

Yadav, V., Marracci, G., Kim, E., Spain, R., Cameron, M., Overs, S., et al. (2016). Low-fat, plant-based diet in multiple sclerosis: A randomized controlled trial. *Multiple Sclerosis and Related Disorders*, *9*, 80–90.

Yu, S., Peng, H. R., Zhang, Y. K., Yin, Y. Q., & Zhou, J. W. (2022). Central dopaminergic control of cell proliferation in the colonic epithelium. *Neuroscience Research*.

Zhang, X., Zhang, D., Jia, H., Feng, Q., Wang, D., Liang, D., et al. (2015). The oral and gut microbiomes are perturbed in rheumatoid arthritis and partly normalized after treatment. *Nature Medicine*, *21*(8), 895–905.

Zhou, C., Zhao, H., Xiao, X. Y., Chen, B. D., Guo, R. J., Wang, Q., et al. (2020). Metagenomic profiling of the pro-inflammatory gut microbiota in ankylosing spondylitis. *Journal of Autoimmunity*, *107*, 102360.

Zhuang, Z. Q., Shen, L. L., Li, W. W., Fu, X., Zeng, F., Gui, L., et al. (2018). Gut Microbiota is Altered in Patients with Alzheimer's Disease. *Journal of Alzheimer's Disease*, *63*(4), 1337–1346.

CHAPTER EIGHT

Interactions between the gut microbiome and ketogenic diet in refractory epilepsy

Ezgi Özcan†, Gregory R. Lum, and Elaine Y. Hsiao*

Department of Integrative Biology & Physiology, University of California, Los Angeles, Los Angeles, CA, United States

*Corresponding author: e-mail address: ehsiao@g.ucla.edu

†Co-corresponding author: e-mail address: ezgiozcan@ucla.edu

Contents

Abstract

Epilepsy is one of the most common neurological diseases globally, afflicting approximately 50 million people worldwide. While many antiepileptic drugs exist, an estimated one-third of individuals do not respond to available medications. The high fat, low carbohydrate ketogenic diet (KD) has been used to treat refractory epilepsy in cases when existing antiepileptic drugs fail. However, there are many variations of the KD, each of which varies greatly in its efficacy and side effects. Increasing evidence suggests that interactions between the KD and gut microbiome may modulate the effects of the diet on host physiology. Herein, we review existing evidence of microbiome differences in epileptic individuals compared to healthy controls. We highlight in particular both clinical and animal studies revealing effects of the KD on the composition and function of the microbiome, as well as proof-of-concept animal studies that implicate the microbiome in the antiseizure effects of the KD. We further synthesize findings suggesting that variations in clinical KD formulations may differentially influence host physiology and discuss the gut microbial interactions with specific dietary factors that may play

International Review of Neurobiology, Volume 167
ISSN 0074-7742
https://doi.org/10.1016/bs.irn.2022.06.002

a role. Overall, understanding interactions between the gut microbiota and specific nutritional components of clinical KDs could reveal foundational mechanisms that underlie the effectiveness, variability, and side effects of different KDs, with the potential to lead to precision nutritional and microbiome-based approaches to treat refractory epilepsy.

1. Introduction to epilepsy

Since the dawn of recorded history, epilepsy has been a burden on human health. Documented in ancient Mesopotamian medical texts as early as 2000 B.C., and later acknowledged by the ancient Egyptians and Greeks, epilepsy was originally attributed to being possessed by an ill-intentioned spirit or deity (World Health Organization, 2005). Despite these early ideas, epilepsy was not completely distanced from its divine beginnings and was not fully accepted as a brain disorder until the early 19th century (Magiorkinis, Sidiropoulou, & Diamantis, 2010).

Presently, epilepsy is the fourth most common neurological disorder worldwide and is broadly defined by recurring epileptic seizures. The general mechanism of an epileptic seizure event originates from unprovoked abnormal or excessive synchronous neuronal activity, commonly originating in the temporal lobe. A patient is considered epileptic if they have two or more unprovoked or reflex seizure events within a span of time greater than 24 h apart (Fisher et al., 2014). Both genetic factors, such as mutations in voltage- or ligand-gated ion channels, and environmental factors, such as brain trauma or infection, contribute to risk for epilepsy (Stafstrom & Carmant, 2015).

Although epileptic seizure events and associated risk factors are well characterized, the fundamental mechanisms underlying epileptogenesis remain poorly understood. The World Health Organization reports over 50 million cases of epilepsy worldwide, accounting for 0.5% of the global health burden (WHO, 2019). Modern antiepileptic drugs (AEDs) are the frontline treatment reducing seizures in approximately 60–70% of cases by blocking sodium or calcium ion channels to decrease the release of the excitatory neurotransmitter glutamate or enhance signaling of the inhibitory neurotransmitter gamma-aminobutyric acid (GABA) (Xia, Ou, & Pan, 2017).

For the remaining one in three epilepsy patients who do not respond to available AEDs, alternative forms of seizure management must be considered, each with their own caveats, or the patient faces a three- to sixfold increase in premature death (Sperling, 2004). First, invasive surgery to resect portions of the brain where seizure activity originates can be used to eliminate seizure activity. However, it must be feasible to accurately locate and safely resect the portion of the brain from which the seizures originate. Success rates vary from 40% to 70% and are highly dependent on seizure etiology (Health Quality Ontario, 2012). Second, refractory (drug-resistant) epilepsy patients can also consider surgical implantation of a neuromodulation device near the vagus nerve or directly in the brain to manage seizures with electrical stimulation. These devices reduce seizures in an estimated 25–50% of refractory epileptic individuals that pursue this intervention (Gooneratne et al., 2016). Lastly, dietary therapies are used as a clinical treatment for refractory epilepsy and are the least invasive alternatives to AEDs. The high fat, low carbohydrate ketogenic diet (KD) is the most well-known dietary therapy for refractory epilepsy. Typically, the KD has between a 22–55% success rate of at least 50% seizure reduction in adults and upwards of 60–75% success rates in infants and children (Castellano & Merlino, 2012; Henderson, Filloux, Alder, Lyon, & Caplin, 2006; Klein, Janousek, Barber, & Weissberger, 2010; Li, Zou, & Ding, 2013; Merrit & Putnam, 1938; Williams & Cervenka, 2017). While the KD can effectively reduce seizures for subsets of individuals, its restrictiveness introduces many side effects including hypoglycemia, metabolic acidosis, and weight loss. These obstacles make it difficult for epilepsy patients to maintain the diet, with about 33% dietary compliance in adults and between 42–60% compliance in children under 10 years old within 6 months of starting the KD (Coppola et al., 2002; Yang et al., 2022). Moreover, the mechanisms of action for the KD are poorly understood. Overall, while several AEDs are available for treating epilepsy, effective and manageable non-invasive treatments are lacking for the remaining one in three individuals with refractory epilepsy. Increased understanding of the mechanisms that enable the efficacy of diet-based therapies for subsets of individuals could uncover new therapeutic targets and approaches to improve the overall success rates, efficacy, and ease of maintenance of dietary treatments.

2. Increasing interest in the microbiome and epilepsy

The microbiome has emerged as a mediator between diet and host, becoming a specific topic of interest to enhance the efficacy of diet-based therapies. The gut microbiota encompasses trillions of native bacteria, viruses, and fungi residing in the host gastrointestinal tract (Proctor et al., 2019). Both gut microbiome composition and function are fluid, rapidly changing when exposed to variables such as medication, stress, and diet but simultaneously influenced by variables such as host genetics (Hasan & Yang, 2019; Turnbaugh et al., 2009). The gut microbiota significantly alters amino acid and carbohydrate metabolism, linked to microglial and astrocyte function, vagal activity, and hippocampal neurotransmitter levels (Lum, Olson, & Hsiao, 2020). Animal studies further reveal that the microbiota is linked to neurobehavior, symptoms of neurological disorders, and brain development (Vuong, Yano, Fung, & Hsiao, 2017). Additionally, alterations in the gut microbiome have been reported across several neuropsychiatric, neurodevelopmental, neurodegenerative disorders, and animal models of epilepsy (Citraro et al., 2021; Olson et al., 2018), but the connection between the microbiome and human epilepsy remains unclear (Fig. 1).

Only a few recent clinical studies have reported differences in baseline fecal microbiota profiles in small cohorts of epileptic individuals compared to healthy controls. One such study analyzed fecal microbiota profiles from 16S rRNA gene sequencing of stool samples collected from 14 infants and toddlers with refractory epilepsy (1–4 years old) and 30 matched healthy controls who did not have any chronic illnesses, metabolic diseases, and had not taken antibiotics 1 month prior to study initiation (Xie et al., 2017). Between healthy and refractory epilepsy groups, there was no significant difference observed in α-diversity measured by the Shannon index for evenness. However, principal component analysis (PCA) of the taxonomic data revealed distinct clustering of refractory epilepsy samples away from the healthy controls, suggesting differences in fecal microbiota β-diversity. Furthermore, Linear Discriminant Analysis Effect Size (LEfSe) indicated that the infant refractory epilepsy fecal microbiota had increased relative abundances of *Bacillota* and *Pseudomonadota* and decreased relative abundances of *Actinomycetota* and *Bacteroidota*. Additional differences in taxonomic membership were notable at the genus level. Refractory epilepsy infants had high levels of *Cronobacter* which was not detected in healthy infant controls. Relative abundances of *Bacteroides, Bifidobacterium,* and *Prevotella* were decreased in epileptic infants compared to healthy controls. Although differences in fecal microbiota profiles were observed between epileptic and healthy infants, it remains unclear whether environmental variables, such as dietary intake and existing medications, may be responsible for the differences reported in this study.

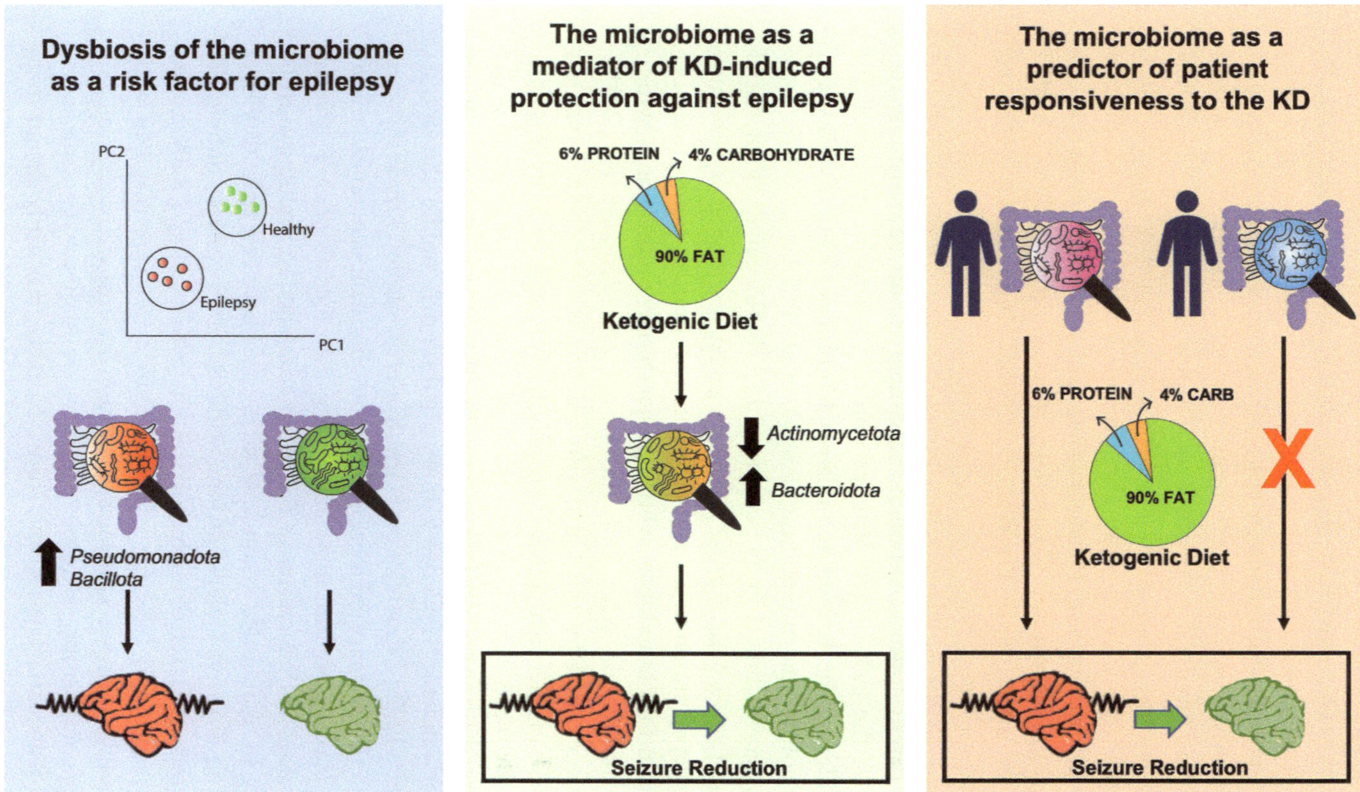

Fig. 1 Potential roles for the gut microbiome in epilepsy. There are many distinct and potentially overlapping ways by which the microbiome could contribute to epilepsy. Dysbiosis in the microbiome could contribute to the pathogenesis or manifestation of epilepsy. The microbiome or changes in the microbiome could contribute to the antiseizure effects of therapeutic treatment, for example with the ketogenic diet. Variations in the gut microbiome could serve as biomarkers for whether a patient responds vs fails to respond to a particular treatment, such as the ketogenic diet.

A second human study profiled the fecal microbiota of 12 children (2–17 years old) with refractory epilepsy and 11 healthy parent controls (Lindefeldt et al., 2019). Shotgun metagenomic sequencing analysis indicated that children with refractory epilepsy exhibited decreased Shannon index α-diversity compared to the healthy parent controls. PCA of fecal microbiome taxonomic profiles displayed distinct clustering of the healthy parent microbiomes away from child refractory epilepsy microbiomes. Children with refractory epilepsy also presented with increased relative abundances of *Bacillota* and *Pseudomonadota* and decreased relative abundances of *Actinomycetota* and *Bacteroidota.* Metagenomic analysis revealed differences in microbiome functional potential, highlighting decreased gene content in the acetyl-CoA pathway from β-hydroxybutyryl-CoA and crotonase in the child refractory epilepsy microbiome compared to the healthy parent control. Although the sample cohort contained household controls, the lack of age-matched controls is a major caveat in this study, considering the known age-related differences in the diversity and the function of the gut microbiome (Nagpal et al., 2018).

Another recent clinical study performed 16S rRNA gene sequencing of stool samples from a cohort of 42 refractory epilepsy individuals, 49 drug-sensitive epilepsy individuals, and 65 matched healthy family member controls, excluding individuals who took antibiotic or probiotics within the past 3 months or who had chronic illness (Peng et al., 2018). α-Diversity measured by the Chao1 diversity index for species richness was increased in the fecal microbiota from refractory epilepsy patients, specifically for those individuals experiencing more than four seizures per year. Weighted principal coordinate analysis (PCoA) revealed no distinct clustering of refractory epilepsy patient microbiomes unlike the distinct clustering seen in the Lindefeldt et al. and Xie et al. studies. Linear discriminant analysis (LDA) reported increased relative abundances of several members of the *Bacillota* phylum including *Coprococcus, Coprobacillus, Roseburia,* and *Ruminococcus* with decreases in *Bacteroides* compared to healthy controls. Interestingly LDA revealed taxonomic differences between refractory epilepsy and drug-sensitive epilepsy groups, with refractory epilepsy individuals exhibiting increased relative abundances of *Akkermansia, Fusobacterium, Methanobrevibacter,* and *Neisseria.* Epileptic individuals were matched with a healthy family member control to decrease the impact of diet on gut microbiome composition. However, there were no dietary controls across the families that participated in this study. Notably, this study design matched age, sex, and exposure to medications across the three representative groups.

Another human study used a cohort of 30 adult patients with idiopathic focal epilepsy and 10 healthy control adults had their stool samples sequenced by 16S rRNA gene sequencing (Şafak, Altunan, Topçu, & Eren Topkaya, 2020). Weighted PCoA revealed distinct clustering between healthy and epileptic patient groups. Epileptic patients were reported to have decreased relative abundance of *Bacillota, Bacteroidota,* and *Actinomycetota.* Similar to the reports from Xie et al. and Lindefeldt et al., Şafak et al. also reports increased relative abundance of *Pseudomonadota* in epileptic individuals compared to the healthy controls. The lack of dietary control and age related differences may explain why the findings from this study do not align with previous reported findings (Lindefeldt et al., 2019; Peng et al., 2018; Xie et al., 2017).

Another study of a cohort of 74 adults, including 24 epilepsy patients and 50 healthy adult controls, had their stool samples sequenced by 16S rRNA gene sequencing (Cui et al., 2022). α-Diversity measured by the Shannon and Simpson indexes showed the healthy controls had significantly increased α-diversity compared to epileptic individuals. Unweighted PCoA to assess β-diversity revealed distinct and separate clustering of healthy control and epileptic individuals. Epileptic individuals were shown to have significantly increased relative abundances of *Pseudomonadota* and *Actinomycetota* and significantly decreased relative abundance of *Bacillota* compared to healthy controls.

A final human clinical study performed 16S rRNA gene sequencing on collected fecal samples from a cohort of 23 children with infantile spasms and a cohort of 21 healthy children (Wan et al., 2021). α-Diversity measured by the Shannon, Simpson, and Chao1 indexes showed no difference between epileptic and healthy individuals. Further assessment of β-diversity by PCA revealed no differences between epileptic and health individuals. LEfSe analysis revealed significantly increased abundances of *Clostridium* and significantly decreased abundances of *Lactobacillus, Roseburia,* and *Lachnospira* compared to healthy individuals.

Across these human studies, alterations in the fecal microbiota profile of refractory epilepsy individuals were observed relative to healthy controls (Cui et al., 2022; Lindefeldt et al., 2019; Peng et al., 2018; Şafak et al., 2020; Xie et al., 2017) with the exception of the Wan et al. study. Several studies reported increased relative abundance of *Bacillota* in refractory epilepsy individuals (Lindefeldt et al., 2019; Peng et al., 2018; Xie et al., 2017). However, microbial signatures differed across studies when considering more resolved taxonomic levels. α-Diversity was a point of contention across these studies with conflicting results. Differences in study design, cohort age, small cohort size, and lack of data on genetic factors and environmental controls each influence the composition and function of the gut microbiome, makes these studies difficult to compare. Differences in taxonomic analysis, by using 16S rRNA gene sequencing compared to shotgun metagenomic sequencing, also complicate study comparisons because shotgun metagenomics affords microbiome functional profiling and strain level specificity while 16S rRNA gene profiling can only capture less resolved levels of taxonomic diversity. The results from these initial studies warrant larger efforts to achieve adequately powered and matched patient and control populations, and to account for variables such as age, genetics, medications, and diet.

Notably, one initial study reported significant differences in gut microbiome composition between drug-responsive and drug-resistant epileptic individuals (Lee, Lee, Lee, & Kim, 2021). As yet, however, strong evidence is lacking for the notion that differences in the microbiota may contribute to the pathogenesis of epilepsy or be used as a biomarker for epilepsies. Several additional studies have since examined the alternative hypothesis that the microbiome could contribute to the efficacy of select interventions for epilepsy (Fig. 1).

3. Ketogenic diet and epilepsy

Disease and diet have shared a close relationship spanning the last 2500 years. Fast-forward to the 1920s, two pivotal observations underlying the modern ketogenic diet were made. In 1921, Woodyatt observed acetone and β-hydroxybutyric acid (BHBA) buildup in a normal fasting paradigm, as well as on a diet with minimal carbohydrates and an over ingestion of fat (Woodyatt, 1921). In the same year, Wilder proposed that ketonemia could provide similar benefits to fasting but could be maintained far longer than fasting (Wilder, 1921). Upon placing patients on this high-fat, low-carbohydrate dietary intervention, he coined the term "ketogenic diet" (KD). The KD was popular as a treatment for seizures throughout the 1920s and 1930s, but with the discovery of the first AED, diphenylhydantoin, in 1938, dietary treatment of epilepsy faded from the mainstream (Merrit & Putnam, 1938). KD was thrust back into the spotlight in 1997 with an episode of NBC's *Dateline* centered on the story of Charlie, a 2-year-old with presenting with refractory epilepsy. Dr. John M. Freeman (Johns Hopkins) treated the boy with KD therapy, and he soon became seizure free.

From this, The Charlie Foundation was formed to provide physicians, dieticians, and parents with informational videos about KD therapy, and interest in the diet began to increase again.

Although the KD is not the first line treatment for epilepsy, it is an effective and non-invasive treatment for refractory epilepsy, especially for those who do not qualify for surgical treatments. However, due to the KD's restrictiveness, patients often find it difficult to maintain on the diet for extended periods of time. For example, a KD patient who wants to maintain ketosis with a caloric intake of 2000 cal per day needs to consume no more than 50 g of carbohydrates per day. A plain bagel equates to 48 g of carbohydrates. Because of this restrictiveness, adults are 50% compliant with KD therapy, seeing successful treatment rates of approximately 30% (Henderson et al., 2006; Klein et al., 2010). Children placed on KD therapy see closer to 90% compliance and successful treatment rates of approximately 40% (Ye, Li, Jiang, Sun, & Liu, 2015). The KD has taken on the role of a niche treatment for certain epilepsy syndromes, including Dravet, Rett, and Lennox-Gastaut syndromes, where it has seen marginally better success rates for reducing seizures (D'Andrea Meira et al., 2019). Exactly how the KD is effective for treating seizures and why it may be more effective for certain individuals over others remains unclear.

Several different mechanisms of action have been proposed, and there are possibly multiple pathways through which the KD works to reduce seizures. Elevated levels of ketone bodies BHBA and acetoacetate, and decreased glucose are hallmarks of the KD's global metabolic effects to increase fatty acid oxidation. As such, the theory that ketone bodies may directly mediate the anti-seizure effects of the KD has been actively pursued, with mixed results in animal models (Kim et al., 2015; Likhodii et al., 2003). Positive results have linked ketone bodies to the modulation of ATP-sensitive potassium channels and neurotransmitter release as a mechanism for seizure reduction (Gasior, Hartman, & Rogawski, 2008; Ma, Berg, & Yellen, 2007; Rho, 2017). However, these are balanced by some studies showing no striking effects of ketone bodies on seizure severity and frequency (Chang et al., 2016; Likhodii et al., 2000).

The KD has also been proposed to confer anti-seizure effects through alterations in neuronal metabolism and mitochondrial activity. Distinct from direct effects of ketone bodies, some studies propose that it is the glucose restriction that protects against seizures. Glucose is transported across the blood brain barrier (BBB) as a key neuronal precursor for neurotransmitter synthesis and ATP to fuel neuronal activity (Greene, Todorova, & Seyfried, 2003). The KD reduces global glucose levels, reducing its bioavailability for neurons in the brain. Studies in animal models have shown the anti-seizure properties of the KD can be abolished with glucose supplementation (Huttenlocher, 1976) and that administration of 2-deoxy-D-glucose, a glycolysis inhibitor, similarly decreases seizure susceptibility (Garriga-Canut et al., 2006). The KD also increases mitochondrial biogenesis and mitochondrial energy reserves in the brain (Devivo, Leckie, Ferrendelli, & McDougal, 1978). This elevates available ATP and ATP-sensitive potassium channel (KATP) function while decreasing neuronal excitability (Bough et al., 2006). As such, the KD may promote seizure protection by limiting the bioavailability of glucose and altering mitochondrial function to a state attenuates seizures.

Beyond core theories regarding ketone body production and glucose restriction, several additional hypotheses exist for the KD. For example, amino acid modulation could also play a role, as the KD is reported to increase levels of GABA in the CFS of epilepsy patients (Wang et al., 2003). KD-induced reductions in levels of aspartate may help to promote GABA synthesis by enhancing glutamate decarboxylase function and facilitating astrocyte conversion of excitatory glutamate to glutamine (Yudkoff, Daikhin, Horyn, Nissim, & Nissim, 2008). In addition, epilepsy has been linked to inflammatory cytokine production, with some evidence suggesting that the KD promotes immunosuppression, reducing IL-1β and other related pro-inflammatory cytokines (Dupuis, Curatolo, Benoist, & Auvin, 2015). Overall, several pathways have been proposed to be important for the KD but the exact mechanism how the diet protects against seizures remains unclear. More research is needed to close this gap and drive the development of more effective KD-based therapies for refractory epilepsy.

The medium-chain triglyceride KD (MCT-KD) is an alternative to the conventional KD used to treat both children and adults with refractory epilepsy with approximately the same success rate (Martin, Jackson, Levy, & Cooper, 2016; Ye et al., 2015). The MCT-KD differs from the conventional KD consisting of long-chain fats that provide 60–80% of dietary energy, are replaced with medium-chain fats, providing closer to 45% of dietary energy (Augustin et al., 2018). This allows a more attractive option as a dietary intervention for epilepsy as it has less restrictive on carbohydrate consumption than classical KD. Ketone body biosynthesis with the MCT-KD is more metabolically efficient because of the increased number of ingested shorter-chained fatty acids. In addition to the mechanisms suggested for the conventional KD (Augustin et al., 2018; Bough et al., 2006; Devivo et al., 1978; Wang et al., 2003), MCT-KD has also been theorized to control seizures through the increase of MCTs specifically. Decanoic acid has been linked to seizure control through inhibition of glutamate receptors (Chang et al., 2016; Citraro et al., 2021). Octanoic acid is also reported to modulate seizures indirectly through adenosine receptors (Chang et al., 2016; Tan, Carrasco-Pozo, McDonald, Puchowicz, & Borges, 2017; Wlaź et al., 2012).

Polyunsaturated fatty acids are another subgroup of fatty acids with purported anti-seizure effects. The polyunsaturated fatty acid KD (PUFA-KD) differs from the conventional KD, switching out the traditional saturated fatty acids for PUFAs such as peanut oil, sunflower oil, butter, or cream to better control seizures. In a study done in Mumbai, India including 50 patients (ages 10 months to 35 years old) previously on conventional KD and not showing further seizure reduction, were switched to PUFA-KD (Nathan, Bailur, Datay, Sharma, & Khedekar Kale, 2019). It was reported that 88% of participants had a > 50% reduction in seizure frequency after 12 months, a significantly increased success rate compared to the conventional KD (Nathan et al., 2019). It is purported that the PUFA-KD through alterations in omega-6 and omega-3 fatty acids levels and proinflammatory cytokine production. An out of balance omega-6 to omega-3 fatty acid ratio has been linked to increased glutamate production, while increased levels of omega-3 fatty acids are linked to increased levels of GABA (Simopoulos, 2002; Taha, Burnham, & Auvin, 2010).

Lastly, the modified Atkins diet (MAD) is another common variation of the conventional KD commonly used to treat refractory epilepsy patients. MAD is a less stringent diet, which is its main advantage over the conventional KD. While on MAD, carbohydrate intake is initially confined to 10 g per day combined with unrestricted intake of calories, fats, and protein. Overall, MAD has been shown to have similar successful treatment rates to the conventional KD across several studies in both children and has been suggested to operate analogously to the conventional KD to provide its anti-seizure effects (Miranda, Mortensen, Povlsen, Nielsen, & Beniczky, 2011; Poorshiri et al., 2021; Porta et al., 2009). Like the conventional KD, several mechanisms of action have been suggested through which the MCT-KD, PUFA-KD, or MAD act to control seizures. However exactly how these alternative diets to the conventional KD provide their anti-seizure effects remains unclear.

4. Ketogenic diet impact on epilepsy via gut microbiome

The gut microbiome is shaped by diet and plays an important role in mediating the effects of diet on host physiology. Evidence is emerging for interactions between the gut microbiome and the KD in epilepsy, following the idea that KD-induced changes in the microbiome may contribute to the seizure protective effects of the diet (Lum et al., 2020). Notably, this hypothesis that the microbiome may mediate seizure protection in response to the KD is distinct from the notion that the microbiome could contribute to epileptogenesis, for which evidence is as yet limited (Fig. 1).

Animal studies have provided proof-of-concept that the microbiome could promote seizure protection. In the 6-Hz model of acute electrically induced seizures, mice fed the KD exhibited increased seizure resistance compared to controls fed a vitamin and mineral matched control diet (Olson et al., 2018). However, mice germ-free or treated with antibiotics to deplete the microbiota failed show increased seizure resistance, indicating that the microbiome is necessary for the anti-seizure effects of the KD. This was similarly seen in the Kcna1 −/− mouse model of sudden unexpected death in epilepsy (SUDEP), where mice treated with the KD displayed decreased frequency and duration of seizures, but treatment with antibiotics abrogated this effect. When KD was administered to mice, the gut microbiota composition was altered, decreasing α-diversity and increasing relative abundances of *Akkermansia muciniphila* and *Parabacteroides* spp. Transplantation of the KD-associated microbiota into naïve mice fed the control diet, or treatment with select KD-associated bacterial taxa, sufficiently conferred seizure protection, further suggesting that the gut microbiota contributes to the anti-seizure effects of the KD in the 6-Hz and Kcna1 −/− mouse models for refractory epilepsy. The KD-associated microbiota was shown to reduce gamma-glutamylation of ketogenic amino acids in the intestinal lumen and serum, and to correlate with increases in bulk GABA relative to glutamate in the hippocampus.

Overall, this initial study implicates the gut microbiota in mediating the anti-seizure effects of the KD in mice. Similarly, in a mouse model with *Scn1a* gene deficiency for Dravet syndrome, increases in *Clostridium* and decreased *Romboutsia* correlated with seizure severity. KD treatment shifted the microbiome by increasing *Bacillota* and decreasing *Bacteroidota,* which was correlated with reduced frequency and duration of motor seizures (Miljanovic & Potschka, 2021). In a rat model of infantile spasms, antibiotic treatment to deplete the gut microbiota improved the effectiveness of KD. In addition, a fecal transplant from KD treated animals into mice fed a conventional diet was effective at mitigating spasms (Mu et al., 2022). Additional experiments in animals and humans are warranted to investigate effects of the KD on the gut microbiota and potential roles for the gut microbiota in the KD.

4.1 Ketogenic diet on the composition of the gut microbiota in epilepsy

A few recent human studies have begun to report associations of the clinical KD with alterations in the composition of the gut microbiota in epileptic individuals (Lindefeldt et al., 2019; Peng et al., 2018; Tagliabue et al., 2017; Xie et al., 2017; Zhang et al., 2018). Several studies of children with refractory epilepsy similarly reported that the KD increased relative abundance of *Bacteroidota* and decreased relative abundance of *Bacillota* and *Actinomycetota* (Lindefeldt et al., 2019; Xie et al., 2017; Zhang et al., 2018). Interestingly, this general pattern was seen regardless of the effectiveness of the diet against seizures or the dietary fat to carbohydrate ratio. Three months of KD treatment in epileptic children ages 2–17, where the KD ratio ranged from 3:1 to 4:1, consistently decreased the relative abundance of *Actinomycetota*, particularly of the members *Bifidobacterium longum* and *Bifidobacterium adolescentis,* in the stool microbiota (Lindefeldt et al., 2019). In another pediatric cohort, 14 patients with refractory epilepsy aged 1–4 years old showed significantly increased *Bacteroides* after 1 week of 2:1 KD therapy (Xie et al., 2017). Similarly, 6 months of treatment with the 4:1 KD in 20 epilepsy patients, aged 1–10 years, was associated with increased relative abundance of *Bacteroidota*, and decreased *Bacillota* and *Actinomycetota* (Zhang et al., 2018). In addition, 6 months of KD treatment (standard protocol by Zeneca, ratio was not indicated) on children ages 2–8 with drug resistant epilepsy was correlated with decreases in *Bifidobacterium*, *Akkermansia*, *Actinomyces* and *Enterococcaceae*, though they did not reach statistical significance when compared to the pre-treatment internal controls (Gong et al., 2021).

While there are some similarities in reported results across independent studies, there are also many differences. In human studies conducted with epileptic children, there were only modest decreases in the α-diversity in KD treated individuals, but changes were not statistically significant (Gong et al., 2021; Lindefeldt et al., 2019; Xie et al., 2017; Zhang et al., 2018). This contrasts studies in mice, where the KD substantially decreased α-diversity compared to controls (Olson et al., 2018). However, in dogs with idiopathic epilepsy, the MCT-KD increased α-diversity compared to baseline diet (Pilla et al., 2020). In addition, despite the similar reports of KD-associated changes in the phylum composition of the microbiota, the specific changes varied greatly at more resolved taxonomic levels. Differential results could be due, at least in part, to heterogeneity in KD regimen, variability in initial microbiome state, differences in patient demographics and medical history, among other relevant variables. Differences in sequencing approach and analysis could also contribute. Metagenomic sequencing of the fecal microbiota revealed that 3 months of KD (3:1 to 4:1) in therapy resistant epileptic children aged 2–17 was linked with increased *Pseudomonadota* (*E. coli*) and reduced *Eubacterium rectale* and *Dialister* (Lindefeldt et al., 2019). In contrast, another study of 14 epileptic infants aged 1–4 who were treated for 1 week with the clinical KD reported diminished levels of *Pseudomonadota* as measured by 16S rRNA gene sequencing (Xie et al., 2017). Increased *Prevotella* and decreased *Coronobacter, Erysipelatoclostridium, Streptococcus, Alistipes, Ruminiclostridium, Barnesiella and Enterococcus* were also reported. Overall, the inconsistencies in KD-associated changes in the composition gut microbiota at species and genus level point to the need to better evaluate factors that could contribute to the variability. Even in healthy individuals the diversity and the abundance of the gut microbiota vary widely, however, the metagenomic carriage stays more stable among individuals despite the variations in community structure (The Human Microbiome Project Consortium, 2012). Therefore, such dietary perturbations call for characterization of metabolic and functional output.

One consideration is whether patient responsiveness to the KD may correlate with the microbiome, and if so, whether the microbiome could serve as a biomarker to identify those subsets of epileptic individuals who would benefit from the KD. The microbiota studies to date have included cohorts of epileptic patients that varied in their responsiveness to the KD, ranging from no improvement to >50% seizure reduction or even complete absence of seizures, while many of the studies did not evaluate microbiota signatures in responders vs non-responders, a few reported some differences. Among 20 patients subjected to the 4:1 KD for 6 months, compared to 10 patients with at least 50% seizure reduction, the non/low-responders (<50% reduction) had significantly increased relative abundances of *Clostridiales, Clostridia, Ruminococcaceae, Lachnospiraceae, Alistipes,* and *Rikenellaceae* (Zhang et al., 2018). In another study, epileptic patients who had positive responses to KD treatment after 6 months exhibited modest elevations in the relative abundances of *Eubacterium* and *Dialister* relative to those who showed poor response to the KD (Gong et al., 2021).

KD-associated taxonomic changes in the microbiota might also be influenced by seizure etiology. Glucose transporter 1 (GLUT1) deficiency syndrome is an early-onset childhood epileptic encephalopathy caused by impaired glucose transport across the BBB. This malfunction largely originates from mutations in the *SLC2A1* gene encoding GLUT1. The KD can effectively treat seizures in individuals with this syndrome (Klepper, 2012). Six patients (three females and three males, aged 8–34 years) with GLUT1 deficiency syndrome were subjected to the KD for 3 months, gradually increased from 1:1 to 4:1. After 90 days of treatment, KD-treated patients showed microbiota profiles with statistically significant increases in *Desulfovibrio* (Tagliabue et al., 2017). Notably, this signature was not observed in other human epilepsy studies of the gut microbiota and KD. In another study of KD-responsive epileptic patients with mutations occurring in the genes *SCN1A*, *NOTCH3,* and *CDKL5,* increases in *Tannerella, Sulfurosprillum* and *Parabacteroides* were reported (Gong et al., 2021). In contrast, patients without these genetic mutations showed an increase in *Hungatella* after KD treatment (Gong et al., 2021).

Some studies are beginning to assess the hypothesis that the effectiveness of KD treatments on epilepsy could be enhanced by prebiotics and/or probiotics. In a mouse model of pentylenetetrazol (PTZ)-induced seizures, co-treatment of the KD together with the probiotic *Lactobacillus fermentum* maintained KD-associated reductions in seizure susceptibility, while also preventing some KD-associated changes in serum lipid profiles and reductions in tight junction protein expression in the BBB. Probiotic treatment was associated with reductions in relative levels of *Bacteroidota* and increases in *Bacillota* and *Defferibacteres* (Eor, Tan, Son, Kwak, & Kim, 2021). In particular, KD treatment was associated with lower abundance of *Actinomycetota* and elevated *Bacteroides* which was prevented by co-treatment with *L. fermentum.* In a separate study of the PTZ seizure model, mice were fed the 4:1 KD for 8 weeks, together with *L. fermentum* MSK 408 and the prebiotic galactooligosaccharide (GOS) (Eor, Son, et al., 2021). Both the KD and co-treatment with synbiotics similarly reduced seizure susceptibility and the diversity of the gut microbiota, with increases in *Bacillota* relative to *Bacteroidota.* This is in contrast with results observed in human studies, where KD was frequently associated with elevations in *Bacteroidota.* In the same study, *Acetatifactor, Anaerotaenia, Escherichia, Flintibacter, Oscillibacter,* and *Erysipelatoclostridium* were significantly higher in KD-treated animals. Notably, while synbiotic treatment had no added effect on KD-induced seizure protection, treatment of mice fed a control diet with the synbiotic sufficiently reduced seizure susceptibility. Together, these animal studies suggest that microbiota-targeted interventions could diminish particular side-effects of the KD and potentially confer seizure protection. More research is needed to further test these ideas in animal models and in human studies.

4.2 Ketogenic diet on the function of the gut microbiome in epilepsy

The limited consistency in microbiota composition across epileptic patients treated with the clinical KD is not surprising considering the heterogeneity of the disorder, the variability in even the healthy human microbiota, the differences in specific nutritional content of the KD, and the several additional technical, medical and lifestyle factors that shape microbial taxonomic profiles. Increasing evidence of functional redundancy across different microbial taxa suggests that common functional signatures could be identified despite variations in taxonomic composition (The Human Microbiome Project Consortium, 2012). In a study that employed metagenomic sequencing of the gut microbiota of 12 children with severe epilepsy that were treated for 3 months with the KD, 29 gene clusters significantly associated with the KD (Lindefeldt et al., 2019). In particular, increases genes related to the hemin transport system and succinate dehydrogenase (which includes fumarate reductase subunits) were elevated with KD treatment, which correlated with *E. coli* and *Bacteroides*, as well as *Eggerthella lenta*, respectively (Lindefeldt et al., 2019). Notably, 26 of the 29 altered subsystems diminished with the KD treatment and seven of them were relevant to carbohydrate metabolism, which aligns with the carbohydrate restriction in the KD. In contrast to these results from a human epilepsy cohort, 4 weeks of KD treatment in the PTZ seizure mouse model yielded microbiomes with inferred enrichment of genes related to mTOR signaling, starch and sucrose metabolism, and PPAR signaling and decreases in genes related to energy production (Eor, Tan, et al., 2021). In particular, the mTOR signaling pathway has gained attention in epilepsy due to its ability to regulate nutrient availability, cell growth, and energy metabolism in response to the KD in rats (Liu et al., 2020; McDaniel, Rensing, Thio, Yamada, & Wong, 2011). The mTOR signaling pathway is also linked to gut microbial metabolism and immunoregulation (Noureldein & Eid, 2018). Moreover, following *L. fermentum* administration in KD-treated PTZ mice, the gut microbiota exhibited inferred enrichment of genes related to the TCA cycle and ketone body synthesis, when compared to that from PTZ mice treated with KD alone and increased microbial genes related to propionate metabolism, PPAR signaling, and pyruvate metabolism compared to PTZ mice with normal diet and same probiotic control (Eor, Tan, et al., 2021). Pathways related to PPAR signaling could be relevant due to prior links to anti-inflammatory activity in the hippocampus of a mouse model for epilepsy (Noureldein & Eid, 2018).

Beyond metagenomic profiling and inferred metagenomic assessments based on taxonomic sequencing data, some microbially modulated metabolites have been implicated as the intermediates in KD-associated host-microbiome interactions (Krautkramer, Fan, & Bäckhed, 2020). SCFAs produced from microbial metabolism of carbohydrates in the gut is of interest due to their potential mediating role in gut microbiota–brain communication (Dalile, Van Oudenhove, Vervliet, & Verbeke, 2019; Koh, De Vadder, Kovatcheva-Datchary, & Bäckhed, 2016). Children with drug resistant epilepsy prior to KD treatment showed a reduced total SCFA content compared to healthy controls (Gong et al., 2021). Although there were no significant differences between prior to KD and after KD groups, the levels of butyric acid from the KD responders were higher compared to non-responders. In contrast, 1 month of treatment with KD (4:1) in 7 epileptic patients (ages 2–46) was associated with significantly decreased total SCFAs, including acetate, butyrate, propionate and isobutyrate, in feces (Ferraris et al., 2021). In the PTZ seizure mouse model, SCFAs were significantly lower in KD-treated mice compared to controls (Eor, Tan, et al., 2021). Concurrent treatment of KD with *L. fermentum* increased acetate and isobutyrate concentrations compared to control treated with KD alone (Eor, Tan, et al., 2021). In a follow-up study, select SCFAs also decreased with KD treatment (Eor, Son, et al., 2021).

Another approach to examine functional outcomes of the gut microbiome is to identify microbiome-dependent metabolomic alterations. While research is needed to determine effects of the KD on microbiome-dependent metabolites in refractory epilepsy, some studies in animal models begin to highlight metabolites of interest. In the study of dogs with idiopathic epilepsy, LC-MS analysis of fecal metabolites revealed increases in classes of long chain triglycerides (LCT) that were associated with KD treatment, as well as 17 unidentified metabolic features that were significantly increased in both MCT and KD treatment compared to typical diet controls (Pilla et al., 2020). This suggests that the KD regardless of type of oil formulation involved may be able to induce shared metabolic responses in epileptic animals. Furthermore, in mice, random forest analysis of metabolomic profiles revealed 30 microbiome-dependent metabolites in intestinal contents and serum that were predictive of seizure protection with >90% accuracy. Notably, many metabolites were relevant to amino acid metabolism, including derivatives of lysine, tyrosine, and threonine (Olson et al., 2018). Widespread decreases in subsets of ketogenic gamma-glutamylated amino acids were observed in both colonic luminal contents and sera from seizure-protected mice compared to seizure-susceptible mice. Gamma glutamylated (GG) amino acids were previously implicated in the KD as sources that modulate GABA-glutamate metabolism in the brain (Calderón, Betancourt, Hernández, & Rada, 2017). Overall, emerging studies in humans and animal models point to a potential for the gut microbiome to be modified by the KD and to in turn, contribute to alterations in host physiology and seizure susceptibility. However, the current studies are few, of limited sample size, and variable in results. Additional well-controlled and -powered studies are required to interrogate potential interactions between the microbiome and KD in epilepsy.

4.3 Ketogenic diet on the gut microbiome in other diseases

Beyond epilepsy, the KD is increasingly being explored for its fundamental effects on host physiology and for its potential to modify risk for symptoms other metabolic, neurological, neurodevelopmental and neurodegenerative diseases. In a study of 33 obese patients enrolled in a weight loss program, treatment with a low calorie KD (2 months with very low calorie KD followed by 2 months of low calorie diet) significantly increased microbial diversity by the end of the dietary intervention (Gutiérrez-Repiso et al., 2019), with reductions in *Pseudomonadota* and increases in *Bacillota*. At the family level, *Enterobacteriaceae*, *Sinobacteraceae*, and *Comamonadacea* decreased, while the abundance of *Ruminococcaceae* and *Mogibacteriaceae* increased. At the genus level, a reduction in *Serratia*, *Erwinia*, and *Citrobacter* abundance was reported, while an increase in *Oscillospira* and *Butyricimonas* abundance was observed. In contrast, 17 obese men between the ages of 18–50 years followed a KD for 4 weeks (Ang et al., 2020). KD was associated with increases in *Bacteroides* and decreases in *Actinomycetota* and *Bacillota*. *Bifidobacterium* showed the greatest decrease in response to KD while no differences in SCFAs were detected in this human cohort. This is consistent with a mouse study reporting that the KD decreased the abundance of

Bifidobacterium. Although the reduction might be concerning due to reported beneficial effects of *Bifidobacterium* in the human gut (O'Callaghan & van Sinderen, 2016), this shift is not surprising given the propensity for Bifidobacteria to rely on carbohydrates metabolism for growth (Milani et al., 2015). In addition, a recent study reported that the ketone body BHBA selectively inhibits bifidobacterial growth, which results in diminished levels of intestinal pro-inflammatory Th17 cells (Ang et al., 2020). Whether these ketone body- and microbiota-driven reductions in intestinal inflammation could play a role in the anti-seizure effects of the KD remains unclear (Agirman, Yu, & Hsiao, 2021). However, there is some evidence that metabolites that are reduced by the KD (Heischmann et al., 2018) and correlated with lower seizure frequency (Zarnowska, 2020) could play a role in the immune response (Cervenka, Agudelo, & Ruas, 2017).

One study evaluated effects of a 6-month treatment with the 1:5:1 (protein:fat:carbs) KD on the fecal microbiota of 25 patients with multiple sclerosis (MS) compared with 14 healthy controls (Swidsinski et al., 2017). Total bacterial loads measured in the fecal microbiota of MS patients were initially decreased upon initiation of the KD, with notable restorations to levels seen in healthy controls beginning at 12 weeks and by week 23–24 of KD treatment. Notably, the KD was associated with decreases in the relative abundance of *Akkermansia*, with contrasts reported effects of the KD in healthy humans (Ang et al., 2020) and epilepsy mouse models (Olson et al., 2018). In a study of 11 individuals diagnosed with cognitive impairment (MCI) and 6 controls with normal cognitive balance 6 weeks of treatment with the Mediterranean ketogenic diet (MMKD) (Nagpal, Neth, Wang, Craft, & Yadav, 2019) yielded no notable differences in α- or β-diversity between fecal microbiota from MCI compared to controls. Regardless of the cognitive status, the MMKD was associated with increased relative abundances of *Enterobacteriaceae*, *Akkermansia*, *Slackia*, *Christensenellaceae* and *Erysipelotriaceae* and reductions in *Bifidobacterium* and *Lachnobacterium*. Inferred metagenomic analysis of the gut microbiota revealed that MMKD treatment was correlated with decreases in microbial gene families said to be related to Alzheimer's disease, diabetes, bacterial toxins and carbohydrate metabolism and increases in gene pathways related to lipid metabolism and steroid biosynthesis.

In young healthy mice, aged 12–14 weeks, 16 weeks of KD treatment decreased microbial diversity and shifted community composition relative to standard diet controls (Ma et al., 2018). KD-treated mice had significant increases in the relative abundances of *A. muciniphila* and *Lactobacillus*. and decreases in *Clostridium* and *Dorea*. *Desulfovibrio* and *Turicibacter* were also significantly and substantially lower in KD-treated mice. In the mouse strain BTBR, which is used to study social behavioral deficits relevant to autism spectrum disorder, mice were fed a standard chow or KD for 10–14 days (Newell et al., 2016). The KD was associated with decreases in *A. muciniphila*, *Methanobrevibacter*, and *Roseburia* in cecal samples and decreased *A. muciniphila*, *Enterobacteriaceae*, and *Lactobacillus* in fecal samples. In another study with that compared KD treatment of C57BL/6 vs BTBR (Klein et al., 2016), the KD similarly reduced α-diversity of the fecal microbiota in both mouse lines, with shared taxonomic alterations that led the authors to conclude that the KD rather than the mouse genetic background was the primary driver for microbial responses. In a study of mice exposed to hypoxia-induced cognitive impairment, the KD was associated with reductions in the relative abundance of *Clostridium cocleatum* and increases in *Bilophila* (Olson et al., 2021). These changes were only seen in mice exposed to hypoxia and on the KD. Altogether these studies reveal the potential context-specificity of microbial responses to the KD, that depend on genotype and exposure to environmental changes, among many other potential factors.

5. Potential mechanisms for microbial interactions with the ketogenic diet

5.1 Microbial effects on host lipid biology

Studies using gnotobiotic animals reveal that the gut microbiome fundamentally regulates lipid biology in the host. In comparing mice reared with a conventional microbiome or with those reared germ-free, the presence of gut microbes elevated levels of hydroxy fatty acids in plasma, small intestine, and colon (Kishino et al., 2013). Similarly, gut microbes modified lipids profiles, including phosphatidylcholines and triglycerides, in the serum, liver and adipose tissue (Velagapudi et al., 2010). Consistent with this, germ free mice transplanted with microbiota from aged mice showed elevated brain phospholipids and monounsaturated fatty acids (MUFAs), with reduced cholesterol and PUFAs, compared to mice transplanted with microbiota from young mice (Albouery et al., 2020). In a separate study, the gut microbiota promoted the synthesis of MUFAs by stearoyl-CoA desaturase 1 and the elongation of PUFAs by fatty acid elongase 5 in the liver and plasma (Kindt et al., 2018). These studies reveal fundamental roles for the gut microbiome in regulating host lipid profiles in the blood, liver, and brain.

Microbial effects on host lipid profiles may be due to their ability to modulate the metabolism of dietary lipids (Fig. 2). In mice fed a high lard diet, for example, the gut microbiota was required for mediating dietary effects on hepatic triglyceride levels and the development of fatty livers (Just et al., 2018). In addition, mice treated with dietary conjugated linoleic acids exhibited increased levels of *Bacteroidota* and decreased levels of *Bacillota*, with elevated microbial production of SCFAs (Marques et al., 2015). The ability of gut microbes to regulate bile acids and SCFAs (Schoeler & Caesar, 2019) could have important implications for diets like the KD, that are based on alterations in dietary fat and carbohydrates (Fig. 2).

Many studies indicate that gut microbes respond to bile acids, modify levels of bile acids, and biotransform bile acids in ways that impact host lipid biology. Bile acids emulsify fat to enable lipid absorption in the small intestine. In mice fed high fat diets containing lard and palm oil, supplementation with bile acids increased weight gain, independently of the fat source (Just et al., 2018). Treatment with bile acids altered the gut microbiota, with increases in *Desulfovibrionaceae, Clostridium lactatifermentans* and *Flintibacter butyricus,* and decreases in *Lachnospiraceae.* Most bile acids are reabsorbed and recirculated to the liver, but bacterially mediated deconjugation of glycine- or taurine-conjugated bile acids reduce their reabsorption. Deconjugated bile acids can be further metabolized to secondary bile acids through dehydrogenation, dehydroxylation, and epimerization by gut bacteria (Wahlström, Sayin, Marschall, & Bäckhed, 2016). In the host, bile acids also act as signaling molecules for glucose and lipid metabolism, by interacting with host bile acid receptors, such as nuclear receptors farnesoid-X-receptor (FXR) and cell membrane receptor Takeda G protein-coupled receptor 5 (TGR5) (Li & Chiang, 2014). Thus, the ability of the gut microbiota to regulate host levels and profiles of specific bile acid species can impact not only lipid absorption, but downstream physiologies that are dependent on bile acid signaling (Yu, Raka, & Adeli, 2019).

SCFAs are the microbial end products from dietary fiber fermentation and play important roles in wide variety of metabolic processes, such as regulation of energy intake, energy harvest, glucose metabolism, lipid metabolism, and adipogenesis, as well as pathophysiology of obesity and related metabolic disorders (Koh et al., 2016). The microbial derived SCFA acetate serves as a precursor for the hepatic synthesis of C16 and C18 fatty acids in mice fed a standard chow diet containing 5% dietary fiber for 2 weeks (Kindt et al., 2018). This was not observed when fiber was replaced with the unfermentable fiber cellulose, indicating a key role for microbial fiber fermentation in regulating host fatty acid synthesis.

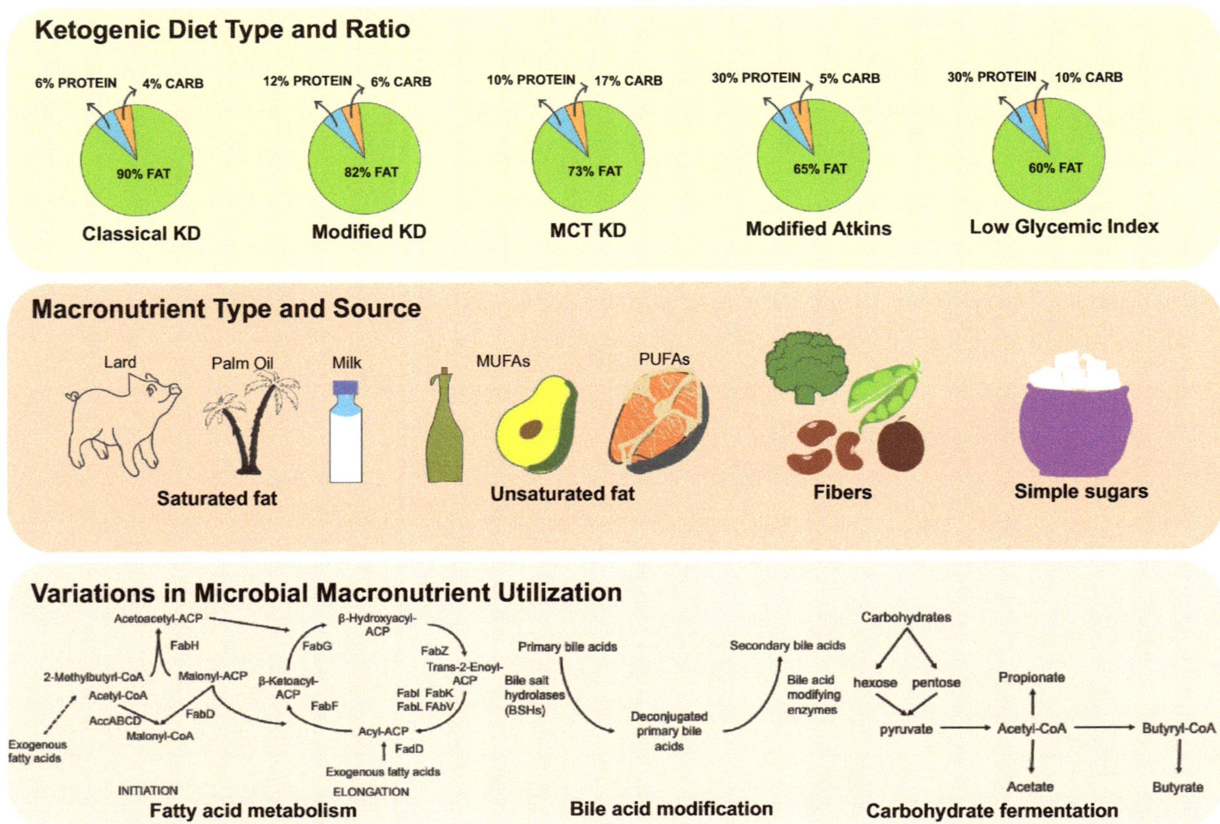

Fig. 2 Nuances of clinical ketogenic diet regimens and human microbiome variation could impact host responses to dietary treatment. There are various types of ketogenic diets that are used clinically and vary in the fat: carbohydrate ratio and/or the fat type and source. Within these types of clinical KDs, specific choices regarding macro- and micro-nutrient types that comprise individual meal plans could differentially impact the gut microbiome and host responses. Variations in the individual patient microbiome could consist of altered compositions and functions that impact dietary metabolism and host metabolites.

In addition, the microbial derived SCFAs butyrate and propionate have been shown to activate PPARγ in vitro (Alex et al., 2013). Consistent with this, dietary butyrate supplementation induced the activation of PPARγ and promoted lipid-mediated energy expenditure in mice fed with high fat diet (Gao et al., 2009). Notably, since many dietary fibers are unfermentable by the host and cannot be directly used as an energy source, they are often discounted as carbohydrates in clinical KD regimen (Rebecca, 2016). Given that the amount of fiber included in the KD treatment will differentially affect microbial metabolism, the production of SCFAs, and downstream consequences on host lipid metabolism, greater attention is needed to consider relative content of host non-fermentable fibers relative to host fermentable fibers and simple sugars in the diet.

5.2 Microbiome responses to variations in dietary fat content and type

Microbiome-dependent effects of dietary lipids on host physiology are influenced by both the amount and the type of the fat used in the diet (Mokkala, Houttu, Cansev, & Laitinen, 2020) (Fig. 2). The composition of gut microbiota differs across mice fed the high fat high sugar diet (HFD), the ketogenic diet, or the low-fat diet. In a study of mice fed the following diets for 5 weeks: (i) ketogenic diet type 1 (KDR, caloric ratio of 89.5% fat, 0.1% carbohydrate, and 10.4% protein), (ii) control normal chow of KDR (NCR, caloric ratio of 10% fat, 70% carbohydrate, and 20% protein), (iii) ketogenic diet type 2 (KDH, caloric ratio of 91.3% fat, 1% carbohydrate, and 7.7% protein), and (iv) control normal chow of KDH (NCH, caloric ratio of 15.5% fat, 64.5% carbohydrate, and 20% protein), 16S rRNA gene sequencing showed that the richness and diversity of the gut microbiota was significantly increased in KDH mice but not in KDR mice compared to their respective control groups (Li et al., 2021). Both KDs increased *Ruminococcaceae*, *Intestinimonas*, and *Lachnospiraceae*. However, these changes were not observed in human cohort studies (Lindefeldt et al., 2019; Peng et al., 2018; Tagliabue et al., 2017; Xie et al., 2017; Zhang et al., 2018). This highlights the need to evaluate whether there are shared functional changes in the microbiome across animal and human studies. In addition to alterations in microbial composition, the KDR but not KDH induced insulin resistance and damaged glucose homeostasis, while KDH increased fat accumulation in mice. The microbial derived tryptophan metabolites 1H-indole-3-propanoic acid (IPA) and indole acetic acid (IAA) were decreased in the KDR and KDH mice, respectively, and were negatively correlated with lipid accumulation. Furthermore, the microbial metabolite *trans*-2-hydroxycinnamic acid, which was elevated only in KDR mice, was positively correlated with glucose intolerance, while the SCFA isobutyrate, which was reduced in KDH mice, was negatively correlated with lipid accumulation.

Furthermore, a customized KD with the same proportions of fat as KDR and the same sources of fat as KDH was tested to evaluate the influences of sources and proportions of fat on the metabolic phenotypes in mice. 14 bacteria taxa were affected by the sources and proportions of fat in the two kinds of KD and linked to alterations in lipid metabolism (Li et al., 2021). Microbial alterations in response to the KDs also led to functional changes in the host, including decreased tryptophan metabolites and SCFAs, and increased deoxycholic acid. KDR mice had significantly higher deoxycholic acid and chenodeoxycholate levels than the KDH mice, suggesting that the differences in levels of these two bile acids may be due to the different fat sources in the KD.

Other diet studies indicate that fat type differentially affects the composition of the gut microbiome. Diets high in saturated fatty acids reduced the diversity and richness of the gut microbiota (Caesar, Tremaroli, Kovatcheva-Datchary, Cani, & Bäckhed, 2015; Devkota et al., 2012), whereas diets high in unsaturated fatty acids exerted the opposite effect (Patterson et al., 2014). In particular, consumption of a diet high in saturated (milk-derived) fat, but not polyunsaturated (safflower oil) fat, led to enrichment of a low-abundance, sulfite-reducing pathobiont, *Bilophila wadsworthia* (Devkota et al., 2012). This was associated with a pro-inflammatory T helper type 1 immune response and increased severity of colitis in genetically susceptible $Il10^{-/-}$ mice. Milk-derived fat promoted the taurine conjugation of hepatic bile acids which increased the availability of sulfur for *B. wadsworthia*. Furthermore, mice fed high fat diets comprised of omega-3 PUFAs showed a different microbial composition than mice saturated HFD or omega-6 PUFA diet (Lam et al., 2015). Hydrogen sulfide-producing bacteria (*Bilophila*) were one of the major groups driving the diet-specific changes in the gut microbiome, with the overall microbial profile being associated with changes in body weight, HOMA-IR, and gut permeability. Besides the saturation levels, the source of the saturated fat (e.g., from lard vs palm oil) differentially alters microbial composition and lipid metabolism (Just et al., 2018). A study of mice fed with either lard or fish oil reported that fat source explained about 24% of the variability in microbiota composition (Caesar et al., 2015). TLR2 and TLR4 were activated by sera from mice fed lard, suggesting that a lard-based diet promotes the influx of microbial antigens into the systemic circulation to drives the inflammatory signatures seen in white adipose tissue (Caesar et al., 2015). Overall, these findings reveal that the specific amounts, sources, and types of fats included in the diet differentially alter the gut microbiome in ways that lead to varied host metabolic and immune outcomes. This highlights the importance of evaluating the effects of specific dietary fats on seizure protection and the gut microbiota in response to clinical KD treatments.

6. Conclusion

The gut microbiome has ability to modify brain activity, behavior, and the severity of symptoms related to a variety of neurological diseases. At the same time, diet is a major determinant of the composition and function of the gut microbiome, with the potential to shape host physiology in ways that depend upon the gut microbiome. These core principles are beginning to be supported by findings from animal studies that implicate the gut microbiome in mediating the anti-seizure effects of KD. More work is needed to extend and validate these animal studies, and to advance translational research that evaluates interactions between the human gut microbiome and clinical KD treatment in refractory epilepsy. Such studies are made challenging by the complexity and heterogeneity of the gut microbiome, refractory epilepsy, and the clinical KD regimens. Exactly what constitutes a "healthy" microbiome is poorly understood, and emerging research suggests that the definition will vary from individual to individual. As such, prospective and longitudinal studies in epileptic patients and well-matched controls before and during dietary treatment are needed to account for baseline and aging-dependent variations in microbiome composition and function. While the KD is widely implemented as a clinical treatment for refractory epilepsy, patient responsiveness to KD treatment varies greatly. This is likely due, at least in part, to the substantial heterogeneity in underlying causes and subtypes of refractory epilepsy, as well as variations in medical history, including anti-epileptic drug treatment. Large studies are needed that include clearly defined cohorts of refractory epilepsy and that account for the many confounding medical and lifestyle factors that have the potential to impact the gut microbiome. The study design, results, and interpretation may also be influenced by the specific type of clinical KD tested (e.g., classical KD, modified KD, MCT, Modified Atkins, low glycemic index), the administered dietary ratios, and the specific macro- and micro-nutritional composition of the individual meal plans. Fundamental research is needed to advance understanding of how specific gut microbes respond to, metabolize, and transform specific dietary components, toward developing predictive tools for using diet to modify the microbiome to achieve targeted functions.

Overall, understanding the mechanistic underpinnings of microbial interactions with the clinical KD will inform the prospect of identifying targeted, personalized, and efficacious approaches for treating refractory epilepsy.

References

Agirman, G., Yu, K. B., & Hsiao, E. Y. (2021). Signaling inflammation across the gut-brain axis. *Science*, *374*(6571), 1087–1092. https://doi.org/10.1126/science.abi6087.

Albouery, M., Buteau, B., Grégoire, S., Cherbuy, C., Pais de Barros, J. P., Martine, L., et al. (2020). Age-related changes in the gut microbiota modify brain lipid composition. *Frontiers in Cellular and Infection Microbiology*, *9*, 1–16. https://doi.org/10.3389/FCIMB.2019.00444.

Alex, S., Lange, K., Amolo, T., Grinstead, J. S., Haakonsson, A. K., Szalowska, E., et al. (2013). Short-chain fatty acids stimulate angiopoietin-like 4 synthesis in human colon adenocarcinoma cells by activating peroxisome proliferator-activated receptor γ. *Molecular and Cellular Biology*, *33*(7), 1303–1316. https://doi.org/10.1128/MCB.00858-12.

Ang, Q. Y., Alexander, M., Newman, J. C., Tian, Y., Cai, J., Upadhyay, V., et al. (2020). Ketogenic diets Alter the gut microbiome resulting in decreased intestinal Th17 cells. *Cell*, *181*(6). https://doi.org/10.1016/j.cell.2020.04.027. 1263-1275.e16.

Augustin, K., Khabbush, A., Williams, S., Eaton, S., Orford, M., Cross, J. H., et al. (2018). Mechanisms of action for the medium-chain triglyceride ketogenic diet in neurological and metabolic disorders. *The Lancet Neurology*, *17*(1), 84–93. https://doi.org/10.1016/S1474-4422(17)30408-8.

Bough, K. J., Wetherington, J., Hassel, B., Pare, J. F., Gawryluk, J. W., Greene, J. G., et al. (2006). Mitochondrial biogenesis in the anticonvulsant mechanism of the ketogenic diet. *Annals of Neurology*, *60*(2), 223–235. https://doi.org/10.1002/ANA.20899.

Caesar, R., Tremaroli, V., Kovatcheva-Datchary, P., Cani, P. D., & Bäckhed, F. (2015). Crosstalk between gut microbiota and dietary lipids aggravates WAT inflammation through TLR signaling. *Cell Metabolism*, *22*(4), 658–668. https://doi.org/10.1016/J.CMET.2015.07.026/ATTACHMENT/1FBBFB7A-EE34-453E-8104-08BF27E25DEA/MMC1.PDF.

Calderón, N., Betancourt, L., Hernández, L., & Rada, P. (2017). A ketogenic diet modifies glutamate, gamma-aminobutyric acid and agmatine levels in the hippocampus of rats: A microdialysis study. *Neuroscience Letters*, *642*, 158–162. https://doi.org/10.1016/j.neulet.2017.02.014.

Castellano, I., & Merlino, A. (2012). γ-Glutamyltranspeptidases: Sequence, structure, biochemical properties, and biotechnological applications. *Cellular and Molecular Life Sciences*, *69*(20), 3381–3394. https://doi.org/10.1007/S00018-012-0988-3/TABLES/4.

Cervenka, I., Agudelo, L. Z., & Ruas, J. L. (2017). Kynurenines: Tryptophan's metabolites in exercise, inflammation, and mental health. *Science*, *357*(6349), 1–8. https://doi.org/10.1126/SCIENCE.AAF9794/ASSET/2178CA51-45B8-4673-A26F-461DCF63B028/ASSETS/GRAPHIC/357_AAF9794_F4.JPEG.

Chang, P., Augustin, K., Boddum, K., Williams, S., Sun, M., Terschak, J. A., et al. (2016). Seizure control by decanoic acid through direct AMPA receptor inhibition. *Brain, 139*(2), 413–443. https://doi.org/10.1093/brain/awv325.

Citraro, R., Lembo, F., De Caro, C., Tallarico, M., Coretti, L., Iannone, L. F., et al. (2021). First evidence of altered microbiota and intestinal damage and their link to absence epilepsy in a genetic animal model, the WAG/Rij rat. *Epilepsia, 62*(2), 529–541. https://doi.org/10.1111/epi.16813.

Coppola, G., Veggiotti, P., Cusmai, R., Bertoli, S., Cardinali, S., Dionisi-Vici, C., et al. (2002). The ketogenic diet in children, adolescents and young adults with refractory epilepsy: An Italian multicentric experience. *Epilepsy Research, 48*(3), 221–227. https://doi.org/10.1016/S0920-1211(01)00315-1.

Cui, G., Liu, S., Liu, Z., Chen, Y., Wu, T., Lou, J., et al. (2022). Gut microbiome distinguishes patients with epilepsy from healthy individuals. *Frontiers in Microbiology, 12*, 1–14. https://doi.org/10.3389/FMICB.2021.696632.

Dalile, B., Van Oudenhove, L., Vervliet, B., & Verbeke, K. (2019). The role of short-chain fatty acids in microbiota–gut–brain communication. *Nature Reviews. Gastroenterology & Hepatology, 16*(8), 461–478. https://doi.org/10.1038/s41575-019-0157-3.

D'Andrea Meira, I., Romão, T. T., Do Prado, H. J. P., Krüger, L. T., Pires, M. E. P., & Da Conceição, P. O. (2019). Ketogenic diet and epilepsy: What we know so far. *Frontiers in Neuroscience, 13*(JAN). https://doi.org/10.3389/FNINS.2019.00005.

Devivo, D. C., Leckie, M. P., Ferrendelli, J. S., & McDougal, D. B. (1978). Chronic ketosis and cerebral metabolism. *Annals of Neurology, 3*(4), 331–337. https://doi.org/10.1002/ANA.410030410.

Devkota, S., Wang, Y., Musch, M. W., Leone, V., Fehlner-Peach, H., Nadimpalli, A., et al. (2012). Dietary-fat-induced taurocholic acid promotes pathobiont expansion and colitis in Il10 −/− mice. *Nature, 487*(7405), 104–108. https://doi.org/10.1038/nature11225.

Dupuis, N., Curatolo, N., Benoist, J. F., & Auvin, S. (2015). Ketogenic diet exhibits anti-inflammatory properties. *Epilepsia, 56*(7), e95–e98. https://doi.org/10.1111/EPI.13038.

Eor, J. Y., Son, Y. J., Kim, J.-Y., Kang, H.-C., Youn, S. E., Kim, J. H., et al. (2021). Neuroprotective effect of both synbiotics and ketogenic diet in a pentylenetetrazol-induced acute seizure murine model. *Epilepsy Research, 174*, 106668. https://doi.org/10.1016/j.eplepsyres.2021.106668.

Eor, J. Y., Tan, P. L., Son, Y. J., Kwak, M. J., & Kim, S. H. (2021). Gut microbiota modulation by both Lactobacillus fermentum MSK 408 and ketogenic diet in a murine model of pentylenetetrazole-induced acute seizure. *Epilepsy Research, 169*, 106506. https://doi.org/10.1016/j.eplepsyres.2020.106506.

Ferraris, C., Meroni, E., Casiraghi, M. C., Tagliabue, A., De Giorgis, V., & Erba, D. (2021). One month of classic therapeutic ketogenic diet decreases short chain fatty acids production in epileptic patients. *Frontiers in Nutrition, 8*, 118. https://doi.org/10.3389/FNUT.2021.613100/BIBTEX.

Fisher, R. S., Acevedo, C., Arzimanoglou, A., Bogacz, A., Cross, J. H., Elger, C. E., et al. (2014). ILAE official report: A practical clinical definition of epilepsy. *Epilepsia, 55*(4), 475–482. https://doi.org/10.1111/EPI.12550.

Gao, Z., Yin, J., Zhang, J., Ward, R. E., Martin, R. J., Lefevre, M., et al. (2009). Butyrate improves insulin sensitivity and increases energy expenditure in mice. *Diabetes, 58*(7), 1509–1517. https://doi.org/10.2337/DB08-1637.

Garriga-Canut, M., Schoenike, B., Qazi, R., Bergendahl, K., Daley, T. J., Pfender, R. M., et al. (2006). 2-deoxy-D-glucose reduces epilepsy progression by NRSF-CtBP-dependent metabolic regulation of chromatin structure. *Nature Neuroscience, 9*(11), 1382–1387. https://doi.org/10.1038/nn1791.

Gasior, M., Hartman, A. L., & Rogawski, M. A. (2008). The anticonvulsant activity of acetone does not depend upon its metabolites. *Epilepsia, 49*(5), 936–937. https://doi.org/10.1111/J.1528-1167.2007.01518_3.X.

Gong, X., Cai, Q., Liu, X., An, D., Zhou, D., Luo, R., et al. (2021). Gut flora and metabolism are altered in epilepsy and partially restored after ketogenic diets. *Microbial Pathogenesis*, *155*, 104899. https://doi.org/10.1016/j.micpath.2021.104899.

Gooneratne, I. K., Green, A. L., Dugan, P., Sen, A., Franzini, A., Aziz, T., et al. (2016). Comparing neurostimulation technologies in refractory focal-onset epilepsy. *Journal of Neurology, Neurosurgery & Psychiatry*, *87*(11), 1174–1182. https://doi.org/10.1136/JNNP-2016-313297.

Greene, A. E., Todorova, M. T., & Seyfried, T. N. (2003). Perspectives on the metabolic management of epilepsy through dietary reduction of glucose and elevation of ketone bodies. *Journal of Neurochemistry*, *86*(3), 529–537. https://doi.org/10.1046/J.1471-4159.2003.01862.X.

Gutiérrez-Repiso, C., Hernández-García, C., Manuel García-Almeida, J., Bellido, D., María Martín-Núñez, G., Sánchez-Alcoholado, L., et al. (2019). Effect of synbiotic supplementation in a very-low-calorie ketogenic diet on weight loss achievement and gut microbiota: A randomized controlled pilot study. *Molecular Nutrition & Food Research*, *63*(19), 1900167. https://doi.org/10.1002/MNFR.201900167.

Hasan, N., & Yang, H. (2019). Factors affecting the composition of the gut microbiota, and its modulation. *PeerJ*, *7*, 1–31. https://doi.org/10.7717/peerj.7502.

Health Quality Ontario. (2012). Epilepsy surgery: An evidence summary. *Ontario Health Technology Assessment Series*, *12*(177), 1–28.

Heischmann, S., Gano, L. B., Quinn, K., Liang, L. P., Klepacki, J., Christians, U., et al. (2018). Regulation of kynurenine metabolism by a ketogenic diet. *Journal of Lipid Research*, *59*(6), 958–966. https://doi.org/10.1194/JLR.M079251.

Henderson, C. B., Filloux, F. M., Alder, S. C., Lyon, J. L., & Caplin, D. A. (2006). Efficacy of the ketogenic diet as a treatment option for epilepsy: Meta-analysis. *Journal of Child Neurology*, *21*(3), 193–198. https://doi.org/10.2310/7010.2006.00044.

Huttenlocher, P. R. (1976). Ketonemia and seizures: Metabolic and anticonvulsant effects of two ketogenic diets in childhood epilepsy. *Pediatric Research*, *10*(5), 536–540. https://doi.org/10.1203/00006450-197605000-00006.

Just, S., Mondot, S., Ecker, J., Wegner, K., Rath, E., Gau, L., et al. (2018). The gut microbiota drives the impact of bile acids and fat source in diet on mouse metabolism. *Microbiome*, *6*(1), 1–18. https://doi.org/10.1186/S40168-018-0510-8/FIGURES/4.

Kim, D. Y., Simeone, K. A., Simeone, T. A., Pandya, J. D., Wilke, J. C., Ahn, Y., et al. (2015). Ketone bodies mediate antiseizure effects through mitochondrial permeability transition. *Annals of Neurology*, *78*(1), 77–87. https://doi.org/10.1002/ANA.24424.

Kindt, A., Liebisch, G., Clavel, T., Haller, D., Hörmannsperger, G., Yoon, H., et al. (2018). The gut microbiota promotes hepatic fatty acid desaturation and elongation in mice. *Nature Communications*, *9*(1), 1–15. https://doi.org/10.1038/S41467-018-05767-4.

Kishino, S., Takeuchi, M., Park, S. B., Hirata, A., Kitamura, N., Kunisawa, J., et al. (2013). Polyunsaturated fatty acid saturation by gut lactic acid bacteria affecting host lipid composition. *Proceedings of the National Academy of Sciences of the United States of America*, *110*(44), 17808–17813. https://doi.org/10.1073/PNAS.1312937110.

Klein, P., Janousek, J., Barber, A., & Weissberger, R. (2010). Ketogenic diet treatment in adults with refractory epilepsy. *Epilepsy & Behavior*, *19*(4), 575–579. https://doi.org/10.1016/j.yebeh.2010.09.016.

Klein, M. S., Newell, C., Bomhof, M. R., Reimer, R. A., Hittel, D. S., Rho, J. M., et al. (2016). Metabolomic modeling to monitor host responsiveness to gut microbiota manipulation in the BTBR T+tf/j mouse. *Journal of Proteome Research*, *15*(4), 1143–1150. https://doi.org/10.1021/acs.jproteome.5b01025.

Klepper, J. (2012). GLUT1 deficiency syndrome in clinical practice. *Epilepsy Research*, *100*(3), 272–277. https://doi.org/10.1016/j.eplepsyres.2011.02.007.

Koh, A., De Vadder, F., Kovatcheva-Datchary, P., & Bäckhed, F. (2016). From dietary fiber to host physiology: Short-chain fatty acids as key bacterial metabolites. *Cell*, *165*(6), 1332–1345. https://doi.org/10.1016/J.CELL.2016.05.041.

Krautkramer, K. A., Fan, J., & Bäckhed, F. (2020). Gut microbial metabolites as multi-kingdom intermediates. *Nature Reviews. Microbiology*, *19*(2), 77–94. https://doi.org/10.1038/s41579-020-0438-4.

Lam, Y. Y., Ha, C. W. Y., Hoffmann, J. M. A., Oscarsson, J., Dinudom, A., Mather, T. J., et al. (2015). Effects of dietary fat profile on gut permeability and microbiota and their relationships with metabolic changes in mice. *Obesity*, *23*(7), 1429–1439. https://doi.org/10.1002/OBY.21122.

Lee, H., Lee, S., Lee, D. H., & Kim, D. W. (2021). A comparison of the gut microbiota among adult patients with drug-responsive and drug-resistant epilepsy: An exploratory study. *Epilepsy Research*, *172*, 1–7. https://doi.org/10.1016/j.eplepsyres.2021.106601.

Li, T., & Chiang, J. Y. L. (2014). Bile acid signaling in metabolic disease and drug therapy. *Pharmacological Reviews*, *66*(4), 948–983. https://doi.org/10.1124/PR.113.008201.

Li, Y., Yang, X., Zhang, J., Jiang, T., Zhang, Z., Wang, Z., et al. (2021). Ketogenic diets induced glucose intolerance and lipid accumulation in mice with alterations in gut microbiota and metabolites. *MBio*, *12*(2), 1–19. https://doi.org/10.1128/MBIO.03601-20.

Li, H., Zou, Y., & Ding, G. (2013). Therapeutic success of the ketogenic diet as a treatment option for epilepsy: A meta-analysis. *Iranian Journal of Pediatrics*, *23*(6), 613–620. Retrieved from https://www.ncbi.nlm.nih.gov/pmc/articles/PMC4025116/.

Likhodii, S. S., Musa, K., Mendonca, A., Dell, C., Burnham, W. M., & Cunnane, S. C. (2000). Dietary fat, ketosis, and seizure resistance in rats on the ketogenic diet. *Epilepsia*, *41*(11), 1400–1410. https://doi.org/10.1111/j.1528-1157.2000.tb00115.x.

Likhodii, S. S., Serbanescu, I., Cortez, M. A., Murphy, P., Snead, O. C., & Burnham, W. M. I. (2003). Anticonvulsant properties of acetone, a brain ketone elevated by the ketogenic diet. *Annals of Neurology*, *54*(2), 219–226. https://doi.org/10.1002/ANA.10634.

Lindefeldt, M., Eng, A., Darban, H., Bjerkner, A., Zetterström, C. K., Allander, T., et al. (2019). The ketogenic diet influences taxonomic and functional composition of the gut microbiota in children with severe epilepsy. *Npj Biofilms and Microbiomes*, *5*(1), 5. https://doi.org/10.1038/s41522-018-0073-2.

Liu, H., Huang, J., Liu, H., Li, F., Peng, Q., & Liu, C. (2020). Effects of ketogenic diet containing medium-chain fatty acids on serum inflammatory factor and mTOR signaling pathway in rats. *Chemical and Biological Technologies in Agriculture*, 7(1), 27. https://doi.org/10.1186/s40538-020-00194-4.

Lum, G. R., Olson, C. A., & Hsiao, E. Y. (2020). Emerging roles for the intestinal microbiome in epilepsy. *Neurobiology of Disease*, *135*, 104576. https://doi.org/10.1016/j.nbd.2019.104576.

Ma, W., Berg, J., & Yellen, G. (2007). Ketogenic diet metabolites reduce firing in central neurons by opening KATP channels. *The Journal of Neuroscience*, 27(14), 3618. https://doi.org/10.1523/JNEUROSCI.0132-07.2007.

Ma, D., Wang, A. C., Parikh, I., Green, S. J., Hoffman, J. D., Chlipala, G., et al. (2018). Ketogenic diet enhances neurovascular function with altered gut microbiome in young healthy mice. *Scientific Reports*, *8*(1), 6670. https://doi.org/10.1038/s41598-018-25190-5.

Magiorkinis, E., Sidiropoulou, K., & Diamantis, A. (2010). Hallmarks in the history of epilepsy: Epilepsy in antiquity. *Epilepsy & Behavior*, *17*(1), 103–108. https://doi.org/10.1016/j.yebeh.2009.10.023.

Marques, T. M., Wall, R., O'Sullivan, O., Fitzgerald, G. F., Shanahan, F., Quigley, E. M., et al. (2015). Dietary trans-10, cis-12-conjugated linoleic acid alters fatty acid metabolism and microbiota composition in mice. *British Journal of Nutrition*, *113*(5), 728–738. https://doi.org/10.1017/S0007114514004206.

Martin, K., Jackson, C. F., Levy, R. G., & Cooper, P. N. (2016). Ketogenic diet and other dietary treatments for epilepsy. *Cochrane Database of Systematic Reviews, 2016*(2), 1–28. https://doi.org/10.1002/14651858.CD001903.pub3.

McDaniel, S. S., Rensing, N. R., Thio, L. L., Yamada, K. A., & Wong, M. (2011). The ketogenic diet inhibits the mammalian target of rapamycin (mTOR) pathway. *Epilepsia, 52*(3), e7. https://doi.org/10.1111/J.1528-1167.2011.02981.X.

Merrit, H. H., & Putnam, T. J. (1938). Sodium diphenyl hydantoinate in the treatment of convulsive disorders. *Journal of the American Medical Association, 111*(12), 1068. https://doi.org/10.1001/jama.1938.02790380010004.

Milani, C., Lugli, G. A., Duranti, S., Turroni, F., Mancabelli, L., Ferrario, C., et al. (2015). Bifidobacteria exhibit social behavior through carbohydrate resource sharing in the gut. *Scientific Reports, 5*(1), 15782. https://doi.org/10.1038/srep15782.

Miljanovic, N., & Potschka, H. (2021). The impact of Scn1a deficiency and ketogenic diet on the intestinal microbiome: A study in a genetic Dravet mouse model. *Epilepsy Research, 178*, 106826. https://doi.org/10.1016/J.EPLEPSYRES.2021.106826.

Miranda, M. J., Mortensen, M., Povlsen, J. H., Nielsen, H., & Beniczky, S. (2011). Danish study of a modified Atkins diet for medically intractable epilepsy in children: Can we achieve the same results as with the classical ketogenic diet? *Seizure, 20*(2), 151–155. https://doi.org/10.1016/j.seizure.2010.11.010.

Mokkala, K., Houttu, N., Cansev, T., & Laitinen, K. (2020). Interactions of dietary fat with the gut microbiota: Evaluation of mechanisms and metabolic consequences. *Clinical Nutrition, 39*(4), 994–1018. https://doi.org/10.1016/J.CLNU.2019.05.003.

Mu, C., Choudhary, A., Mayengbam, S., Barrett, K. T., Rho, J. M., Shearer, J., et al. (2022). Seizure modulation by the gut microbiota and tryptophan-kynurenine metabolism in an animal model of infantile spasms. *eBioMedicine, 76*, 103833. https://doi.org/10.1016/J.EBIOM.2022.103833.

Nagpal, R., Mainali, R., Ahmadi, S., Wang, S., Singh, R., Kavanagh, K., et al. (2018). Gut microbiome and aging: Physiological and mechanistic insights. *Nutrition and Healthy Aging, 4*(4), 267. https://doi.org/10.3233/NHA-170030.

Nagpal, R., Neth, B. J., Wang, S., Craft, S., & Yadav, H. (2019). Modified Mediterranean-ketogenic diet modulates gut microbiome and short-chain fatty acids in association with Alzheimer's disease markers in subjects with mild cognitive impairment. *eBioMedicine, 47*, 529–542. https://doi.org/10.1016/j.ebiom.2019.08.032.

Nathan, J., Bailur, S., Datay, K., Sharma, S., & Khedekar Kale, D. (2019). A switch to polyunsaturated fatty acid based ketogenic diet improves seizure control in patients with drug-resistant epilepsy on the mixed fat ketogenic Diet: A retrospective open label trial. *Cureus, 11*(12), 1–12. https://doi.org/10.7759/cureus.6399.

Newell, C., Bomhof, M. R., Reimer, R. A., Hittel, D. S., Rho, J. M., & Shearer, J. (2016). Ketogenic diet modifies the gut microbiota in a murine model of autism spectrum disorder. *Molecular Autism*, 7(1), 37. https://doi.org/10.1186/s13229-016-0099-3.

Noureldein, M. H., & Eid, A. A. (2018). Gut microbiota and mTOR signaling: Insight on a new pathophysiological interaction. *Microbial Pathogenesis, 118*, 98–104. https://doi.org/10.1016/J.MICPATH.2018.03.021.

O'Callaghan, A., & van Sinderen, D. (2016). Bifidobacteria and their role as members of the human gut microbiota. *Frontiers in Microbiology*, 7, 925. https://doi.org/10.3389/fmicb.2016.00925.

Olson, C. A., Iñiguez, A. J., Yang, G. E., Fang, P., Pronovost, G. N., Jameson, K. G., et al. (2021). Alterations in the gut microbiota contribute to cognitive impairment induced by the ketogenic diet and hypoxia. *Cell Host & Microbe, 29*(9), 1378-1392.e6. https://doi.org/10.1016/J.CHOM.2021.07.004.

Olson, C. A., Vuong, H. E., Yano, J. M., Liang, Q. Y., Nusbaum, D. J., & Hsiao, E. Y. (2018). The gut microbiota mediates the anti-seizure effects of the ketogenic diet. *Cell, 173*(7), 1728-1741.e13. https://doi.org/10.1016/j.cell.2018.04.027.

Patterson, E., O'Doherty, R. M., Murphy, E. F., Wall, R., O'Sullivan, O., Nilaweera, K., et al. (2014). Impact of dietary fatty acids on metabolic activity and host intestinal microbiota composition in C57BL/6J mice. *British Journal of Nutrition, 111*(11), 1905–1917. https://doi.org/10.1017/S0007114514000117.

Peng, A., Qiu, X., Lai, W., Li, W., Zhang, L., Zhu, X., et al. (2018). Altered composition of the gut microbiome in patients with drug-resistant epilepsy. *Epilepsy Research, 147*, 102–107. https://doi.org/10.1016/j.eplepsyres.2018.09.013.

Pilla, R., Law, T. H., Pan, Y., Zanghi, B. M., Li, Q., Want, E. J., et al. (2020). The effects of a ketogenic medium-chain triglyceride diet on the feces in dogs with idiopathic epilepsy. *Frontiers in Veterinary Science*, 7, 1102. https://doi.org/10.3389/fvets.2020.541547.

Poorshiri, B., Barzegar, M., Tahmasebi, S., Shiva, S., Raeisi, S., & Ebadi, Z. (2021). The efficacy comparison of classic ketogenic diet and modified Atkins diet in children with refractory epilepsy: A clinical trial. *Acta Neurologica Belgica, 121*(2), 483–487. https://doi.org/10.1007/s13760-019-01225-0.

Porta, N., Vallée, L., Boutry, E., Fontaine, M., Dessein, A. F., Joriot, S., et al. (2009). Comparison of seizure reduction and serum fatty acid levels after receiving the ketogenic and modified Atkins diet. *Seizure, 18*(5), 359–364. https://doi.org/10.1016/j.seizure.2009.01.004.

Proctor, L. M., Creasy, H. H., Fettweis, J. M., Lloyd-Price, J., Mahurkar, A., Zhou, W., et al. (2019). The integrative human microbiome project. *Nature, 569*(7758), 641–648. https://doi.org/10.1038/s41586-019-1238-8.

Rebecca, R. (2016). The ketogenic diet for epilepsy—Learn about the diet, the medical conditions it's used to treat, and its mechanism of action. *Today's Dietitian, 18*(5), 1–20.

Rho, J. M. (2017). How does the ketogenic diet induce anti-seizure effects? *Neuroscience Letters, 637*, 4–10. https://doi.org/10.1016/J.NEULET.2015.07.034.

Şafak, B., Altunan, B., Topçu, B., & Eren Topkaya, A. (2020). The gut microbiome in epilepsy. *Microbial Pathogenesis, 139*, 103853. https://doi.org/10.1016/J.MICPATH.2019.103853.

Schoeler, M., & Caesar, R. (2019). Dietary lipids, gut microbiota and lipid metabolism. *Reviews in Endocrine and Metabolic Disorders, 20*(4), 461–472. https://doi.org/10.1007/s11154-019-09512-0.

Simopoulos, A. P. (2002). The importance of the ratio of omega-6/omega-3 essential fatty acids. *Biomedicine and Pharmacotherapy, 56*(8), 365–379. https://doi.org/10.1016/S0753-3322(02)00253-6.

Sperling, M. R. (2004). The consequences of uncontrolled epilepsy. *CNS Spectrums, 9*(2), 98–109. https://doi.org/10.1017/S1092852900008464.

Stafstrom, C. E., & Carmant, L. (2015). Seizures and epilepsy: An overview for neuroscientists. *Cold Spring Harbor Perspectives in Medicine, 5*(6), 1–19. https://doi.org/10.1101/CSHPERSPECT.A022426.

Swidsinski, A., Dörffel, Y., Loening-Baucke, V., Gille, C., Göktas, Ö., Reißhauer, A., et al. (2017). Reduced mass and diversity of the colonic microbiome in patients with multiple sclerosis and their improvement with ketogenic diet. *Frontiers in Microbiology, 8*(JUN), 1–9. https://doi.org/10.3389/FMICB.2017.01141.

Tagliabue, A., Ferraris, C., Uggeri, F., Trentani, C., Bertoli, S., de Giorgis, V., et al. (2017). Short-term impact of a classical ketogenic diet on gut microbiota in GLUT1 deficiency syndrome: A 3-month prospective observational study. *Clinical Nutrition ESPEN, 17*, 33–37. https://doi.org/10.1016/j.clnesp.2016.11.003.

Taha, A. Y., Burnham, W. M. I., & Auvin, S. (2010). Polyunsaturated fatty acids and epilepsy. *Epilepsia, 51*(8), 1348–1358. https://doi.org/10.1111/j.1528-1167.2010.02654.x.

Tan, K. N., Carrasco-Pozo, C., McDonald, T. S., Puchowicz, M., & Borges, K. (2017). Tridecanoin is anticonvulsant, antioxidant, and improves mitochondrial function. *Journal of Cerebral Blood Flow and Metabolism, 37*(6), 2035–2048. https://doi.org/10.1177/0271678X16659498.

The Human Microbiome Project Consortium. (2012). Structure, function and diversity of the healthy human microbiome. *Nature*, *486*(7402), 207–214. https://doi.org/10.1038/nature11234.

Turnbaugh, P. J., Hamady, M., Yatsunenko, T., Cantarel, B. L., Duncan, A., Ley, R. E., et al. (2009). A core gut microbiome in obese and lean twins. *Nature*, *457*(7228), 480–484. https://doi.org/10.1038/nature07540.

Velagapudi, V. R., Hezaveh, R., Reigstad, C. S., Gopalacharyulu, P., Yetukuri, L., Islam, S., et al. (2010). The gut microbiota modulates host energy and lipid metabolism in mice. *Journal of Lipid Research*, *51*(5), 1101–1112. https://doi.org/10.1194/jlr.M002774.

Vuong, H. E., Yano, J. M., Fung, T. C., & Hsiao, E. Y. (2017). The microbiome and host behavior. *Annual Review of Neuroscience*, *40*(1), 21–49. https://doi.org/10.1146/annurev-neuro-072116-031347.

Wahlström, A., Sayin, S. I., Marschall, H. U., & Bäckhed, F. (2016). Intestinal crosstalk between bile acids and microbiota and its impact on host metabolism. *Cell Metabolism*, *24*(1), 41–50. https://doi.org/10.1016/J.CMET.2016.05.005.

Wan, L., Yang, G., Zhang, S., Sun, Y., Li, Z., Wang, J., et al. (2021). Investigation of the association between imbalance of the intestinal flora and infantile spasms: A pilot case-control study. *Translational Pediatrics*, *10*(4), 819–833. https://doi.org/10.21037/TP-20-384.

Wang, Z. J., Bergqvist, C., Hunter, J. V., Jin, D., Wang, D. J., Wehrli, S., et al. (2003). In vivo measurement of brain metabolites using two-dimensional double-quantum MR spectroscopy—Exploration of GABA levels in a ketogenic diet. *Magnetic Resonance in Medicine*, *49*(4), 615–619. https://doi.org/10.1002/MRM.10429.

WHO. (2019). *Epilepsy*. Retrieved September 30, 2021, from https://www.who.int/news-room/fact-sheets/detail/epilepsy.

Wilder, R. M. (1921). The effects of ketonemia on the course of epilepsy. *Mayo Clinic Proceedings*, *2*, 307.

Williams, T. J., & Cervenka, M. C. (2017). The role for ketogenic diets in epilepsy and status epilepticus in adults. *Clinical Neurophysiology Practice*, *2*, 154–160. https://doi.org/10.1016/J.CNP.2017.06.001.

Wlaź, P., Socała, K., Nieoczym, D., Łuszczki, J. J., Arnowska, I., Arnowski, T., et al. (2012). Anticonvulsant profile of caprylic acid, a main constituent of the medium-chain triglyceride (MCT) ketogenic diet, in mice. *Neuropharmacology*, *62*(4), 1882–1889. https://doi.org/10.1016/j.neuropharm.2011.12.015.

Woodyatt, R. T. (1921). Objects and methods of diet adjustment in diabetes. *Archives of Internal Medicine*, *28*(2), 125–141.

World Health Organization. (2005). *Atlas: Epilepsy care in the world*. World Health Organization.

Xia, L., Ou, S., & Pan, S. (2017). Initial response to antiepileptic drugs in patients with newly diagnosed epilepsy as a predictor of long-term outcome. *Frontiers in Neurology*, *8*(DEC), 658. https://doi.org/10.3389/FNEUR.2017.00658.

Xie, G., Zhou, Q., Qiu, C.-Z., Dai, W.-K., Wang, H.-P., Li, Y.-H., et al. (2017). Ketogenic diet poses a significant effect on imbalanced gut microbiota in infants with refractory epilepsy. *World Journal of Gastroenterology*, *23*(33), 6164–6171. https://doi.org/10.3748/wjg.v23.i33.6164.

Yang, R., Wen, J., Wei, W., Chen, H., Cao, D., Chen, L., et al. (2022). Improving the effects of ketogenic diet therapy in children with drug-resistant epilepsy. *Seizure*, *94*, 183–188. https://doi.org/10.1016/J.SEIZURE.2021.10.021.

Ye, F., Li, X. J., Jiang, W. L., Sun, H. B., & Liu, J. (2015). Efficacy of and patient compliance with a ketogenic diet in adults with intractable epilepsy: A meta-analysis. *Journal of Clinical Neurology*, *11*(1), 26–31. https://doi.org/10.3988/JCN.2015.11.1.26.

Yu, Y., Raka, F., & Adeli, K. (2019). The role of the gut microbiota in lipid and lipoprotein metabolism. *Journal of Clinical Medicine*, *8*(12), 2227. https://doi.org/10.3390/JCM8122227.

Yudkoff, M., Daikhin, Y., Horyn, O., Nissim, I., & Nissim, I. (2008). Ketosis and brain handling of glutamate, glutamine and GABA. *Epilepsia*, *49*(Suppl. 8), 73. https://doi.org/10.1111/J.1528-1167.2008.01841.X.

Zarnowska, I. M. (2020). Therapeutic use of the ketogenic diet in refractory epilepsy: What we know and what still needs to be learned. *Nutrients*, *12*(9), 2616. https://doi.org/10.3390/NU12092616.

Zhang, Y., Zhou, S., Zhou, Y., Yu, L., Zhang, L., & Wang, Y. (2018). Altered gut microbiome composition in children with refractory epilepsy after ketogenic diet. *Epilepsy Research*, *145*, 163–168. https://doi.org/10.1016/j.eplepsyres.2018.06.015.

CHAPTER NINE

Traumatic spinal cord injury and the contributions of the post-injury microbiome

Adam M. Hamilton and Timothy R. Sampson*
Department of Cell Biology, Emory University School of Medicine, Atlanta, GA, United States
*Corresponding author: e-mail address: trsamps@emory.edu

Contents

Abstract

Spinal cord injuries are an enormous burden on injured individuals and their caregivers. The pathophysiological effects of injury are not limited to the spine and limb function, but affect numerous body systems. Growing observations in human studies and experimental models suggest that the gut microbiome is altered following spinal cord injury. Given the importance of signals derived from the gut microbiome for host physiology, it is possible that injury-triggered dysbiosis subsequently affects aspects of recovery. Here, we review emerging literature on the role of the microbiome following spinal cord injury. Specifically, we highlight findings from both human and experimental studies that correlate taxonomic changes to aspects of injury recovery. Examination of both observational and emerging interventional studies supports the notion that future therapeutic avenues for spinal cord injury pathologies may lie at the interface of the host and indigenous microbes.

International Review of Neurobiology, Volume 167
ISSN 0074-7742
https://doi.org/10.1016/bs.irn.2022.06.003

1. Introduction

Spinal cord injuries (SCI) are traumatic occurrences with an annual incidence of roughly 15,000 new cases in the United States, and almost half a million world-wide (Lee, Cripps, Fitzharris, & Wing, 2014). Due largely to injuries resultant from ground transportation incidents or falls (Lee et al., 2014), SCI leads to life-long impairment across numerous body systems. It further imposes an enormous mental, physical, and financial burden on the injured individual and their caregivers. In the US, estimates of financial costs that individuals and their caregivers will incur over the post-injury lifetime is $1.2–$5 million per person, excluding lost wages, in conjunction with significant decreases in overall quality of life (Lee et al., 2014).

Injuries themselves are highly variable in their etiology (e.g., contusion, transection, etc.) severity (e.g., depth of contusion, duration of compression/distraction, completeness of severance) and level of injury (e.g., cervical, thoracic, lumbar). These etiological factors lead to a wide range of both primary injury phenotypes and secondary sequelae. Certainly, para- and tetraplegia are visible, well-known, and debilitating indications of SCI, however dysfunction in the gastrointestinal (GI), genitourinary, and cardiac systems are also apparent both acutely and chronically following SCI (Ahuja et al., 2017; Alizadeh, Dyck, & Karimi-Abdolrezaee, 2019; Simpson et al., 2012). Further, long term effects on mood disorders and somatosensation (i.e., nociception) arise (Craig, Guest, Tran, & Middleton, 2017; Jurkiewicz, Crawley, Verrier, Fehlings, & Mikulis, 2006; Khazaeipour, Taheri-Otaghsara, & Naghdi, 2015).

Given the broad and systemic effects of SCI across physiological systems, emerging studies in the last 5–6 years have begun to examine potential contributions of the indigenous microbiome to SCI recovery. As the microbiome has established capacity to modulate inflammatory and metabolic responses (Fung, Olson, & Hsiao, 2017), two critical players in setting the tone of SCI recovery, it is likely that microbiome-dependent activities impart effects on recovery. In fact, a number of recent experimental studies establish concrete associations with microbiome-mediated contributions to recovery after SCI in rodent models.

Additionally, a wide array of intrinsic GI dysfunctions (e.g., neuropathy) are observed acutely following injury (Holmes & Blanke, 2019). We, and others, predict that these acute alterations to GI physiology drive changes to the structure of the microbiome. Across both animal models and human incidences of SCI, shifts in particular microbial taxa of the gut microbiome have been observed (Bannerman, Douchant, Sheth, & Ghasemlou, 2021;

Bannerman, Douchant, Segal, et al., 2021; Bazzocchi et al., 2021; Cheng et al., 2021; Doelman et al., 2021; Du et al., 2021; Gungor et al., 2016; Jing et al., 2019; Jing, Bai, & Yu, 2021; Jing, Yu, et al., 2021; Jing et al., 2022; Jogia & Ruitenberg, 2020; Kigerl et al., 2016; Kigerl, Mostacada, & Popovich, 2018; Kigerl, Zane, Adams, Sullivan, & Popovich, 2020; Li et al., 2020, 2021; Li, Morrow, McLain, Womack, & Yarar-Fisher, 2022; Lin et al., 2020; Musleh-Vega, Ojeda, & Vidal, 2022; Myers et al., 2019; O'Connor et al., 2018; Panther, Dodd, Clark, & Lucke-Wold, 2022; Rong et al., 2021; Schmidt et al., 2020, 2021; Schmidt, Raposo, Madsen, et al., 2021; Valido et al., 2022; Wallace et al., 2019; Yu et al., 2021; Yuan, Lu, & Wu, 2021; Zhang et al., 2018, 2019). This dysbiosis (a detrimental disruption to the microbiome community) has the capacity to feed into the physiological pathways that impact recovery following injury. From these emerging observational and experimental studies, a more complete picture of SCI recovery is being developed. The drastic and acute, whole-body, physiological dysfunctions that arise following SCI trigger disturbances in the intestinal environment, influencing both the composition and functionality of the microbiome. As microbiome structure and activities change, altered detrimental signals derived from this dysbiotic community differentially modulate inflammatory activation and metabolic input across all body systems, leading to effects on recovery both within and outside the spinal cord.

2. Remodeling of the gut microbiome after SCI

Growing evidence indicates that the gut microbiome undergoes significant alteration following SCI and other traumatic injuries. Effects of medical treatment (i.e., antibiotics, steroids, analgesics), as well as secondary injuries, confound assigning primacy of SCI itself to those microbiome shifts observed in humans. However, experimental models which can limit these confounding variables also note substantial injury-dependent effects on microbiome structure.

To date, eight independent cohorts of persons with SCI have had microbiome assessments performed, with six of these studies publishing detailed genus-level gut microbiome analysis. While true metanalysis of taxonomic alterations is difficult at this juncture, owing to the low number of individuals (relative to the larger variability of the human microbiome), as well as technical considerations (region of 16S sequenced, amplicon database used, etc.), injury-dependent taxonomic alterations can be qualitatively combined (Table 1). Collapsing these different variables across studies

Table 1 Microbiome alterations in incidence of human spinal cord injury.

Phylum	2021 Yu	2016 Gungor	2018/19 Zhang	2021 Bazzocchi	2020 Lin	2020 Li
(A)synergistetes/ Synergistota	⇑	N/A	=	N/A	N/A	N/A
Actinobacteria	⇑	N/A	⇑	N/A	N/A	N/A
Bacteroidetes	↓	N/A	=	↓	N/A	N/A
Cyanobacteria	↓	N/A	=	N/A	N/A	N/A
Firmicutes	N/A	N/A	↓	=	↓	N/A
Fusobacteria	N/A	N/A	⇑	N/A	N/A	N/A
Proteobacteria/ Pseudomonadota	↓	N/A	⇑	N/A	N/A	N/A
Saccharibacteria	N/A	N/A	=	N/A	N/A	N/A
Tenericutes	N/A	N/A	=	N/A	N/A	N/A
Verrucomicrobia	N/A	N/A	⇑	N/A	N/A	N/A

Genus	2021 Yu	2016 Gungor	2018/19 Zhang	2021 Bazzocchi	2020 Lin	2020 Li
Akkermansia	N/A	⇑	⇑	⇑	N/A	N/A
Alistipes	N/A	=	⇑	N/A	⇑	⇑
Anaerostipes	N/A	=	N/A	N/A	N/A	N/A
Anaerotruncus	N/A	=	N/A	N/A	⇑	⇑
Bacteroides	N/A	=	⇑	↓	N/A	N/A
Barnesiella	N/A	N/A	N/A	N/A	⇑	N/A
Bifidobacterium	⇑	⇑	⇑	N/A	N/A	N/A
Bilophila	N/A	N/A	N/A	N/A	⇑	N/A
Blautia	N/A	=	⇑	N/A	N/A	N/A
Butyricicoccus	↓	N/A	N/A	N/A	N/A	N/A
Campylobacter	N/A	N/A	N/A	N/A	N/A	↓

Table 1 Microbiome alterations in incidence of human spinal cord injury.—cont'd

Genus	2021 Yu	2016 Gungor	2018/19 Zhang	2021 Bazzocchi	2020 Lin	2020 Li
Christensenella	⥣	N/A	N/A	N/A	⥣	N/A
Cloacibacillus	⥣	N/A	N/A	N/A	N/A	N/A
Clostridium	⥣	=	N/A	↓	↓	N/A
Collinsella	N/A	⥣	N/A	N/A	N/A	N/A
Coprobacillus	⥣	N/A	N/A	N/A	⥣	N/A
Coprococcus	↓	↓	N/A	↓	N/A	N/A
Coriobacteriaceae	⥣	N/A	N/A	⥣	N/A	N/A
Desulfovibrio	↓	N/A	N/A	N/A	N/A	N/A
Dialister	↓	↓	↓	↓	↓	N/A
Dorea	⥣	=	N/A	N/A	N/A	N/A
Eggerthella	⥣	N/A	N/A	N/A	⥣	⥣
Eisenbergiella	N/A	N/A	N/A	N/A	N/A	⥣
Enterococcus	N/A	⥣	N/A	⥣	N/A	N/A
Erysipelatoclostridium	N/A	N/A	N/A	N/A	N/A	⥣
Escherichia	N/A	⥣	⥣	N/A	N/A	N/A
Eubacterium	⥣	=	↓	N/A	N/A	⥣
Faecalibacterium	↓	↓	↓	↓	↓	N/A
Flavonifractor	N/A	=	N/A	N/A	⥣	N/A
Gemmiger	↓	N/A	N/A	N/A	N/A	N/A
Gordonibacter	N/A	N/A	N/A	N/A	⥣	N/A
Haemophilus	N/A	N/A	N/A	N/A	↓	N/A
Holdemania	N/A	N/A	N/A	N/A	⥣	N/A
Intestinimonas	N/A	N/A	N/A	N/A	⥣	N/A
Klebsiella	N/A	N/A	⥣	⥣	N/A	N/A

Continued

Table 1 Microbiome alterations in incidence of human spinal cord injury.—cont'd

Genus	2021 Yu	2016 Gungor	2018/19 Zhang	2021 Bazzocchi	2020 Lin	2020 Li
Lachnoclostridium	N/A	N/A	⇑	N/A	N/A	⇑
Lachnospira	N/A	N/A	N/A	N/A	↓	N/A
Lachnospiraceae (various genera)	↓	↓	N/A	↓	↓	N/A
Lactobacillus	N/A	⇑	⇑	⇑	N/A	N/A
Leuconostoc	N/A	N/A	N/A	N/A	↓	N/A
Marvinbryantia	N/A	↓	N/A	N/A	N/A	N/A
Megamonas	N/A	↓	↓	N/A	↓	N/A
Megasphaera	N/A	N/A	⇑	N/A	↓	N/A
Methanobrevibacter	N/A	N/A	N/A	⇑	N/A	N/A
Odoribacter	N/A	N/A	N/A	N/A	N/A	⇑
Oscillospira	N/A	=	N/A	↓	N/A	N/A
Oscillibacter	N/A	N/A	N/A	N/A	N/A	⇑
Parabacteroides	N/A	N/A	N/A	N/A	⇑	N/A
Pasteurellaceae	↓	N/A	N/A	N/A	N/A	N/A
Phascolarctobacterium	N/A	N/A	⇑	N/A	⇑	⇑
Prevotella	↓	↓	↓	↓	N/A	N/A
Pseudobutyrivibrio	N/A	↓	N/A	N/A	↓	N/A
Pseudoramibacter	⇑	N/A	N/A	N/A	N/A	N/A
Rhodococcus	N/A	N/A	N/A	N/A	↓	N/A
Roseburia	↓	↓	=	N/A	↓	N/A
Ruminococcus	N/A	=	=	↓	↓	N/A
Sarcina	N/A	⇑	N/A	N/A	N/A	N/A
Streptococcus	N/A	=	N/A	⇑	N/A	N/A
Subdoligranulum	N/A	↓	↓	N/A	↓	N/A

Table 1 Microbiome alterations in incidence of human spinal cord injury.—cont'd

Genus	2021 Yu	2016 Gungor	2018/19 Zhang	2021 Bazzocchi	2020 Lin	2020 Li
Sutterella	↓	N/A	N/A	N/A	N/A	⇑
Thalassospira	N/A	=	N/A	N/A	N/A	N/A
Veillonella	N/A	N/A	N/A	N/A	↓	N/A

Key

Higher in Controls/Lower in SCI (↓)
Lower in Controls/Higher in SCI (⇑)
Data Not Available/Not Reported (N/A)
Data Reported, No Different Between Groups (=)

Phyla Higher in SCI (⇑), in 2 or more studies
Actinobacteria

Genera Higher in SCI (⇑), in 2 or more studies
Akkermansia
Alistipes
Anaerotruncus
Bifidobacterium
Christensenella
Coprobacillus
Eggerthella
Enterococcus
Escherichia
Klebsiella
Lachnoclostridium
Lactobacillus
Phascolarctobacterium

Phyla Lower in SCI (↓), in 2 or more studies
Bacteroidetes
Firmicutes

Genera Lower in SCI (↓), in 2 or more studies
Coprococcus
Dialister
Faecalibacterium
Lachnospiraceae (genus: N/A)
Megamonas
Prevotella
Pseudobutyrivibrio
Roseburia
Ruminococcus
Subdoligranulum.

(anatomical level of injury, time post-injury) to assess those taxa altered by injury reveals some consistent differences. Common taxa changes associated with SCI are increases in the genera *Akkermansia*, *Alistipes*, *Bifidobacterium*, *Lactobacillus*, and *Phascolarctobacterium* and decreases in the genera *Coprococcus, Dialister, Faecalibacterium, Lachnospiraceae, Megamonas, Prevotella, Roseburia,* and *Subdoligranulum*. Interestingly, many of those taxa that are depleted are those which ferment complex polysaccharides to beneficial short-chain fatty acids (SCFAs) and appear vulnerable to perturbation in other inflammatory neurological conditions such as Parkinson's disease (Sampson, 2020), as well as non-traumatic SCI, including spina bifida (Furuta et al., 2021).

Animal models of SCI allow for greater control over injury and recovery conditions, enabling investigators to elucidate not only temporal changes in microbiome composition, but those that are specific to a predetermined injury level and severity. However, the reported taxa altered post-injury have not been consistent across studies, likely due to differences in methodologies, timing, and facility-level variations in murine gut microbiomes prior to perturbation/injury. Generally, the most consistent alterations to the post-SCI microbiome are characterized by an increase in the genus *Bacteroides* (5/9 studies identified with genus-level data) and a decrease in the genus *Lactobacillus* (observed in 6/9 studies) (Table 2).

Within studies, control for initial microbiome composition (comparing pre- and post-injury status) has indicated progressive changes to the microbiome that appear acutely after injury and progressively become more pronounced. The earliest significant alterations that have been reported occur 3 days post-injury, that become more engrained at later stages, between 7 and 40 days post-injury. (Bannerman, Douchant, Segal, et al., 2021; Kigerl et al., 2016; Myers et al., 2019; Schmidt, Raposo, Torres-Espin, Fenrich, & Fouad, 2021). However, some studies have reported a recovery of the microbiome, with the most significant changes occurring in the acute phase post-injury, and a subsequent restoration of specific taxa to their pre-injury abundances (Bannerman, Douchant, Segal, et al., 2021; Doelman et al., 2021; Schmidt, Raposo, Madsen, et al., 2021; Schmidt et al., 2020). Differences in injury severity and level may account for discrepancies between those that observe progressive dysbiosis and those studies which observe a microbiome recovery in conjunction with injury recovery. In one recent study, early (3 days post-injury) increases were observed in the orders Bifidobacteriales, Lactobacillales, Erysipelotrichales, and Verrucomicrobiales, with significant decreases in the genus *Bifidobacterium* (Bannerman, Douchant, Segal, et al., 2021). Conversely, a loss of the genus *Akkermansia* appeared later at ~13 days post-injury. While the alterations of these organisms reached statistical significance at the acute timepoints, they were qualitatively more reduced throughout recovery following injury, in a more severely injured group of animals, and showed longitudinal changes in magnitude out to 41 days post-injury (Bannerman, Douchant, Segal, et al., 2021). Collectively, these data indicate not only temporal alterations, but also a "dose-response" to injury severity with more severe injuries leading to greater impacts on the composition of the gut microbiome.

Table 2 Alterations to the gut microbiome in experimental spinal cord injury.

Phylum	2018 O'Connor	2020 Schmidt	2021 Schmidt	2021 Jing	2019 Jing	2021 Cheng	2021 Du	2021 Rong	2021 Bannerman	2021 Doelman
Actinobacteria	N/A	N/A	N/A	N/A	N/A	N/A	⇑	N/A	N/A	N/A
Bacteroidetes	N/A	N/A	=	⇑	N/A	N/A	N/A	⇑	N/A	⇑ (acute)
Cyanobacteria	N/A	N/A	⇑	N/A	N/A	N/A	N/A	N/A	N/A	↓
Epsilonbacteraeota	N/A	N/A	N/A	N/A	N/A	N/A	N/A	N/A	N/A	↓ (acute)
Firmicutes	N/A	N/A	=	↓	N/A	↓	↓	↓	N/A	⇑ (acute)
Proteobacteria/ Pseudomonadota	N/A	N/A	↓	N/A	N/A	⇑	N/A	⇑	N/A	↓
Tenericutes	N/A	N/A	N/A	N/A	N/A	N/A	N/A	N/A	N/A	↓ (acute)
Spirochaetes	N/A	N/A	N/A	N/A	N/A	N/A	N/A	N/A	N/A	⇑
Genus	**2018 O'Connor**	**2020 Schmidt**	**2021 Schmidt**	**2021 Jing**	**2019 Jing**	**2021 Cheng**	**2021 Du**	**2021 Rong**	**2021 Bannerman**	**2021 Doelman**
Akkermansia	N/A	N/A	=	⇑	N/A	N/A	N/A	N/A	↓	N/A
Alistipes	N/A	N/A	=	N/A	⇑	N/A	N/A	N/A	N/A	N/A
Allobaculum	N/A	N/A	N/A	N/A	N/A	⇑	N/A	N/A	N/A	N/A
Alloprevotella	N/A	N/A	N/A	N/A	N/A	↓	N/A	N/A	N/A	N/A
Anaerostipes	N/A	N/A	N/A	↓	N/A	N/A	N/A	N/A	N/A	N/A
Anaerotruncus	N/A	N/A	N/A	N/A	N/A	N/A	N/A	N/A	N/A	N/A
Azospirillum sp. 47_25	N/A	N/A	⇑	N/A	N/A	N/A	N/A	N/A	N/A	N/A

Continued

Table 2 Alterations to the gut microbiome in experimental spinal cord injury.—cont'd

Genus	2018 O'Connor	2020 Schmidt	2021 Schmidt	2021 Jing	2019 Jing	2021 Cheng	2021 Du	2021 Rong	2021 Bannerman	2021 Doelman
Bacteroides	N/A	N/A	=	⇑	⇑	⇑	⇑	⇑	N/A	N/A
Bifidobacterium	⇑	↓	N/A	N/A	⇑	N/A	N/A	N/A	↓	N/A
Bilophila	N/A	N/A	N/A	⇑	N/A	N/A	N/A	N/A	N/A	N/A
Blautia	N/A	↓	N/A	↓	N/A	⇑	N/A	⇑	N/A	N/A
Butyricicoccus	N/A	↓	⇑	N/A	N/A	N/A	N/A	N/A	N/A	N/A
Butyricimonas	N/A	N/A	N/A	↓	N/A	N/A	N/A	N/A	N/A	N/A
Christensenellaceae	N/A	N/A	↓ (R-7 Group)	↓	N/A	N/A	N/A	N/A	N/A	N/A
Clostridium	N/A	↓	N/A	N/A	N/A	N/A	↓	N/A	N/A	N/A
Coprococcus	N/A	↓	N/A	N/A	N/A	N/A	N/A	N/A	N/A	N/A
Dialister	N/A	↓	N/A	N/A	N/A	N/A	N/A	N/A	N/A	N/A
Dorea	N/A	N/A	N/A	N/A	N/A	⇑	N/A	N/A	N/A	N/A
Eggerthella	N/A	↓	N/A	N/A	N/A	N/A	N/A	N/A	N/A	N/A
Enterobacter	N/A	N/A	N/A	N/A	N/A	N/A	N/A	⇑	N/A	N/A
Enterococcus	N/A	↓	N/A	N/A	N/A	N/A	N/A	N/A	N/A	N/A
Escherichia-Shigella	N/A	↓	N/A	N/A	N/A	N/A	N/A	⇑	N/A	N/A
Eubacterium	N/A	⇑	N/A	N/A	N/A	N/A	⇑	N/A	N/A	N/A
Faecalibacterium	N/A	↓	N/A	N/A	N/A	N/A	N/A	N/A	N/A	N/A
Flavonifractor	N/A	N/A	N/A	N/A	N/A	N/A	⇑	N/A	N/A	N/A

Helicobacter	N/A	N/A	N/A	N/A	N/A	↓	N/A	N/A	N/A	N/A
Intestinimonas	N/A	N/A	↓	N/A	N/A	N/A	N/A	N/A	N/A	N/A
Klebsiella	N/A	N/A	N/A	↓	N/A	N/A	N/A	⇑	N/A	N/A
Lachnoclostridium	N/A	N/A	N/A	⇑	N/A	⇑	N/A	N/A	N/A	N/A
Lachnospiraceae/ Lachnospira	N/A	↓	↓ (ND3007)	↓	⇑	N/A	⇑	N/A	N/A	N/A
Lactobacillus	N/A	N/A	↓	↓	↓	↓	↓	↓	N/A	N/A
Lactococcus	N/A	⇑	N/A	N/A	N/A	N/A	N/A	N/A	N/A	N/A
Megamonas	N/A	N/A	N/A	N/A	N/A	N/A	N/A	N/A	N/A	N/A
Mogibacterium	N/A	⇑	N/A	N/A	N/A	N/A	N/A	N/A	N/A	N/A
Muribaculaceae (fam for CAG)	N/A	N/A	N/A	↓	N/A	N/A	N/A	N/A	N/A	N/A
Oscillibacter	N/A	N/A	⇑	N/A	N/A	N/A	N/A	N/A	N/A	N/A
Parabacteroides	N/A	↓	N/A	⇑	N/A	N/A	N/A	N/A	N/A	N/A
Prevotella	N/A	N/A	↓	N/A	N/A	N/A	↓ (CAG-1031)	N/A	N/A	N/A
Romboutsia	N/A	N/A	⇑	N/A	N/A	N/A	N/A	N/A	N/A	N/A
Roseburia	N/A	↓	N/A	N/A	N/A	N/A	N/A	N/A	N/A	N/A
Ruminiclostridium 6	N/A	N/A	⇑	N/A	N/A	N/A	N/A	N/A	N/A	N/A
Ruminococcaceae UCG-005	N/A	N/A	↓	N/A	N/A	N/A	N/A	N/A	N/A	N/A

Continued

Table 2 Alterations to the gut microbiome in experimental spinal cord injury.—cont'd

Genus	2018 O'Connor	2020 Schmidt	2021 Schmidt	2021 Jing	2019 Jing	2021 Cheng	2021 Du	2021 Rong	2021 Bannerman	2021 Doelman
Ruminococcus 1	N/A	N/A	⇑	N/A	N/A	N/A	N/A	N/A	N/A	N/A
Slackia	N/A	⇑	N/A	N/A	N/A	N/A	⇑	N/A	N/A	N/A
Turicibacter	↓	N/A	N/A	N/A	N/A	N/A	↓	N/A	N/A	N/A
Weissella	N/A	⇑	N/A	N/A	N/A	N/A	⇑	N/A	N/A	N/A

Key
Higher in Controls/Lower in SCI (↓)
Lower in Controls/Higher in SCI (⇑)
Data Not Available/Not Reported (N/A)
Data Reported, No Different Between Groups (=)

Phyla Higher in SCI (⇑), in 2 or more studies
Bacteroidetes

Genera Higher in SCI (⇑), in 2 or more studies
Bacteroides
Eubacterium
Lachnoclostridium
Slackia
Weissella

Phyla Lower in SCI (↓), in 2 or more studies
Firmicutes

Genera Lower in SCI (↓), in 2 or more studies
Christensenellaceae
Clostridium
Lactobacillus
Prevotella
Turicibacter.

Overall, the published data to date show that traumatic SCI results in a progressive shift in the gut microbiome in experimental models and suggest injury-dependent alterations in human incidences of SCI as well. However, the precise mechanisms that alter the intestinal environment and mediate impacts on the microbial community are not yet known. Robust physiological dysfunctions, such as hypoxia of intestinal tissues and increased inflammatory signaling are known to occur acutely post-injury, and likely lead to diminished intestinal motility and increased permeability (Ahuja et al., 2017; Holmes & Blanke, 2019). These acute pathologies likely give rise to an altered intestinal environment that promotes selection of certain organisms, which through competition, depletes others. These effects may become more pronounced with time post-injury (Ahuja et al., 2017; Kigerl et al., 2016; Stone, Nino-Murcia, Wolfe, & Perkash, 1990), suggesting the possibility that these altered and dysbiotic microbiomes become engrained in a feed-forward loop of dysregulation. Given the substantial roles for certain members of the microbiome in inflammatory and metabolic responses (Fung et al., 2017), it is likely then that this cycle of dysfunction ultimately contributes to impaired recovery of numerous organ systems.

3. Interactions between the gut microbiome and SCI-induced neurogenic bowel

The progressive shift in the gut microbial community following SCI is likely the result of multiple factors. Certainly, loss of skeletal muscle control will lead to varied effects on dietary choices and food intake (Bigford & Nash, 2017; Thibault-Halman, Casha, Singer, & Christie, 2011), inducing taxonomic alterations due to nutritional changes (David et al., 2014; Dey et al., 2015). Further, depending on injury level, SCI may limit abdominal and rectal control of fecal evacuation, impairing overall intestinal motility, and impacting the intestinal environment (Brading & Ramalingam, 2006; Krassioukov, Eng, Claxton, Sakakibara, & Shum, 2010; Margolis, Cryan, & Mayer, 2021; Quigley, 2011). These, and other, post-SCI effects on the gut including enteric neuropathy and hypoperfusion (Holmes & Blanke, 2019), lead to neurogenic bowel dysfunction (NBD), a post-injury condition characterized by constipation and/or fecal incontinence (Hakim, Gaglani, & Cash, 2022), that also likely impacts the microbiome (Singal, Rosman, Bauman, & Korsten, 2006; Stiens, Bergman, & Goetz, 1997; Stone et al., 1990).

Across numerous surveys, bowel dysfunction has proven to be one of the most detrimental aspects of SCI, in terms of quality of life (Simpson et al., 2012). Despite this, its exploration and therapeutic avenues remain understudied compared to other, more visible, components of SCI recovery. While the microbiome is likely influenced by neurogenic bowel, growing evidence indicates that microbiome signals promote intestinal motility through neurogenesis and hormonal signaling (Anitha, Vijay-Kumar, Sitaraman, Gewirtz, & Srinivasan, 2012; De Vadder et al., 2018; Margolis et al., 2021; Quigley, 2011; Vicentini et al., 2021; Yarandi et al., 2020; Zhao & Yu, 2016). However, evaluation of the efficacy of microbiome-based interventions to ameliorate SCI-associated NBD in human injuries has not yet occurred.

To our knowledge, no studies to date have established cause-effect relationships between microbiome composition and SCI-triggered NBD. Some, however, have correlated certain microbial taxa with aspects of NBD and/or have modulated outcomes with microbiome manipulations (Cheng et al., 2021; Gungor, Adiguzel, Gursel, Yilmaz, & Gursel, 2016; Jing et al., 2019; Jing, Yu, et al., 2021; Yu et al., 2021; Zhang et al., 2019, 2018). Comparisons between severity of injury, bowel dysfunction, and gut microbiome dysbiosis has revealed that these attributes are positively correlated (Yu et al., 2021). One observation of a human cohort compared microbiome compositions between individuals with upper motor neuron (UMN) bowel dysfunction and those with lower motor neuron (LMN) bowel dysfunction following SCI (Gungor et al., 2016). UMN bowel dysfunction is characterized by an inability to defecate due to increased sphincter and colonic tonus, but a preservation of reflex circuitry between the ENS and the spinal cord (Singal et al., 2006; Stiens et al., 1997). In contrast, LMN bowel dysfunction is characterized by a loss of reflex circuitry, inhibiting peristalsis, and a lack of tonus within the anal sphincter and colon (Singal et al., 2006; Stiens et al., 1997). These deficits are associated with either constipation or constipation with incontinence, respectively. Significant alterations were observed in both UMN and LMN groups compared to uninjured individuals, for instance a decrease in SCFA-producing microbes. Both groups of injured persons displayed significantly reduced levels of the genus *Pseudobutyrivibrio*, that produces lactate, butyrate, and formate, as well as reduced levels of the genus *Megamonas*, that produces acetate and propionate (Gungor et al., 2016). Interestingly, microbiome composition was observed to differ based on injury level in this study, a trend seen by others as well (Bazzocchi et al., 2021; Lin et al., 2020; Yu et al., 2021; Zhang et al., 2019, 2018). Those with UMN (constipation alone) had significantly decreased *Marvinbryantia* and *Dialister* species, while those with LMN (constipation with fecal incontinence) had significantly decreased *Roseburia* (Gungor et al., 2016). These findings suggest that SCI remodels the gut microbiome in both an injury-dependent and injury-level dependent manner, reflective of the physiological status of the gut.

It is interesting to note that there is a relatively small body of pre-clinical observations of NBD in experimental models of SCI (Holmes & Blanke, 2019), and even more limited associations with the gut microbiome at this time. Rodent models of SCI recapitulate enteric neuropathy and diminished total intestinal transit that appears in human injuries (Fealey, Szurszewski, Merritt, & DiMagno, 1984; Jing, Yu, et al., 2021; White & Holmes, 2018; White, Werner, & Holmes, 2020; Williams et al., 2012), and microbiome contributions to intestinal motility post-injury were recently described (Jing, Yu, et al., 2021). Following a fecal microbiome transplant (FMT) from a healthy, uninjured mouse, mice with a midthoracic contusion SCI showed improved parameters of intestinal motility (Jing, Yu, et al., 2021). However, in this study, FMT did not alter intestinal motility in uninjured controls, indicating the microbiome contribution to this deficit may be specific to animals with SCI. Additional studies to specifically investigate cause-effect relationships between the microbiome and SCI-induced NBD are needed.

4. SCI-triggered local and systemic immune responses

While trauma to the spinal cord results in significant inflammatory responses at the site of injury, an ever-growing body of literature indicates that SCI also results in drastic inflammatory effects across numerous body systems (Hellenbrand et al., 2021; Schwab, Zhang, Kopp, Brommer, & Popovich, 2014; Zhang, Yin, Xu, Wu, & Chen, 2012). This includes a range of peripheral inflammatory responses that are quite physically distant from the local injury originating within the cord (Hausmann, 2003). In human incidences, examination of acute peripheral inflammation is largely confounded by the fact that SCI does not often occur alone and instead arises in conjunction with trauma at other body sites (Yue et al., 2017). Experimental rodent models of injury have allowed for the isolation of peripheral effects after injury, solely due to the injury within the spinal cord (Cheriyan et al., 2014). Even when isolated to a single injury event within the cord itself, SCI results in dramatic changes to the inflammatory status at peripheral body sites (Bank et al., 2015; Hausmann, 2003; Hellenbrand et al., 2021; Schwab et al., 2014; Zhang et al., 2012).

The timing and magnitude of both systemic and CNS-resident inflammation in humans and experimental models' post-injury is dynamic and is dependent on the level, completeness, and severity of injury. In general, an acute increase in pro-inflammatory cytokine signals and infiltration of circulating monocytes is observed locally within the cord, as well as systemically. Following SCI, immune cells are recruited to the site of injury and infiltrate the cord, in an injury severity-dependent fashion (Bannerman, Douchant, Segal, et al., 2021). This was exemplified in mice through experimental comparison between a spinal cord contusion and a more severe contusion with spinal compression. Substantial differences in pro-inflammatory cytokine expression were observed in the spinal cord based on injury severity, with a more robust response observed in the more severely injured group (Bannerman, Douchant, Segal, et al., 2021). Acutely following injury, both IL-1β and TNF expression were elevated in the compression/contusion group, compared to contusion alone. Both injury groups displayed increased lymphoid and monocyte/macrophage infiltration into spinal cord than sham controls. However, spinal cords of the more severe contusion/compression injury group had significantly more monocytes/macrophages (particularly $Ly6C^+$ infiltrating cells), as well as a substantial increase in total and phagocytic ($F4/80^+$) microglia than those with less severe injury and sham controls. During chronic points of recovery, >40 days post-injury, in the dorsal root ganglia of animals with more severe contusion/compression injuries, but not those with less severe injury, $Ly6C^+$ macrophages remained present.

In higher-level injury, this acute local induction is followed by a dramatic decrease in peripheral immune function, SCI-induced immune depression syndrome (SCI-IDS) (Pruss et al., 2017; Riegger et al., 2007, 2009; Schwab et al., 2014; Zhang et al., 2013). This loss of immune inflammatory capacity and increase in regulatory immune cells is thought to be beneficial to prevent autoimmunity against CNS-derived antigens released following traumatic injury (Schwab et al., 2014). However, the significant immune impairment from SCI-IDS leads to increased infection susceptibility, ultimately hampering recovery. Conversely, longer-term increased inflammation can be detrimental to multi-organ systems including the nervous system (pain hypersensitivity, mood impairments) and gastrointestinal tract (neurogenic bowel) (Fenn et al., 2014; Luedtke et al., 2014; Maldonado-Bouchard et al., 2016; Schwab et al., 2014).

Therefore, understanding the molecular contributions to these distal inflammatory events outside the spinal cord is critical for furthering an understanding of secondary effects of injury and limiting their impact on recovery outcomes.

5. Contributions of the microbiome to SCI-associated inflammation in humans

The gut microbiome has a critical role in the development and dynamics of the immune system, not only in the intestinal tract, but the periphery, and CNS (Round & Mazmanian, 2009; Wu & Wu, 2012; Yoo & Mazmanian, 2017). This modulation of inflammatory responses is critical for mediating recovery from infections, predisposing to neuroinflammatory insults, and prevention of autoimmunity. It is therefore not surprising to envision that these microbiome-dependent activities may also influence the inflammatory signals that arise following SCI and other traumatic CNS injuries. Indeed, inflammatory outcomes after injury have been associated with microbiome composition in both humans and in experimental models.

To the best of our knowledge (and at the time of this writing), only one published human dataset sought to directly correlate microbiome composition to inflammatory outcomes, after SCI. This recent observational study compared groups of individuals with SCI who were either pre- or type 2-diabetic (P/DM) to those with SCI but normal glucose tolerance (NGT). Interestingly, none of the participants had previously been diagnosed with diabetes prior to this study, suggesting that specific pathophysiology arising after SCI may trigger metabolic inflammation and glucose intolerance. It was observed that those with SCI and post-injury P/DM status had an increase in circulating inflammatory cytokines, including IL-6 and TNF compared to individuals with SCI but NGT (Li et al., 2021). This is notable as these cytokines are known to be particularly elevated in individuals with type 2 diabetes (Dinh et al., 2009; Goyal, Faizy, Siddiqui, & Singhai, 2012; Mirza et al., 2012).

Since the composition of the gut microbiome has the capacity to differentially modulate lipid and glucose metabolism (Li et al., 2020), it is possible that SCI-induced microbiome disruptions contribute directly to these metabolic and inflammatory outcomes and that the microbiome alterations arising following SCI are themselves responsible for triggering metabolic inflammation and glucose intolerance (Brown & Hazen, 2015). Examination of the microbiome by 16S rRNA profiling revealed substantial differences in both alpha and beta diversity of the fecal microbiome between the P/DM and NGT groups. Notably, individuals with SCI and P/DM also had a lower abundance of the order Clostridiales and a higher abundance of the genus *Akkermansia* (Li et al., 2021). These observations are somewhat contradictory to prior studies linking microbiome alterations to metabolic inflammation (Tilg, Zmora, Adolph, & Elinav, 2020). Others have instead observed increased Clostridiales abundance following SCI in experimental models (Table 2) (Kigerl et al., 2016; Li et al., 2020). Additionally, *Akkermansia* is highly correlated with beneficial outcomes in many metabolic disorders, including obesity and diabetes (Dao et al., 2016; Liu et al., 2017; Plovier et al., 2017). This organism has been experimentally shown to be beneficial against metabolic syndrome in both animal models and human trials of these disorders (Depommier et al., 2019; Schneeberger et al., 2015), and generally presents in lower abundance in patients with metabolic disorders (Hasani et al., 2021). Given the role of *Akkermansia* in metabolizing intestinal mucosa and modulating intestinal integrity (Desai et al., 2016), and its response to both dietary fiber and antibiotics (Desai et al., 2016; Dubourg et al., 2013), understanding how this and other organisms function in the distinct context of SCI (with altered host intestinal physiology and microbiome structures) is necessary. Indeed, many other microbially-impacted metabolic parameters are altered following SCI. Traumatic SCI is associated with decreased levels of circulating high-density lipoprotein (HDL) (Zhang et al., 2019), and within those with chronic SCI, glucose, HDL, and creatinine have been correlated with microbiome composition (Yu et al., 2021; Zhang et al., 2019, 2018), including positive association with the phylum Actinobacteria. Further, both studies report an increase in the genus *Bifidobacterium* and decreases in *Prevotella*, *Dialister*, and *Faecalibacterium* genera (Yu et al., 2021; Zhang et al., 2019, 2018), suggesting potential metabolism-mediating roles of these taxa. Thus, use of experimental systems will be critical toward understanding how these organisms contribute to metabolic-inflammation and delineating their roles in recovery from SCI.

6. Microbiome contributions to SCI-associated inflammation in experimental models

While associations between SCI-triggered inflammation and microbiome composition in humans are more limited, numerous experimental animal studies have begun linking these intrinsic factors. By assessing the inflammatory changes, arising both systemically and locally within the spinal cord, in conjunction with changes to microbiome composition that arise after injury, one may begin identifying key players in the microbial community that impart beneficial or detrimental effects on recovery. These may not only reveal new, underlying biology of systemic interactions following traumatic CNS injury, but more coercively, establish potential probiotic organisms with therapeutic benefit towards recovery.

A recent study examining inflammatory outcomes in the rat model of thoracic SCI identified a strong correlation between three specific cytokines (IL-1β, IL-12, and MIP-2) and specific gut microbial taxa (O'Connor et al., 2018). The strongest correlations were found between IL-1β and 23 distinct operational taxonomic units (OTUs). Eight OTUs that were able to be characterized to the species level were each negatively correlated with circulating IL-1β levels, suggesting that these particular organisms may modulate levels of this inflammatory cytokine. One of these specific taxa was *Faecalibacterium prausnitzii*, which was negatively correlated with IL-1β. This organism is well established as an SCFA-producing member of the microbiome, converting dietary fiber preferentially to butyrate (Lopez-Siles, Duncan, Garcia-Gil, & Martinez-Medina, 2017). This is notable because one common occurrence in the SCI-associated microbiome is the general loss of SCFA-producing bacteria (see Tables 1 and 2). *Faecalibacterium* is of particular note as it was found to be decreased in 5/6 studies of humans with SCI, that reported genus-level data (Table 1). Since butyrate potently dampens inflammatory activation (Lenoir et al., 2020; Segain et al., 2000), the loss of *F. prausnitzii* and other SCFA-producing organisms, suggests that the SCI-associated microbiome has decreased capacity to limit detrimental inflammatory responses. In fact, three distinct studies have evaluated the efficacy of SCFA-based therapeutics for recovery from SCI, and all three studies found improvements to SCI-associated inflammation in animal models provided increased SCFAs (Filippone et al., 2020; Lanza et al., 2019; Qian, Zhu, Lu, Ni, & Yang, 2017).

While peripheral inflammation arises following injury, emerging data indicate the microbiome has capacity to modulate immune responses within the CNS itself, including the spinal cord. For instance, recent experimental studies have identified components of signaling pathways activated by microbes, as modulators of spinal cord inflammation following SCI. One receptor, Toll-like receptor 4 (TLR4) has garnered recent attention. This pattern recognition receptor detects and responds to bacterial lipopolysaccharide

(LPS) derived from Gram-negative bacterial cell envelopes, leading to significant inflammatory responses. Following SCI in mice, a robust activation of the TLR4 signaling cascade was observed to occur in both intestinal tissue and the spinal cord (Rong et al., 2021). This includes, the TLR4 adaptor protein MyD88, and the p65 and IκBα signaling molecules, which lead to increased NFκB activation, the transcription factor for numerous inflammatory cytokines and chemokines. While only an association, this implicates a role for microbial triggers of post-SCI inflammation *via* a TLR4-dependent pathway.

FMT allows the ability to directly assess the contribution of particular microbiomes to outcomes of interest, such as TLR4-mediated inflammation. Using this approach, microbiomes derived from either mice with SCI or sham injury were introduced into conventionally-colonized mice with SCI (Rong et al., 2021). Injured mice treated with SCI-derived microbiomes displayed increased inflammatory cytokines relative to injured mice without an FMT. Further, mice treated with sham-derived FMT showed reduced inflammation, strongly suggesting that the microbiome that arises post-SCI is sufficient to exacerbate injury-induced inflammation and that the two differential microbiome communities (sham-derived or SCI-derived) have discrete impacts on inflammatory signaling (Jing et al., 2022; Rong et al., 2021). While unable to elucidate specific bacterial populations that directly associate with changes in the inflammatory status (likely due to an insufficient sample size) TLR4-signaling was further implicated by the observation that those molecular components downstream of the TLR4 activation cascade were concurrently upregulated. Thus, suggesting (although not directly testing) that the SCI-associated microbiomes have enhanced capacity to stimulate TLR4 and exacerbate inflammatory signaling.

Additional implication of microbial-stimulated TLR4 activation following SCI has come from the study of cAMP-specific phosphodiesterase 4B (PDE4B) activation on SCI recovery. PDE4B is a downstream and parallel component of TLR4 activation and promotes cytokine activation (Jin & Conti, 2002; Jin, Lan, Zoudilova, & Conti, 2005; Myers et al., 2019). Comparison of SCI outcomes between wild-type and PDE4B null mice revealed a substantial decrease in inflammatory markers including GFAP, CD11b, Iba1, and Cox2 in injured mice lacking PDE4B, implicating this signaling pathway in post-SCI inflammation (Myers et al., 2019). Intriguingly, PDE4B null mice also showed improved locomotor recovery following SCI and did not present with other SCI-triggered pathophysiologies, such as gut microbiome alterations, bacterial overgrowth, or endotoxemia, which the wild-type animals robustly displayed. In fact, wild-type mice displayed a substantial increase in TLR4-stimulating

Proteobacteria following SCI, perhaps suggesting that the bloom of these microbes potentiates post-injury inflammation (Myers et al., 2019). These data further support a broader model, whereby acute inflammatory signaling triggers microbiome alterations, which then lead to a chronic feed-forward cycle and exacerbation of systemic inflammation and neurological impairment, leading to additional dysbiosis and endotoxemia. Rolipram, a non-specific PDE4 inhibitor, is widely available and has been used to treat a variety of human conditions (Zhu, Mix, & Winblad, 2001), and has shown efficacy for recovery in preclinical models of SCI (Costa et al., 2013; Iannotti et al., 2011; Schaal et al., 2012).

7. SCI microbiome association with gut permeability after injury

One largely embraced theory surrounding how gut bacteria may lead to detrimental systemic and neuro- inflammatory responses, relevant to SCI, is through disruption of intestinal epithelial tight junctions and subsequent increased translocation of pro-inflammatory bacteria and/or their metabolites. Not just observed in traumatic neurological injury, various physiological disturbances can give rise to impaired barrier function and increased bacterial translocation from the gut, leading to an exacerbated inflammatory response throughout the body after injury. These ultimately result in systemic effects that can have drastic impacts on recovery from injury. SCI has long been established to trigger increased gut permeability and bacterial translocation, leading to quantifiable levels of bacterial endotoxin in the blood within 24h, and within various tissues and organ systems within 48h of SCI (Liu et al., 2004). Recent studies have further validated and characterized these observations with combinations of both functional assessments (such as direct measurement of bacterial burden in peripheral tissues following injury, and absorption of typically non-absorbable molecules—e.g., FITC-dextrin) and molecular characterizations (e.g., expression and production of tight junction proteins in the intestinal epithelium). Overall, across experimental models, SCI results in an acute disruption of the intestinal epithelial barrier with diminished barrier function, increased permeability, enhanced bacterial translocation, and increased circulating endotoxin, all within 7days post-injury (Jing et al., 2019; Jing, Yu, et al., 2021; Kigerl et al., 2016; Myers et al., 2019).

To combat this, prior recommendations entailed immediate administration of broad-spectrum antibiotics after injury. However, as our understanding of the complex interactions between host and microbiome after traumatic injury have evolved, it is becoming apparent that a more targeted approach is essential. A number of recent studies have indicated that broad-spectrum antibiotics prior to, or immediately post-injury, can adversely affect recovery in experimental models (Jing et al., 2019; Kigerl et al., 2016). Therefore, while broad antibiotic treatment may prevent translocation, as well as limit secondary infections that are a significant risk post-injury, these benefits may be outweighed by their effects on the native microbiome. Rather than using broad-spectrum antibiotics to rescue the negative effects of bacterial translocation, a number of recent studies have evaluated microbiome-based approaches to restore inflammatory responses by shifting the composition of the microbiome to a less disrupted state.

8. Microbiome manipulations with therapeutic potential for SCI recovery

Recovery of locomotor function is perhaps the most visible and most targeted component of SCI recuperation. To a greater extent than human injuries, rodents with SCI are able to recover limb functions following a period of paralysis. With a set of highly standardized behavioral assays (e.g., Basso, Beattie, Bresnahan [BBB] and Basso Mouse Scale [BMS]) to measure injury-induced locomotor dysfunctions, experimental rodent models of injury provide a dynamic system to measure how limb function can be restored or impaired following SCI (Basso, Beattie, & Bresnahan, 1995; Basso et al., 2006).

Through its impacts on inflammation (perhaps *via* modulation of gut permeability), the gut microbiome represents a potential therapeutic target for SCI recovery. The gut microbiome is readily accessible and potentially modifiable through a number of different methodologies. Such methods range from whole microbiome FMT that reconstitute a complex microbial community, to probiotic therapy with specified organisms, or prebiotic treatments that select for certain taxa or induce a specific function. Further advances are also being made in the field of "postbiotics," in which microbial-derived metabolites are introduced directly to mediate host and/or microbiome physiologies. A wide variety of emerging studies have begun to explore these potential avenues for benefit following SCI.

8.1 Fecal microbiome transplants

Fecal microbiome transplants are notable as they have the capacity to elucidate specific microbiome-mediated effects on various host physiologies in experimental settings, to determine potential key contributors to desired outcomes. A recent study investigated the efficacy of a healthy donor-derived FMT following a T10 contusive SCI in mice (Jing, Yu, et al., 2021). Injured mice, when provided a daily, oral FMT, displayed significantly decreased levels of NFκB in the spinal cord and colon 4 weeks after injury, compared to injured animals provided a vehicle gavage. Corresponding to this inflammatory marker, SCI animals receiving FMT also displayed improved locomotor scores (Jing, Yu, et al., 2021). In conjunction with previous described studies, this suggests that microbiome transplantation with a healthy microbiome may have beneficial effects on inflammatory responses post SCI. Conversely in a milder rat injury model, FMT from an uninjured donor did not result in significant improvement to locomotor recovery in the animal with SCI (Schmidt et al., 2020). However, FMT did reduce anxiety-like behaviors in recipient animals (Schmidt et al., 2020), suggesting microbiome-dependent effects beyond inflammatory and locomotor recovery and emphasizing a need for careful selection of FMT donors (Schmidt, Raposo, Madsen, et al., 2021).

FMT has also been observed to improve motor evoked potentials (MEP) in the gastrocnemius muscle after SCI (Jing, Yu, et al., 2021). FMT was able to restore the amplitude and latency of the MEP when compared to untreated mice with SCI, suggesting a capacity of microbiome-derived signals to modulate post-injury neuronal physiology beneficially and at a highly refined level. In fact, FMT treated animals with SCI also displayed greater grip strength and higher numbers of motoneurons, synapses, and neuronal axons in spinal cord sections compared to those untreated animals with SCI (Jing, Yu, et al., 2021). This was further validated with observations that FMT from uninjured animals improved outcomes in white matter sparing, neurotrophic factors, and other critical physiological parameters in the cord, following SCI in mice (Jing et al., 2022). Thus, not only does the gut microbiome contribute to central aspects of recovery from SCI, but modulation of the community as a whole with FMT (to restore to a pre-injured taxonomic structure), has capacity to promote therapeutic outcomes following SCI.

8.2 Probiotic therapeutics

While FMT has garnered a great deal of attention due to its efficacy and ability to broadly impact the entirety of the gut community, efforts to identify specific microbial contributors may reveal keystone taxa that are sufficient to drive beneficial immune responses after SCI. One such study, by Kigerl et al., noted a significant increase in inflammatory immune cell activation within the gut-associated lymphoid tissues (GALT) after a T9 contusion SCI (Kigerl et al., 2016). A shift in immune cell activation was observed, with increased B-cells, macrophages, $CD3^+$ T-cells, and dendritic cells within GALT following injury. This coincided with an increased expression of cytokines, both pro- and anti-inflammatory, including IL-10, TGFβ, TNF, and IL-1β after injury. Administration of the probiotic cocktail VSL#3 (now Visbiome, consisting of *Streptococcus thermophilus, Bifidobacterium breve, Bifidobacterium longum, Bifidobacterium infantis, Lactobacillus acidophilus, Lactobacillus plantarum, Lactobacillus paracasei*, and *Lactobacillus deslbrueckii* subsp. *bulgaricus*) significantly diminished this immune activation in the GALT following SCI, suggesting a therapeutic benefit from certain organisms not necessarily requiring a full FMT.

These beneficial effects of select bacterial taxa were not confined to the intestinal inflammatory environment. Those animals receiving VSL#3 also displayed substantially improved inflammation in the spinal cord, with fewer infiltrating macrophages, and increased dendritic cells and $FOXP3^+$ regulatory T-cells (Kigerl et al., 2016). Notably, antibiotic treatment to alter the gut microbiome prior to SCI, resulted in increased markers of inflammation in the cord compared to SCI alone, including macrophages and $CD3^+$ T-cells. These inflammatory parameters correlated directly with survival of white matter in the injured cord tissues, with VSL#3 treated animals having substantially more white matter spared following injury. Locomotor recovery and the percentage of spared white matter in mice receiving VSL#3 probiotics were also significantly improved. While these data strongly implicate the immune response in the cord and periphery with recovery, they also further support the notion of a gut microbiome contribution to SCI recovery. In fact, VSL#3 has recently shown efficacy in promoting functional recovery following traumatic peripheral nerve injury, in contrast to impaired recovery following antibiotic pre-treatment (Rodenhouse et al., 2022). These observations together indicate that specific microbial taxa can be beneficial not solely in the local intestinal environment, but within the spinal cord itself at the site of injury and even the peripheral nervous system. Thus, association with specific microbes may have a tremendous impact on long-term neurological recovery following SCI.

8.3 Selective antibiotic treatment

Administration of a broad-spectrum cocktail of antibiotics *prior to* injury has been shown to greatly diminish recovery after T9 contusion SCI in mice—including impaired hindlimb function, increased inflammation, and greater lesion volume at the site of injury (Jing et al., 2019; Kigerl et al., 2016). However, the use of selective antibiotics post-injury may provide a mechanism to prevent pathobionts from blooming and allow for beneficial microbes to become enriched. Certain antibiotics have shown promise for reducing inflammation and improving outcomes in a wide range of neuro-inflammatory diseases, although whether their activities are dependent on an intact microbiome (as opposed to "off-target effects" on host factors) is not fully established.

One such antibiotic is minocycline, which has been shown to confer anti-inflammatory and neuroprotective effects in rodents (Festoff et al., 2006; Lee et al., 2003; Shultz & Zhong, 2017; Wells, Hurlbert, Fehlings, & Yong, 2003), as well as modest motor improvement in humans (Casha et al., 2012), after SCI. The therapeutic value of minocycline was tested in a rat injury model of unilateral cervical (C5) SCI (Schmidt, Raposo, Torres-Espin, et al., 2021). At this cervical lesion level, but not at lower thoracic injury sites such as those described in prior sections, a significant immune depression occurs, with decreases in abundance of a number of cytokine and chemokines (Pruss et al., 2017; Riegger et al., 2007, 2009; Schwab et al., 2014; Zhang et al., 2013). Irrespective of minocycline treatment, these injured animals all displayed significantly depressed cytokine responses compared to their baseline at 5 days post-injury (Schmidt, Raposo, Torres-Espin, et al., 2021). However, minocycline treatment post-injury rescued immune suppression by 28 days following injury, restoring cytokine profiles to pre-injury levels. Rats with a C5 SCI, but no minocycline treatment remained in a pro-longed immune suppressed state. Not surprisingly, minocycline treatment resulted in acute, although transient, changes to the microbiome that were more robust than those alterations induced by injury alone. These microbiome alterations were characterized predominantly by a significant decrease in the Firmicutes/Bacteroidetes ratio (Schmidt, Raposo, Torres-Espin, et al., 2021). These acute changes to the microbiome preceded the effect of minocycline on cytokines and chemokines, implicating the microbiome in minocycline-dependent immune regulation following SCI. However, examination of minocycline's effects in germ-free animals or reconstitution of minocycline-shifted microbiomes would help to provide evidence towards causality, rather than correlation of microbiome alterations during treatment. It is important to note that this more mild, unilateral injury model did not observe an effect on locomotor recovery following minocycline treatment. Minocycline has additional impacts on neuroprotection and immune regulation in cell culture systems (Elewa, Hilali, Hess, Machado, & Fagan, 2006; Giuliani, Hader, & Yong, 2005), suggesting that its beneficial effects *in vivo* may be more complex than modifying the microbiome alone.

8.4 Beneficial microbiome-related metabolites

While FMT, probiotic, and antibiotic experiments provide evidence that the presence or absence of select microbes (or the disruption of the community) can modulate SCI outcomes, small molecules derived from these gut communities (so-called postbiotics) may allow more precise tuning of physiological responses during SCI recovery. One such class of beneficial metabolites are the short chain fatty acids (SCFAs). Following therapeutic FMT in a mouse T10 contusion model of SCI, roughly half of the microbial taxa that were depleted post-injury returned to normal levels (Jing, Yu, et al., 2021). Notably, these included the genera *Blautia*, *Anaerostipes*, and *Butyricimonas* all of which are capable of producing the beneficial SCFA, butyrate (Keshavarzian et al., 2015; Parada Venegas et al., 2019; Riviere, Selak, Lantin, Leroy, & De Vuyst, 2016). Given its broad roles in energy homeostasis, anti-inflammatory signaling, and epigenetic modification (through histone deacetylase modulation), butyrate and other SCFAs represent a physiologically beneficial microbiome metabolite (Chang, Hao, Offermanns, & Medzhitov, 2014; Lin, de Zoete, van Putten, & Strijbis, 2015; Soret et al., 2010; Yao et al., 2022). Post-SCI, many SCFAs are reduced in the gut, including propionate, butyrate, isobutyrate, and caproic acid- with butyrate the most robustly affected of the SCFAs (depleted >50%) (Jing, Yu, et al., 2021). Despite the drastic effects of SCI on SCFA levels, FMT from an uninjured donor led to a > 40% increase in butyrate levels (Jing, Yu, et al., 2021). Interestingly, SCFA levels were only altered following FMT into mice with SCI and not when introduced intro uninjured controls. This may indicate a homeostatic effect of SCFAs in a healthy microbiome, with beneficial impacts only observed in the presence of an existing deficit (i.e., SCI). SCFAs have also been shown to promote barrier integrity in both the intestine and at the blood-brain barrier (Braniste et al., 2014; Silva, Bernardi, & Frozza, 2020). Following FMT, tight junction protein expression in the gut and pericyte coverage in the spinal cord were restored. Gut permeability and blood-spinal cord-barrier integrity were concurrently improved, suggesting a potential mechanism by which FMTs ameliorate inflammatory outcomes in SCI (Jing et al., 2022, 2021). These early data help to support a notion that butyrate and other SCFAs may be providing benefit to recovery post-injury, and provide a foundation for uncovering specific, modifiable microbial factors as therapeutics.

Other SCFA-derived pharmaceuticals have also shown efficacy in reducing inflammation and improving locomotor recovery (Filippone et al., 2020; Lanza et al., 2019; Qian et al., 2017). For instance, D-β-hydroxybutyrate (DBHB) treatment resulted in improved locomotor function, reduced microglial and glial activation, suppressed NLRP3 inflammasome activation, and decreased protein levels of IL-1β and IL-18 in the spinal cord of injured mice (Qian et al., 2017). Treatment with other SCFAs, including sodium butyrate (Lanza et al., 2019) or sodium propionate (Filippone et al., 2020), has also been observed to greatly improve locomotor outcomes in rodent models of SCI in a dose-dependent fashion. These SCFAs additionally reduced inflammation in the spinal cord, with both diminished proinflammatory cytokine levels and cellular markers of immune activation (Filippone et al., 2020; Lanza et al., 2019).

Postbiotics, such as SCFAs, may themselves serve as modulators of therapeutic physiologies, however changing the functional output of the microbiome also represents an easily accessible pharmaceutical target—i.e., drugging the microbiome. This avenue has shown promise in models of cardiovascular and Parkinson's disease, with small molecules developed to act directly on bacterial metabolic activities within the microbiome which then affect host functions (Maini Rekdal, Bess, Bisanz, Turnbaugh, & Balskus, 2019; Wang et al., 2015). In the context of SCI, one such example is melatonin, a signaling molecule with activities both on the host and on the microbiome (Erland & Saxena, 2017). Melatonin is produced primarily by enterochromaffin cells of the GI tract and has capacity to influence the composition of the gut microbiome as well as the integrity of the intestinal barrier (Chen, Fichna, Bashashati, Li, & Storr, 2011; Esteban-Zubero et al., 2017; Park et al., 2020). A small number of studies in various models of traumatic injury have observed beneficial neuroprotective effects of melatonin administration (Naseem & Parvez, 2014; Samantaray et al., 2009; Xu, He, & Li, 2022; Yang et al., 2016). In a mouse model of midthoracic SCI, intraperitoneal administration of melatonin was found to reduce colonic inflammation following injury (Jing et al., 2019). SCI-triggered increases in colonic concentrations of the proinflammatory cytokines IL-17, IFNγ, and MCP-1 were all reversed by melatonin treatment, indicating its anti-inflammatory benefits. Correspondingly, systemic administration of melatonin also resulted in some restoration of SCI-induced gut microbiome alterations (Jing et al., 2019). Consistent with other studies (see Tables 1 and 2), following SCI, representatives of the Lactobacillales and Bifidobacterium orders were substantially decreased, while members of the order Clostridiales were increased. Intriguingly, levels of the cytokine MCP-1 positively correlated with abundance of Clostridiales, and negatively correlated with abundances of Lactobacillales, suggesting an association between these taxa and gut inflammatory responses. Members of the order Lactobacillales are known to produce lactocepin, a potent anti-inflammatory protease (von Schillde et al., 2012), which others have identified as decreased after experimental SCI (Du et al., 2021). Increased inflammatory state was also associated with a disrupted and more permeable intestinal barrier, suggesting that increased translocation of gut bacterial components may drive systemic inflammation. In fact, increased gut permeability was correlated with levels of Clostridiales but inversely correlated with levels of Lactobacillales implicating select taxa in this physiological dysfunction. Melatonin treatment restored some components of the gut microbiome, including increased levels of Lactobacillales and decreased levels of Clostridiales, but had no impact on Bifidobacterium (Jing et al., 2019). Intestinal tight junction protein expression and intestinal barrier integrity were also restored following melatonin administration post-SCI, as was intestinal motility. Overall these data indicate that melatonin is capable of limiting SCI-induced dysbiosis as well as molecular markers of intestinal function, either by mediating host intestinal physiology or through direct action on the microbiome.

The beneficial effects of melatonin post-SCI also corresponded with substantially improved locomotor recovery in melatonin treated mice. Notably, co-administration of melatonin and broad-spectrum antibiotics in mice with SCI partially negated the protective effects of melatonin treatment on locomotor recovery, suggesting, in part, the necessity of the microbiome for melatonin-based improvements in many aspects of SCI recovery. The association between melatonin-mediated benefit to SCI recovery and microbiome alterations may therefore be directly related. However, it is unclear if melatonin acts independently on inflammatory pathways or if this activity requires an intact microbiome.

While still in early pre-clinical, experimental studies, these emerging investigations provide the foundational proof-of-concept for microbiome-related modulation of SCI recovery. Through FMT, select probiotic organisms, or microbiome-derived metabolites, these may provide more accessible avenues for SCI therapy and uncover new biological pathways involved in traumatic neurological injury. Further understanding of diet-microbiome interactions may also provide new direction for methods to improve microbiome resilience in the face of injury-induced perturbation and prevent the feed-forward cycle of inflammatory dysfunction that occurs following SCI.

9. Looking broadly into the future at microbiome effects on SCI

SCI results in life-long dysregulation of numerous body systems, much more than the visible impacts on limb function. As discussed in prior sections, gastrointestinal dysfunction is a major quality of life issue for those individuals recovering from SCI (Simpson et al., 2012; Wheeler et al., 2018). In addition, rates of anxiety, depression, and other affective disorders are considerably higher in persons with SCI than in the able-bodied population (Craig et al., 2017; Craig, Tran, & Middleton, 2009; Khazaeipour et al., 2015). While the mechanisms to explain this phenomenon are likely highly multifactorial and complex, a plethora of data indicate contributions of indigenous microbes to mood disorders in both humans and experimental models (Appleton, 2018; Clapp et al., 2017; Cryan, O'Riordan, Sandhu, Peterson, & Dinan, 2020; Huang et al., 2019; Vuong, Yano, Fung, & Hsiao, 2017). Emerging data in experimental models of SCI, evaluating the onset of anxiety-like behaviors have suggested that microbiome manipulations post-injury (for instance FMT or minocycline antibiotics) can rescue various rodent behaviors that mirror anxiety (Schmidt, Raposo, Madsen, et al., 2021; Schmidt, Raposo, Torres-Espin, et al., 2021; Schmidt et al., 2020). Given broad, whole-body impacts of microbiome manipulation, it is hopeful that future microbiome-targeted therapeutics will alleviate some of the multi-system physiological burdens experienced by those with SCI and their caregivers. Since the microbiome is readily accessible for modulation, translation to clinical studies are highly tempting with the growing body of pre-clinical data. In fact, an ongoing, randomized clinical trial has early reported findings demonstrating a restoration of the SCI-associated gut microbiome with a low carbohydrate/high protein (LC/HP) diet, including an increase in SCFA-producing bacterial species, suggesting the viability and potential efficacy of such therapeutic avenues (Li et al., 2022; Yarar-Fisher et al., 2018).

Deeper understanding of the large-scale microbial community alterations that occur following injury will be necessary to generate further hypotheses regarding specific organisms that may contribute to dysfunction following SCI. Metagenomic analysis of the gut microbial community has revealed further insights into SCI-triggered dysregulation of not only the well-studied bacterial taxa, but also viral organisms present in the gut (Du et al., 2021), and the ability to begin to assess eukaryotic microbes as well. The use of metagenomics, in combination with metabolomics studies are needed to begin to draw direct lines between the metabolic and immunologic output of the microbiome and the microbiome composition itself. Current studies inferring genetic content and metabolic output from 16S sequencing lack the specificity and depth that metagenomics provides.

Most importantly however, experimental validation in gnotobiotic models will be key to addressing both the contribution of certain gut microbes to SCI-relevant physiologies and to establish direct functional connections between specific organisms and their beneficial outcomes. By combining gnotobiotic techniques with experimental models of SCI, and informed by growing human microbiome datasets, the microbiome is poised to be a valuable therapeutic target for recovery of traumatic SCI.

Acknowledgments

The authors apologize to any authors whose contributions was not able to be included in this review. Spinal cord injury research related to the work in this review was supported by an NIH T32 (NS096050-24) to A.M.H and by the Craig H. Neilsen Foundation (642928) to T.R.S.

References

Ahuja, C. S., Wilson, J. R., Nori, S., Kotter, M. R. N., Druschel, C., Curt, A., et al. (2017). Traumatic spinal cord injury. *Nature Reviews. Disease Primers*, *3*, 17018. https://doi.org/10.1038/nrdp.2017.18.

Alizadeh, A., Dyck, S. M., & Karimi-Abdolrezaee, S. (2019). Traumatic spinal cord injury: An overview of pathophysiology, models and acute injury mechanisms. *Frontiers in Neurology*, *10*, 282. https://doi.org/10.3389/fneur.2019.00282.

Anitha, M., Vijay-Kumar, M., Sitaraman, S. V., Gewirtz, A. T., & Srinivasan, S. (2012). Gut microbial products regulate murine gastrointestinal motility via toll-like receptor 4 signaling. *Gastroenterology*, *143*(4). https://doi.org/10.1053/j.gastro.2012.06.034. 1006–1016.e1004.

Appleton, J. (2018). The gut-brain axis: Influence of microbiota on mood and mental health. *Integr Med (Encinitas)*, *17*(4), 28–32. Retrieved from https://www.ncbi.nlm.nih.gov/pubmed/31043907.

Bank, M, Stein, A., Sison, C., Glazer, A., Jassal, N., McCarthy, D., et al. (2015). Elevated circulating levels of the pro-inflammatory cytokine macrophage migration inhibitory factor in individuals with acute spinal cord injury. *Archives of Physical Medicine and Rehabilitation*, *96*(4), 633–644. https://doi.org/10.1016/j.apmr.2014.10.021.

Bannerman, C. A., Douchant, K., Segal, J. P., Knezic, M., Mack, A. E., Lundell-Creagh, C., et al. (2021). Spinal cord injury in mice impacts central and peripheral pathology in a

severity-dependent manner. *Pain, 163*, 1172–1185. https://doi.org/10.1097/j.pain.0000000000002471.
Bannerman, C. A., Douchant, K., Sheth, P. M., & Ghasemlou, N. (2021). The gut-brain axis and beyond: Microbiome control of spinal cord injury pain in humans and rodents. *Neurobiol Pain, 9*, 100059. https://doi.org/10.1016/j.ynpai.2020.100059.
Basso, D. M., Beattie, M. S., & Bresnahan, J. C. (1995). A sensitive and reliable locomotor rating scale for open field testing in rats. *Journal of Neurotrauma, 12*(1), 1–21. https://doi.org/10.1089/neu.1995.12.1.
Basso, D. M., Fisher, L. C., Anderson, A. J., Jakeman, L. B., McTigue, D. M., & Popovich, P. G. (2006). Basso mouse scale for locomotion detects differences in recovery after spinal cord injury in five common mouse strains. *Journal of Neurotrauma, 23*(5), 635–659. https://doi.org/10.1089/neu.2006.23.635.
Bazzocchi, G., Turroni, S., Bulzamini, M. C., D'Amico, F., Bava, A., Castiglioni, M., et al. (2021). Changes in gut microbiota in the acute phase after spinal cord injury correlate with severity of the lesion. *Scientific Reports, 11*(1), 12743. https://doi.org/10.1038/s41598-021-92027-z.
Bigford, G., & Nash, M. S. (2017). Nutritional health considerations for persons with spinal cord injury. *Topics in Spinal Cord Injury Rehabilitation, 23*(3), 188–206. https://doi.org/10.1310/sci2303-188.
Brading, A. F., & Ramalingam, T. (2006). Mechanisms controlling normal defecation and the potential effects of spinal cord injury. *Progress in Brain Research, 152*, 345–358. https://doi.org/10.1016/S0079-6123(05)52023-5.
Braniste, V., Al-Asmakh, M., Kowal, C., Anuar, F., Abbaspour, A., Toth, M., et al. (2014). The gut microbiota influences blood-brain barrier permeability in mice. *Science Translational Medicine, 6*(263), 263ra158. https://doi.org/10.1126/scitranslmed.3009759.
Brown, J. M., & Hazen, S. L. (2015). The gut microbial endocrine organ: Bacterially derived signals driving cardiometabolic diseases. *Annual Review of Medicine, 66*, 343–359. https://doi.org/10.1146/annurev-med-060513-093205.
Casha, S., Zygun, D., McGowan, M. D., Bains, I., Yong, V. W., & Hurlbert, R. J. (2012). Results of a phase II placebo-controlled randomized trial of minocycline in acute spinal cord injury. *Brain, 135*(Pt. 4), 1224–1236. https://doi.org/10.1093/brain/aws072.
Chang, P. V., Hao, L., Offermanns, S., & Medzhitov, R. (2014). The microbial metabolite butyrate regulates intestinal macrophage function via histone deacetylase inhibition. *Proceedings of the National Academy of Sciences of the United States of America, 111*(6), 2247–2252. https://doi.org/10.1073/pnas.1322269111.
Chen, C. Q., Fichna, J., Bashashati, M., Li, Y. Y., & Storr, M. (2011). Distribution, function and physiological role of melatonin in the lower gut. *World Journal of Gastroenterology, 17*(34), 3888–3898. https://doi.org/10.3748/wjg.v17.i34.3888.
Cheng, J., Li, W., Wang, Y., Cao, Q., Ni, Y., Zhang, W., et al. (2021). Electroacupuncture modulates the intestinal microecology to improve intestinal motility in spinal cord injury rats. *Microbial Biotechnology, 15*, 862–873. https://doi.org/10.1111/1751-7915.13968.
Cheriyan, T., Ryan, D. J., Weinreb, J. H., Cheriyan, J., Paul, J. C., Lafage, V., et al. (2014). Spinal cord injury models: A review. *Spinal Cord, 52*(8), 588–595. https://doi.org/10.1038/sc.2014.91.
Clapp, M., Aurora, N., Herrera, L., Bhatia, M., Wilen, E., & Wakefield, S. (2017). Gut microbiota's effect on mental health: The gut-brain axis. *Clinics and Practice, 7*(4), 987. https://doi.org/10.4081/cp.2017.987.
Costa, L. M., Pereira, J. E., Filipe, V. M., Magalhaes, L. G., Couto, P. A., Gonzalo-Orden, J. M., et al. (2013). Rolipram promotes functional recovery after contusive thoracic spinal cord injury in rats. *Behavioural Brain Research, 243*, 66–73. https://doi.org/10.1016/j.bbr.2012.12.056.

Craig, A., Guest, R., Tran, Y., & Middleton, J. (2017). Depressive mood in adults with spinal cord injury as they transition from an inpatient to a community setting: Secondary analyses from a clinical trial. *Spinal Cord*, *55*(10), 926–934. https://doi.org/10.1038/sc.2017.41.

Craig, A., Tran, Y., & Middleton, J. (2009). Psychological morbidity and spinal cord injury: A systematic review. *Spinal Cord*, *47*(2), 108–114. https://doi.org/10.1038/sc.2008.115.

Cryan, J. F., O'Riordan, K. J., Sandhu, K., Peterson, V., & Dinan, T. G. (2020). The gut microbiome in neurological disorders. *Lancet Neurology*, *19*(2), 179–194. https://doi.org/10.1016/S1474-4422(19)30356-4.

Dao, M. C., Everard, A., Aron-Wisnewsky, J., Sokolovska, N., Prifti, E., Verger, E. O., et al. (2016). Akkermansia muciniphila and improved metabolic health during a dietary intervention in obesity: Relationship with gut microbiome richness and ecology. *Gut*, *65*(3), 426–436. https://doi.org/10.1136/gutjnl-2014-308778.

David, L. A., Maurice, C. F., Carmody, R. N., Gootenberg, D. B., Button, J. E., Wolfe, B. E., et al. (2014). Diet rapidly and reproducibly alters the human gut microbiome. *Nature*, *505*(7484), 559–563. https://doi.org/10.1038/nature12820.

De Vadder, F., Grasset, E., Manneras Holm, L., Karsenty, G., Macpherson, A. J., Olofsson, L. E., et al. (2018). Gut microbiota regulates maturation of the adult enteric nervous system via enteric serotonin networks. *Proceedings of the National Academy of Sciences of the United States of America*, *115*(25), 6458–6463. https://doi.org/10.1073/pnas.1720017115.

Depommier, C., Everard, A., Druart, C., Plovier, H., Van Hul, M., Vieira-Silva, S., et al. (2019). Supplementation with Akkermansia muciniphila in overweight and obese human volunteers: A proof-of-concept exploratory study. *Nature Medicine*, *25*(7), 1096–1103. https://doi.org/10.1038/s41591-019-0495-2.

Desai, M. S., Seekatz, A. M., Koropatkin, N. M., Kamada, N., Hickey, C. A., Wolter, M., et al. (2016). A dietary Fiber-deprived gut microbiota degrades the colonic mucus barrier and enhances pathogen susceptibility. *Cell*, *167*(5), 1339–1353.e1321. https://doi.org/10.1016/j.cell.2016.10.043.

Dey, N., Wagner, V. E., Blanton, L. V., Cheng, J., Fontana, L., Haque, R., et al. (2015). Regulators of gut motility revealed by a gnotobiotic model of diet-microbiome interactions related to travel. *Cell*, *163*(1), 95–107. https://doi.org/10.1016/j.cell.2015.08.059.

Dinh, W., Futh, R., Nickl, W., Krahn, T., Ellinghaus, P., Scheffold, T., et al. (2009). Elevated plasma levels of TNF-alpha and interleukin-6 in patients with diastolic dysfunction and glucose metabolism disorders. *Cardiovascular Diabetology*, *8*, 58. https://doi.org/10.1186/1475-2840-8-58.

Doelman, A., Tigchelaar, S., McConeghy, B., Sinha, S., Keung, M. S., Manouchehri, N., et al. (2021). Characterization of the gut microbiome in a porcine model of thoracic spinal cord injury. *BMC Genomics*, *22*(1), 775. https://doi.org/10.1186/s12864-021-07979-3.

Du, J., Zayed, A. A., Kigerl, K. A., Zane, K., Sullivan, M. B., & Popovich, P. G. (2021). Spinal cord injury changes the structure and functional potential of gut bacterial and viral communities. *mSystems*, *6*(3), e01356–20. https://doi.org/10.1128/mSystems.01356-20.

Dubourg, G., Lagier, J. C., Armougom, F., Robert, C., Audoly, G., Papazian, L., et al. (2013). High-level colonisation of the human gut by Verrucomicrobia following broad-spectrum antibiotic treatment. *International Journal of Antimicrobial Agents*, *41*(2), 149–155. https://doi.org/10.1016/j.ijantimicag.2012.10.012.

Elewa, H. F., Hilali, H., Hess, D. C., Machado, L. S., & Fagan, S. C. (2006). Minocycline for short-term neuroprotection. *Pharmacotherapy*, *26*(4), 515–521. https://doi.org/10.1592/phco.26.4.515.

Erland, L. A., & Saxena, P. K. (2017). Melatonin natural health products and supplements: Presence of serotonin and significant variability of melatonin content. *Journal of Clinical Sleep Medicine, 13*(2), 275–281. https://doi.org/10.5664/jcsm.6462.

Esteban-Zubero, E., Lopez-Pingarron, L., Alatorre-Jimenez, M. A., Ochoa-Moneo, P., Buisac-Ramon, C., Rivas-Jimenez, M., et al. (2017). Melatonin's role as a co-adjuvant treatment in colonic diseases: A review. *Life Sciences, 170*, 72–81. https://doi.org/10.1016/j.lfs.2016.11.031.

Fealey, R. D., Szurszewski, J. H., Merritt, J. L., & DiMagno, E. P. (1984). Effect of traumatic spinal cord transection on human upper gastrointestinal motility and gastric emptying. *Gastroenterology, 87*(1), 69–75. Retrieved from https://www.ncbi.nlm.nih.gov/pubmed/6724276.

Fenn, A. M., Gensel, J. C., Huang, Y., Popovich, P. G., Lifshitz, J., & Godbout, J. P. (2014). Immune activation promotes depression 1 month after diffuse brain injury: A role for primed microglia. *Biological Psychiatry, 76*(7), 575–584. https://doi.org/10.1016/j.biopsych.2013.10.014.

Festoff, B. W., Ameenuddin, S., Arnold, P. M., Wong, A., Santacruz, K. S., & Citron, B. A. (2006). Minocycline neuroprotects, reduces microgliosis, and inhibits caspase protease expression early after spinal cord injury. *Journal of Neurochemistry, 97*(5), 1314–1326. https://doi.org/10.1111/j.1471-4159.2006.03799.x.

Filippone, A., Lanza, M., Campolo, M., Casili, G., Paterniti, I., Cuzzocrea, S., et al. (2020). Protective effect of sodium propionate in Abeta1-42 -induced neurotoxicity and spinal cord trauma. *Neuropharmacology, 166*, 107977. https://doi.org/10.1016/j.neuropharm.2020.107977.

Fung, T. C., Olson, C. A., & Hsiao, E. Y. (2017). Interactions between the microbiota, immune and nervous systems in health and disease. *Nature Neuroscience, 20*(2), 145–155. https://doi.org/10.1038/nn.4476.

Furuta, A., Suzuki, Y., Takahashi, R., Jakobsen, B. P., Kimura, T., Egawa, S., et al. (2021). Effects of Transanal irrigation on gut microbiota in pediatric patients with spina bifida. *Journal of Clinical Medicine, 10*(2), 224. https://doi.org/10.3390/jcm10020224.

Giuliani, F., Hader, W., & Yong, V. W. (2005). Minocycline attenuates T cell and microglia activity to impair cytokine production in T cell-microglia interaction. *Journal of Leukocyte Biology, 78*(1), 135–143. https://doi.org/10.1189/jlb.0804477.

Goyal, R., Faizy, A. F., Siddiqui, S. S., & Singhai, M. (2012). Evaluation of TNF-alpha and IL-6 levels in obese and non-obese diabetics: Pre- and postinsulin effects. *North American Journal of Medical Sciences, 4*(4), 180–184. https://doi.org/10.4103/1947-2714.94944.

Gungor, B., Adiguzel, E., Gursel, I., Yilmaz, B., & Gursel, M. (2016). Intestinal microbiota in patients with spinal cord injury. *PLoS One, 11*(1), e0145878. https://doi.org/10.1371/journal.pone.0145878.

Hakim, S., Gaglani, T., & Cash, B. D. (2022). Neurogenic Bowel dysfunction: The impact of the central nervous system in constipation and fecal incontinence. *Gastroenterology Clinics of North America, 51*(1), 93–105. https://doi.org/10.1016/j.gtc.2021.10.006.

Hasani, A., Ebrahimzadeh, S., Hemmati, F., Khabbaz, A., Hasani, A., & Gholizadeh, P. (2021). The role of Akkermansia muciniphila in obesity, diabetes and atherosclerosis. *Journal of Medical Microbiology, 70*(10), 001435. https://doi.org/10.1099/jmm.0.001435.

Hausmann, O. N. (2003). Post-traumatic inflammation following spinal cord injury. *Spinal Cord, 41*(7), 369–378. https://doi.org/10.1038/sj.sc.3101483.

Hellenbrand, D. J., Quinn, C. M., Piper, Z. J., Morehouse, C. N., Fixel, J. A., & Hanna, A. S. (2021). Inflammation after spinal cord injury: A review of the critical timeline of signaling cues and cellular infiltration. *Journal of Neuroinflammation, 18*(1), 284. https://doi.org/10.1186/s12974-021-02337-2.

Holmes, G. M., & Blanke, E. N. (2019). Gastrointestinal dysfunction after spinal cord injury. *Experimental Neurology, 320*, 113009. https://doi.org/10.1016/j.expneurol.2019.113009.

Huang, T. T., Lai, J. B., Du, Y. L., Xu, Y., Ruan, L. M., & Hu, S. H. (2019). Current understanding of gut microbiota in mood disorders: An update of human studies. *Frontiers in Genetics*, *10*, 98. https://doi.org/10.3389/fgene.2019.00098.

Iannotti, C. A., Clark, M., Horn, K. P., van Rooijen, N., Silver, J., & Steinmetz, M. P. (2011). A combination immunomodulatory treatment promotes neuroprotection and locomotor recovery after contusion SCI. *Experimental Neurology*, *230*(1), 3–15. https://doi.org/10.1016/j.expneurol.2010.03.010.

Jin, S. L., & Conti, M. (2002). Induction of the cyclic nucleotide phosphodiesterase PDE4B is essential for LPS-activated TNF-alpha responses. *Proceedings of the National Academy of Sciences of the United States of America*, *99*(11), 7628–7633. https://doi.org/10.1073/pnas.122041599.

Jin, S. L., Lan, L., Zoudilova, M., & Conti, M. (2005). Specific role of phosphodiesterase 4B in lipopolysaccharide-induced signaling in mouse macrophages. *Journal of Immunology*, *175*(3), 1523–1531. https://doi.org/10.4049/jimmunol.175.3.1523.

Jing, Y., Bai, F., Wang, L., Yang, D., Yan, Y., Wang, Q., et al. (2022). Fecal microbiota transplantation exerts neuroprotective effects in a mouse spinal cord injury model by modulating the microenvironment at the lesion site. *Microbiology Spectrum*, e0017722. https://doi.org/10.1128/spectrum.00177-22.

Jing, Y., Bai, F., & Yu, Y. (2021). Spinal cord injury and gut microbiota: A review. *Life Sciences*, *266*, 118865. https://doi.org/10.1016/j.lfs.2020.118865.

Jing, Y., Yang, D., Bai, F., Zhang, C., Qin, C., Li, D., et al. (2019). Melatonin treatment alleviates spinal cord injury-induced gut Dysbiosis in mice. *Journal of Neurotrauma*, *36*(18), 2646–2664. https://doi.org/10.1089/neu.2018.6012.

Jing, Y., Yu, Y., Bai, F., Wang, L., Yang, D., Zhang, C., et al. (2021). Effect of fecal microbiota transplantation on neurological restoration in a spinal cord injury mouse model: Involvement of brain-gut axis. *Microbiome*, *9*(1), 59. https://doi.org/10.1186/s40168-021-01007-y.

Jogia, T., & Ruitenberg, M. J. (2020). Traumatic spinal cord injury and the gut microbiota: Current insights and future challenges. *Frontiers in Immunology*, *11*, 704. https://doi.org/10.3389/fimmu.2020.00704.

Jurkiewicz, M. T., Crawley, A. P., Verrier, M. C., Fehlings, M. G., & Mikulis, D. J. (2006). Somatosensory cortical atrophy after spinal cord injury: A voxel-based morphometry study. *Neurology*, *66*(5), 762–764. https://doi.org/10.1212/01.wnl.0000201276.28141.40.

Keshavarzian, A., Green, S. J., Engen, P. A., Voigt, R. M., Naqib, A., Forsyth, C. B., et al. (2015). Colonic bacterial composition in Parkinson's disease. *Movement Disorders*, *30*(10), 1351–1360. https://doi.org/10.1002/mds.26307.

Khazaeipour, Z., Taheri-Otaghsara, S. M., & Naghdi, M. (2015). Depression following spinal cord injury: Its relationship to demographic and socioeconomic indicators. *Topics in Spinal Cord Injury Rehabilitation*, *21*(2), 149–155. https://doi.org/10.1310/sci2102-149.

Kigerl, K. A., Hall, J. C., Wang, L., Mo, X., Yu, Z., & Popovich, P. G. (2016). Gut dysbiosis impairs recovery after spinal cord injury. *The Journal of Experimental Medicine*, *213*(12), 2603–2620. https://doi.org/10.1084/jem.20151345.

Kigerl, K. A., Mostacada, K., & Popovich, P. G. (2018). Gut microbiota are disease-modifying factors after traumatic spinal cord injury. *Neurotherapeutics*, *15*(1), 60–67. https://doi.org/10.1007/s13311-017-0583-2.

Kigerl, K. A., Zane, K., Adams, K., Sullivan, M. B., & Popovich, P. G. (2020). The spinal cord-gut-immune axis as a master regulator of health and neurological function after spinal cord injury. *Experimental Neurology*, *323*, 113085. https://doi.org/10.1016/j.expneurol.2019.113085.

Krassioukov, A., Eng, J. J., Claxton, G., Sakakibara, B. M., & Shum, S. (2010). Neurogenic bowel management after spinal cord injury: A systematic review of the evidence. *Spinal Cord*, *48*(10), 718–733. https://doi.org/10.1038/sc.2010.14.

Lanza, M., Campolo, M., Casili, G., Filippone, A., Paterniti, I., Cuzzocrea, S., et al. (2019). Sodium butyrate exerts neuroprotective effects in spinal cord injury. *Molecular Neurobiology, 56*(6), 3937–3947. https://doi.org/10.1007/s12035-018-1347-7.

Lee, B. B., Cripps, R. A., Fitzharris, M., & Wing, P. C. (2014). The global map for traumatic spinal cord injury epidemiology: Update 2011, global incidence rate. *Spinal Cord, 52*(2), 110–116. https://doi.org/10.1038/sc.2012.158.

Lee, S. M., Yune, T. Y., Kim, S. J., Park, D. W., Lee, Y. K., Kim, Y. C., et al. (2003). Minocycline reduces cell death and improves functional recovery after traumatic spinal cord injury in the rat. *Journal of Neurotrauma, 20*(10), 1017–1027. https://doi.org/10.1089/089771503770195867.

Lenoir, M., Martin, R., Torres-Maravilla, E., Chadi, S., Gonzalez-Davila, P., Sokol, H., et al. (2020). Butyrate mediates anti-inflammatory effects of Faecalibacterium prausnitzii in intestinal epithelial cells through Dact3. *Gut Microbes, 12*(1), 1–16. https://doi.org/10.1080/19490976.2020.1826748.

Li, J., Morrow, C., Barnes, S., Wilson, L., Womack, E. D., McLain, A., et al. (2021). Gut microbiome composition and serum metabolome profile among individuals with spinal cord injury and Normal glucose tolerance or prediabetes/type 2 diabetes. *Archives of Physical Medicine and Rehabilitation, 103*, 702–710. https://doi.org/10.1016/j.apmr.2021.03.043.

Li, J., Morrow, C., McLain, A., Womack, E. D., & Yarar-Fisher, C. (2022). Effects of a low-carbohydrate/high-protein diet on gut microbiome composition in insulin resistant individuals with chronic spinal cord injury: Preliminary results from a randomized controlled trial. *Archives of Physical Medicine and Rehabilitation*. https://doi.org/10.1016/j.apmr.2022.03.014.

Li, J., Van Der Pol, W., Eraslan, M., McLain, A., Cetin, H., Cetin, B., et al. (2020). Comparison of the gut microbiome composition among individuals with acute or long-standing spinal cord injury vs. able-bodied controls. *The Journal of Spinal Cord Medicine, 45*, 1–9. https://doi.org/10.1080/10790268.2020.1769949.

Lin, M. Y., de Zoete, M. R., van Putten, J. P., & Strijbis, K. (2015). Redirection of epithelial immune responses by short-chain fatty acids through inhibition of histone deacetylases. *Frontiers in Immunology, 6*, 554. https://doi.org/10.3389/fimmu.2015.00554.

Lin, R., Xu, J., Ma, Q., Chen, M., Wang, L., Wen, S., et al. (2020). Alterations in the fecal microbiota of patients with spinal cord injury. *PLoS One, 15*(8), e0236470. https://doi.org/10.1371/journal.pone.0236470.

Liu, J., An, H., Jiang, D., Huang, W., Zou, H., Meng, C., et al. (2004). Study of bacterial translocation from gut after paraplegia caused by spinal cord injury in rats. *Spine (Phila Pa 1976), 29*(2), 164–169. https://doi.org/10.1097/01.BRS.0000107234.74249.CD.

Liu, R., Hong, J., Xu, X., Feng, Q., Zhang, D., Gu, Y., et al. (2017). Gut microbiome and serum metabolome alterations in obesity and after weight-loss intervention. *Nature Medicine, 23*(7), 859–868. https://doi.org/10.1038/nm.4358.

Lopez-Siles, M., Duncan, S. H., Garcia-Gil, L. J., & Martinez-Medina, M. (2017). Faecalibacterium prausnitzii: From microbiology to diagnostics and prognostics. *The ISME Journal, 11*(4), 841–852. https://doi.org/10.1038/ismej.2016.176.

Luedtke, K., Bouchard, S. M., Woller, S. A., Funk, M. K., Aceves, M., & Hook, M. A. (2014). Assessment of depression in a rodent model of spinal cord injury. *Journal of Neurotrauma, 31*(12), 1107–1121. https://doi.org/10.1089/neu.2013.3204.

Maini Rekdal, V., Bess, E. N., Bisanz, J. E., Turnbaugh, P. J., & Balskus, E. P. (2019). Discovery and inhibition of an interspecies gut bacterial pathway for levodopa metabolism. *Science, 364*(6445), eaau6323. https://doi.org/10.1126/science.aau6323.

Maldonado-Bouchard, S., Peters, K., Woller, S. A., Madahian, B., Faghihi, U., Patel, S., et al. (2016). Inflammation is increased with anxiety- and depression-like signs in a rat model of spinal cord injury. *Brain, Behavior, and Immunity*, *51*, 176–195. https://doi.org/10.1016/j.bbi.2015.08.009.

Margolis, K. G., Cryan, J. F., & Mayer, E. A. (2021). The microbiota-gut-brain axis: From motility to mood. *Gastroenterology*, *160*(5), 1486–1501. https://doi.org/10.1053/j.gastro.2020.10.066.

Mirza, S., Hossain, M., Mathews, C., Martinez, P., Pino, P., Gay, J. L., et al. (2012). Type 2-diabetes is associated with elevated levels of TNF-alpha, IL-6 and adiponectin and low levels of leptin in a population of Mexican Americans: A cross-sectional study. *Cytokine*, *57*(1), 136–142. https://doi.org/10.1016/j.cyto.2011.09.029.

Musleh-Vega, S., Ojeda, J., & Vidal, P. M. (2022). Gut microbiota-brain axis as a potential modulator of psychological stress after spinal cord injury. *Biomedicine*, *10*(4), 847. https://doi.org/10.3390/biomedicines10040847.

Myers, S. A., Gobejishvili, L., Saraswat Ohri, S., Garrett Wilson, C., Andres, K. R., Riegler, A. S., et al. (2019). Following spinal cord injury, PDE4B drives an acute, local inflammatory response and a chronic, systemic response exacerbated by gut dysbiosis and endotoxemia. *Neurobiology of Disease*, *124*, 353–363. https://doi.org/10.1016/j.nbd.2018.12.008.

Naseem, M., & Parvez, S. (2014). Role of melatonin in traumatic brain injury and spinal cord injury. *ScientificWorldJournal*, *2014*, 586270. https://doi.org/10.1155/2014/586270.

O'Connor, G., Jeffrey, E., Madorma, D., Marcillo, A., Abreu, M. T., Deo, S. K., et al. (2018). Investigation of microbiota alterations and intestinal inflammation post-spinal cord injury in rat model. *Journal of Neurotrauma*, *35*(18), 2159–2166. https://doi.org/10.1089/neu.2017.5349.

Panther, E. J., Dodd, W., Clark, A., & Lucke-Wold, B. (2022). Gastrointestinal microbiome and neurologic injury. *Biomedicine*, *10*(2), 500. https://doi.org/10.3390/biomedicines10020500.

Parada Venegas, D., De la Fuente, M. K., Landskron, G., Gonzalez, M. J., Quera, R., Dijkstra, G., et al. (2019). Short chain fatty acids (SCFAs)-mediated gut epithelial and immune regulation and its relevance for inflammatory bowel diseases. *Frontiers in Immunology*, *10*, 277. https://doi.org/10.3389/fimmu.2019.00277.

Park, Y. S., Kim, S. H., Park, J. W., Kho, Y., Seok, P. R., Shin, J. H., et al. (2020). Melatonin in the colon modulates intestinal microbiota in response to stress and sleep deprivation. *Intestinal Research*, *18*(3), 325–336. https://doi.org/10.5217/ir.2019.00093.

Plovier, H., Everard, A., Druart, C., Depommier, C., Van Hul, M., Geurts, L., et al. (2017). A purified membrane protein from Akkermansia muciniphila or the pasteurized bacterium improves metabolism in obese and diabetic mice. *Nature Medicine*, *23*(1), 107–113. https://doi.org/10.1038/nm.4236.

Pruss, H., Tedeschi, A., Thiriot, A., Lynch, L., Loughhead, S. M., Stutte, S., et al. (2017). Spinal cord injury-induced immunodeficiency is mediated by a sympathetic-neuroendocrine adrenal reflex. *Nature Neuroscience*, *20*(11), 1549–1559. https://doi.org/10.1038/nn.4643.

Qian, J., Zhu, W., Lu, M., Ni, B., & Yang, J. (2017). D-beta-hydroxybutyrate promotes functional recovery and relieves pain hypersensitivity in mice with spinal cord injury. *British Journal of Pharmacology*, *174*(13), 1961–1971. https://doi.org/10.1111/bph.13788.

Quigley, E. M. (2011). Microflora modulation of motility. *Journal of Neurogastroenterology and Motility*, *17*(2), 140–147. https://doi.org/10.5056/jnm.2011.17.2.140.

Riegger, T., Conrad, S., Liu, K., Schluesener, H. J., Adibzahdeh, M., & Schwab, J. M. (2007). Spinal cord injury-induced immune depression syndrome (SCI-IDS). *The European Journal of Neuroscience*, *25*(6), 1743–1747. https://doi.org/10.1111/j.1460-9568.2007.05447.x.

Riegger, T., Conrad, S., Schluesener, H. J., Kaps, H. P., Badke, A., Baron, C., et al. (2009). Immune depression syndrome following human spinal cord injury (SCI): A pilot study. *Neuroscience*, *158*(3), 1194–1199. https://doi.org/10.1016/j.neuroscience.2008.08.021.

Riviere, A., Selak, M., Lantin, D., Leroy, F., & De Vuyst, L. (2016). Bifidobacteria and butyrate-producing colon bacteria: Importance and strategies for their stimulation in the human gut. *Frontiers in Microbiology*, 7, 979. https://doi.org/10.3389/fmicb.2016.00979.

Rodenhouse, A., Talukder, M. A. H., Lee, J. I., Govindappa, P. K., O'Brien, M., Manto, K. M., et al. (2022). Altered gut microbiota composition with antibiotic treatment impairs functional recovery after traumatic peripheral nerve crush injury in mice: Effects of probiotics with butyrate producing bacteria. *BMC Research Notes*, *15*(1), 80. https://doi.org/10.1186/s13104-022-05967-8.

Rong, Z., Huang, Y., Cai, H., Chen, M., Wang, H., Liu, G., et al. (2021). Gut microbiota disorders promote inflammation and aggravate spinal cord injury through the TLR4/MyD88 signaling pathway. *Frontiers in Nutrition*, *8*, 702659. https://doi.org/10.3389/fnut.2021.702659.

Round, J. L., & Mazmanian, S. K. (2009). The gut microbiota shapes intestinal immune responses during health and disease. *Nature Reviews. Immunology*, *9*(5), 313–323. https://doi.org/10.1038/nri2515.

Samantaray, S., Das, A., Thakore, N. P., Matzelle, D. D., Reiter, R. J., Ray, S. K., et al. (2009). Therapeutic potential of melatonin in traumatic central nervous system injury. *Journal of Pineal Research*, *47*(2), 134–142. https://doi.org/10.1111/j.1600-079X.2009.00703.x.

Sampson, T. (2020). The impact of indigenous microbes on Parkinson's disease. *Neurobiology of Disease*, *135*, 104426. https://doi.org/10.1016/j.nbd.2019.03.014.

Schaal, S. M., Garg, M. S., Ghosh, M., Lovera, L., Lopez, M., Patel, M., et al. (2012). The therapeutic profile of rolipram, PDE target and mechanism of action as a neuroprotectant following spinal cord injury. *PLoS One*, 7(9), e43634. https://doi.org/10.1371/journal.pone.0043634.

Schmidt, E. K. A., Raposo, P. J. F., Madsen, K. L., Fenrich, K. K., Kabarchuk, G., & Fouad, K. (2021). What makes a successful donor? Fecal transplant from anxious-like rats does not prevent spinal cord injury-induced dysbiosis. *Biology (Basel)*, *10*(4), 254. https://doi.org/10.3390/biology10040254.

Schmidt, E. K. A., Raposo, P. J. F., Torres-Espin, A., Fenrich, K. K., & Fouad, K. (2021). Beyond the lesion site: Minocycline augments inflammation and anxiety-like behavior following SCI in rats through action on the gut microbiota. *Journal of Neuroinflammation*, *18*(1), 144. https://doi.org/10.1186/s12974-021-02123-0.

Schmidt, E. K. A., Torres-Espin, A., Raposo, P. J. F., Madsen, K. L., Kigerl, K. A., Popovich, P. G., et al. (2020). Fecal transplant prevents gut dysbiosis and anxiety-like behaviour after spinal cord injury in rats. *PLoS One*, *15*(1), e0226128. https://doi.org/10.1371/journal.pone.0226128.

Schneeberger, M., Everard, A., Gomez-Valades, A. G., Matamoros, S., Ramirez, S., Delzenne, N. M., et al. (2015). Akkermansia muciniphila inversely correlates with the onset of inflammation, altered adipose tissue metabolism and metabolic disorders during obesity in mice. *Scientific Reports*, *5*, 16643. https://doi.org/10.1038/srep16643.

Schwab, J. M., Zhang, Y., Kopp, M. A., Brommer, B., & Popovich, P. G. (2014). The paradox of chronic neuroinflammation, systemic immune suppression, autoimmunity after traumatic chronic spinal cord injury. *Experimental Neurology*, *258*, 121–129. https://doi.org/10.1016/j.expneurol.2014.04.023.

Segain, J. P., Raingeard de la Bletiere, D., Bourreille, A., Leray, V., Gervois, N., Rosales, C., et al. (2000). Butyrate inhibits inflammatory responses through NFkappaB inhibition: Implications for Crohn's disease. *Gut*, *47*(3), 397–403. https://doi.org/10.1136/gut.47.3.397.

Shultz, R. B., & Zhong, Y. (2017). Minocycline targets multiple secondary injury mechanisms in traumatic spinal cord injury. *Neural Regeneration Research*, *12*(5), 702–713. https://doi.org/10.4103/1673-5374.206633.

Silva, Y. P., Bernardi, A., & Frozza, R. L. (2020). The role of short-chain fatty acids from gut microbiota in gut-brain communication. *Front Endocrinol (Lausanne)*, *11*, 25. https://doi.org/10.3389/fendo.2020.00025.

Simpson, L. A., Eng, J. J., Hsieh, J. T., Wolfe, D. L., & Spinal Cord Injury Rehabilitation Evidence Scire Research, T. (2012). The health and life priorities of individuals with spinal cord injury: A systematic review. *Journal of Neurotrauma*, *29*(8), 1548–1555. https://doi.org/10.1089/neu.2011.2226.

Singal, A. K., Rosman, A. S., Bauman, W. A., & Korsten, M. A. (2006). Recent concepts in the management of bowel problems after spinal cord injury. *Advances in Medical Sciences*, *51*, 15–22. Retrieved from https://www.ncbi.nlm.nih.gov/pubmed/17357271.

Soret, R., Chevalier, J., De Coppet, P., Poupeau, G., Derkinderen, P., Segain, J. P., et al. (2010). Short-chain fatty acids regulate the enteric neurons and control gastrointestinal motility in rats. *Gastroenterology*, *138*(5), 1772–1782. https://doi.org/10.1053/j.gastro.2010.01.053.

Stiens, S. A., Bergman, S. B., & Goetz, L. L. (1997). Neurogenic bowel dysfunction after spinal cord injury: Clinical evaluation and rehabilitative management. *Archives of Physical Medicine and Rehabilitation*, *78*(3 Suppl), S86–102. https://doi.org/10.1016/s0003-9993(97)90416-0.

Stone, J. M., Nino-Murcia, M., Wolfe, V. A., & Perkash, I. (1990). Chronic gastrointestinal problems in spinal cord injury patients: A prospective analysis. *The American Journal of Gastroenterology*, *85*(9), 1114–1119. Retrieved from https://www.ncbi.nlm.nih.gov/pubmed/2389723.

Thibault-Halman, G., Casha, S., Singer, S., & Christie, S. (2011). Acute management of nutritional demands after spinal cord injury. *Journal of Neurotrauma*, *28*(8), 1497–1507. https://doi.org/10.1089/neu.2009.1155.

Tilg, H., Zmora, N., Adolph, T. E., & Elinav, E. (2020). The intestinal microbiota fuelling metabolic inflammation. *Nature Reviews. Immunology*, *20*(1), 40–54. https://doi.org/10.1038/s41577-019-0198-4.

Valido, E., Bertolo, A., Frankl, G. P., Itodo, O. A., Pinheiro, T., Pannek, J., et al. (2022). Systematic review of the changes in the microbiome following spinal cord injury: Animal and human evidence. *Spinal Cord*, *60*, 288–300. https://doi.org/10.1038/s41393-021-00737-y.

Vicentini, F. A., Keenan, C. M., Wallace, L. E., Woods, C., Cavin, J. B., Flockton, A. R., et al. (2021). Intestinal microbiota shapes gut physiology and regulates enteric neurons and glia. *Microbiome*, *9*(1), 210. https://doi.org/10.1186/s40168-021-01165-z.

von Schillde, M. A., Hormannsperger, G., Weiher, M., Alpert, C. A., Hahne, H., Bauerl, C., et al. (2012). Lactocepin secreted by Lactobacillus exerts anti-inflammatory effects by selectively degrading proinflammatory chemokines. *Cell Host & Microbe*, *11*(4), 387–396. https://doi.org/10.1016/j.chom.2012.02.006.

Vuong, H. E., Yano, J. M., Fung, T. C., & Hsiao, E. Y. (2017). The microbiome and host behavior. *Annual Review of Neuroscience*, *40*, 21–49. https://doi.org/10.1146/annurev-neuro-072116-031347.

Wallace, D. J., Sayre, N. L., Patterson, T. T., Nicholson, S. E., Hilton, D., & Grandhi, R. (2019). Spinal cord injury and the human microbiome: Beyond the brain-gut axis. *Neurosurgical Focus*, *46*(3), E11. https://doi.org/10.3171/2018.12.FOCUS18206.

Wang, Z., Roberts, A. B., Buffa, J. A., Levison, B. S., Zhu, W., Org, E., et al. (2015). Non-lethal inhibition of gut microbial trimethylamine production for the treatment of atherosclerosis. *Cell*, *163*(7), 1585–1595. https://doi.org/10.1016/j.cell.2015.11.055.

Wells, J. E., Hurlbert, R. J., Fehlings, M. G., & Yong, V. W. (2003). Neuroprotection by minocycline facilitates significant recovery from spinal cord injury in mice. *Brain*, *126*(Pt. 7), 1628–1637. https://doi.org/10.1093/brain/awg178.

Wheeler, T. L., Bowel, Bladder Workshop, P, de Groat, W., Eisner, K., Emmanuel, A., French, J., et al. (2018). Translating promising strategies for bowel and bladder management in spinal cord injury. *Experimental Neurology*, *306*, 169–176. https://doi.org/10.1016/j.expneurol.2018.05.006.

White, A. R., & Holmes, G. M. (2018). Anatomical and functional changes to the colonic neuromuscular compartment after experimental spinal cord injury. *Journal of Neurotrauma*, *35*(9), 1079–1090. https://doi.org/10.1089/neu.2017.5369.

White, A. R., Werner, C. M., & Holmes, G. M. (2020). Diminished enteric neuromuscular transmission in the distal colon following experimental spinal cord injury. *Experimental Neurology*, *331*, 113377. https://doi.org/10.1016/j.expneurol.2020.113377.

Williams, R. E., 3rd, Bauman, W. A., Spungen, A. M., Vinnakota, R. R., Farid, R. Z., Galea, M., et al. (2012). SmartPill technology provides safe and effective assessment of gastrointestinal function in persons with spinal cord injury. *Spinal Cord*, *50*(1), 81–84. https://doi.org/10.1038/sc.2011.92.

Wu, H. J., & Wu, E. (2012). The role of gut microbiota in immune homeostasis and autoimmunity. *Gut Microbes*, *3*(1), 4–14. https://doi.org/10.4161/gmic.19320.

Xu, C., He, Z., & Li, J. (2022). Melatonin as a potential neuroprotectant: Mechanisms in subarachnoid hemorrhage-induced early brain injury. *Frontiers in Aging Neuroscience*, *14*, 899678. https://doi.org/10.3389/fnagi.2022.899678.

Yang, L., Yao, M., Lan, Y., Mo, W., Sun, Y. L., Wang, J., et al. (2016). Melatonin for spinal cord injury in animal models: A systematic review and network meta-analysis. *Journal of Neurotrauma*, *33*(3), 290–300. https://doi.org/10.1089/neu.2015.4038.

Yao, Y., Cai, X., Fei, W., Ye, Y., Zhao, M., & Zheng, C. (2022). The role of short-chain fatty acids in immunity, inflammation and metabolism. *Critical Reviews in Food Science and Nutrition*, *62*(1), 1–12. https://doi.org/10.1080/10408398.2020.1854675.

Yarandi, S. S., Kulkarni, S., Saha, M., Sylvia, K. E., Sears, C. L., & Pasricha, P. J. (2020). Intestinal bacteria maintain adult enteric nervous system and nitrergic neurons via toll-like receptor 2-induced neurogenesis in mice. *Gastroenterology*, *159*(1), 200–213.e208. https://doi.org/10.1053/j.gastro.2020.03.050.

Yarar-Fisher, C., Kulkarni, A., Li, J., Farley, P., Renfro, C., Aslam, H., et al. (2018). Evaluation of a ketogenic diet for improvement of neurological recovery in individuals with acute spinal cord injury: A pilot, randomized safety and feasibility trial. *Spinal Cord Series And Cases*, *4*, 88. https://doi.org/10.1038/s41394-018-0121-4.

Yoo, B. B., & Mazmanian, S. K. (2017). The enteric network: Interactions between the immune and nervous systems of the gut. *Immunity*, *46*(6), 910–926. https://doi.org/10.1016/j.immuni.2017.05.011.

Yu, B., Qiu, H., Cheng, S., Ye, F., Li, J., Chen, S., et al. (2021). Profile of gut microbiota in patients with traumatic thoracic spinal cord injury and its clinical implications: A case-control study in a rehabilitation setting. *Bioengineered*, *12*(1), 4489–4499. https://doi.org/10.1080/21655979.2021.1955543.

Yuan, B., Lu, X. J., & Wu, Q. (2021). Gut microbiota and acute central nervous system injury: A new target for therapeutic intervention. *Frontiers in Immunology*, *12*, 800796. https://doi.org/10.3389/fimmu.2021.800796.

Yue, J. K., Winkler, E. A., Rick, J. W., Deng, H., Partow, C. P., Upadhyayula, P. S., et al. (2017). Update on critical care for acute spinal cord injury in the setting of polytrauma. *Neurosurgical Focus*, *43*(5), E19. https://doi.org/10.3171/2017.7.FOCUS17396.

Zhang, Y., Guan, Z., Reader, B., Shawler, T., Mandrekar-Colucci, S., Huang, K., et al. (2013). Autonomic dysreflexia causes chronic immune suppression after spinal cord injury. *The Journal of Neuroscience*, *33*(32), 12970–12981. https://doi.org/10.1523/JNEUROSCI.1974-13.2013.

Zhang, C., Jing, Y., Zhang, W., Zhang, J., Yang, M., Du, L., et al. (2019). Dysbiosis of gut microbiota is associated with serum lipid profiles in male patients with chronic traumatic cervical spinal cord injury. *American Journal of Translational Research*, *11*(8), 4817–4834. Retrieved from https://www.ncbi.nlm.nih.gov/pubmed/31497202.

Zhang, N., Yin, Y., Xu, S. J., Wu, Y. P., & Chen, W. S. (2012). Inflammation & apoptosis in spinal cord injury. *Indian Journal of Medical Research*, *135*, 287–296. Retrieved from https://www.ncbi.nlm.nih.gov/pubmed/22561613.

Zhang, C., Zhang, W., Zhang, J., Jing, Y., Yang, M., Du, L., et al. (2018). Gut microbiota dysbiosis in male patients with chronic traumatic complete spinal cord injury. *Journal of Translational Medicine*, *16*(1), 353. https://doi.org/10.1186/s12967-018-1735-9.

Zhao, Y., & Yu, Y. B. (2016). Intestinal microbiota and chronic constipation. *Springerplus*, *5*(1), 1130. https://doi.org/10.1186/s40064-016-2821-1.

Zhu, J., Mix, E., & Winblad, B. (2001). The antidepressant and antiinflammatory effects of rolipram in the central nervous system. *CNS Drug Reviews*, 7(4), 387–398. https://doi.org/10.1111/j.1527-3458.2001.tb00206.x.

FSC
www.fsc.org
MIX
Papier | Fördert gute Waldnutzung
FSC® C083411

Kolibri 360 GmbH
Ferdinand-Jühlke-Straße 7
99095 Erfurt, Deutschland
produktsicherheit@kolibri360.de